D0871938

ADAM'S ANCESTORS

Medicine, Science, and Religion in Historical Context

Ronald L. Numbers, Consulting Editor

DAVID N. LIVINGSTONE

ADAM'S ANCESTORS

Race, Religion, and the
Politics of Human Origins

THE JOHNS HOPKINS UNIVERSITY PRESS   *Baltimore*

9   8   7   6   5   4   3   2   1

The Johns Hopkins University Press
2715 North Charles Street
Baltimore, Maryland 21218-4363
www.press.jhu.edu

Library of Congress Cataloging-in-Publication Data

Livingstone, David N., 1953–
    Adam's ancestors : race, religion, and the politics of human origins /
David N. Livingstone.
        p.   cm. — (Medicine, science, and religion in historical context)
    Includes bibliographical references and index.
    ISBN-13: 978-0-8018-8813-7 (hardcover : alk. paper)
    ISBN-10: 0-8018-8813-1 (hardcover : alk. paper)
    1. Ethnology—Religious aspects. 2. Theological anthropology.
3. Human beings—Origin. 4. Human evolution. I. Title.
    BL256.L57 2008
    202'.2—dc22   2007033706

A catalog record for this book is available from the British Library.

*Special discounts are available for bulk purchases of this book. For more
information, please contact Special Sales at 410-516-6936 or specialsales@
press.jhu.edu.*

for
Frances

CONTENTS

PREFACE

The idea for this book was born one afternoon in March 1999 at the University of California at Berkeley. I got into a conversation with Ronald Numbers, who mentioned an exploratory essay I had published a few years earlier on the idea of humans before Adam. Ron suggested that I might return to that theme to try to flesh out the story and write a full history of the scheme. The journey on which I embarked that afternoon has taken me to many intellectual destinations, some strange, some familiar, all fascinating. And while I am sure that I have not said everything that could be said about the idea of pre-adamic humanity, I am greatly indebted to Ron for encouraging me to embark on this expedition. Along the way I have benefited enormously from his continuing interest and support and from the help I have derived from many friends and scholars.

I owe an immense debt to Colin Kidd both for making his *Forging of Races* available to me prior to publication and for numerous bibliographical leads. The stimulus of his scholarship on the history of ethnic identities has been inspirational. Another afternoon conversation, this time with Andrew Holmes, proved to be invaluable in helping me sort out a coherent structure for the entire book. The fact that he also persistently drew my attention to numerous fugitive publications has only placed me more deeply in his debt. The care that Nicolaas Rupke took in reading the entire manuscript and saving me from some embarrassing errors is both typical of his erudition and scholarship and a mark of his valued friendship. I have benefited greatly, too, from many enlightening conversations with good friends such as Diarmid Finnegan, Frank Gourley, Nuala Johnson, Mark Noll, and Stephen Williams, some of whom took the time and trouble to read portions of the manuscript and offer the best of advice. Luke Harlow willingly provided me with useful bibliographical leads on some American dimensions of the subject; Simon Schaffer directed me to some important seventeenth-century work; Philip Orr brought his dramatist's eye to bear on the entire text and offered excellent counsel; Jeremy Crampton awakened me to dimensions of the story with which I was unfamiliar and shared with me some of his unpublished work; and Martin Rudwick read several parts of the manuscript, offering sage and salient advice on several points. I am also extremely grateful to Gill Alexander and Maura Pringle for their skill in working with illustrations and to Elizabeth Gratch for patient and careful copyediting.

To all these colleagues and friends I record my appreciation in the certain knowledge that, whatever its imperfections, *Adam's Ancestors* is the better for the help they have willingly given. But my greatest debt is, as always, to Frances, Emma, and Justin, who have had to share too many dinners with the ghosts of Adam's ancestors.

ADAM'S ANCESTORS

1 ℬ BEGINNINGS

Questioning the Mosaic Record

In the beginning God created the heaven and the earth. And the earth was without form and void; and darkness was upon the face of the deep. And the Spirit of God moved upon the face of the waters. And God said, Let there be light, and there was light. And God saw the light, that it was good: and God divided the light from the darkness. And God called the light Day, and the darkness he called Night. And the evening and the morning were the first day . . . So God created man in his own image, in the image of God created he him; male and female created he them . . . And the Lord God took the man, and put him into the garden of Eden to dress it and to keep it . . . And out of the ground the Lord God formed every beast of the field, and every fowl of the air; and brought them to Adam to see what he would call them: and whatsoever Adam called every living creature, that was the name thereof. And Adam gave names to all cattle and to the fowl of the air, and to every beast of the field.

E VER SINCE 1611, when the King James Bible first appeared, these words have introduced Bible readers to Adam, the father of the human race. With sonorous majesty these words give voice to a doctrine that has circulated since ancient times, stretching back through the Geneva Bible and Miles Coverdale's translation, Saint Jerome's medieval Latin Vulgate and the Greek Septuagint, to the Aramaic Targums of the early Hebrews. It is the doctrine of Creation, the story of beginnings—of the heavens and the earth, of light and dark, of sun and moon, of plants and animals. It is a chronicle that positions humankind at the pinnacle of the narrative. It tells how we came to be here, what life was like in the morning of the earth, why human beings come in male and female forms, where evil originated, and how ugliness entered the world. And here, in the cradle of creation, stands one individual, Adam, the first man.

To leave nothing to the imagination, pictures in Bibles would soon give visual expression to the world's first couple, newly minted, fresh, pristine, unspoiled. Take, for instance, the magnificent Bible published by the Edinburgh-born John Ogilby (1600–1676), sometime dancing master, theatrical impresario, translator, bookseller, publisher, and cartographer. The author of a series of folio travel books on various countries—Britain, Japan, and Africa among them—and a road atlas of England and Wales under the title *Britannia* in 1675, when he was in his seventies, Ogilby reissued in 1660 the large folio Bible published by John Field the previous year.[1] This hugely expensive text, illustrated with what were called "chorographical sculps," included a plate engraved by Pierre Lombard depicting Adam and Eve in the Garden at the moment of their fall from grace (fig. 1). The lavish abundance of Eden's original perfection, not to mention its portrayal of both the harmonious and the fabulous, only served to anticipate the colossal cost of their banishment from its glories.

Other illustrations reinforced this picture of the world's first occupants enjoying the glorious surroundings of the Garden of Eden, with its lush vegetation, peaceful river, and tree of life. In 1615, for example, Jan Breughel the elder (1568–1625) placed Adam and Eve in a tropical paradise surrounded by bountiful plant and animal life. In his *Historie of the Perfect-Cursed-Blessed Man* of 1628 Joseph Fletcher showed them strolling peacefully among camels, elephants, and lions. In 1629 John Parkinson, himself an apothecary with an extensive physic garden, used an illustration of a superabundant paradise superintended by Adam and Eve as the title page of his work on plant cultivation in flower gardens, kitchen gardens, and orchards (fig. 2).[2] Such portrayals, of course, were the latest expression of a long-standing convention in Western religious art: Adam and Eve were to be found in stained-glass windows, in large medieval maps known as the *mappaemundi,* and in tempera wall paintings. Later they would feature in works of natural history such as Johann Jakob Scheuchzer's *Sacred Physics* (1731–33), in which the cycle of creation was pictorially depicted using the best available scientific evidence and culminating in "Homo ex Humo"—the creation of man, the "most noble of all creatures, the Microcosm or epitome of all this great World," from the dust of the earth in a world "clothed with trees and shrubs, and ornamented with flowers and fruits."[3] To this Zurich medical practitioner and

FIG. I. Plate from John Ogilby's 1660 Bible illustrating the Garden of Eden at the point of the fall from grace.

firm believer in the Bible, the glories of the creation existed in anticipation of its future human inhabitants.

All of this was entirely in keeping with the image of the unspoiled Eden conjured up in the King James Bible. After all, numbered among those charged with the task of translating the opening chapters of Genesis was John Layfield (d. 1617), chaplain to the earl of Cumberland, explorer and chronicler of the English coming to the New World, who wrote of the forests of Dominica that "the trees doe continually maintain themselves in a greene-good liking."[4] Evidently, the verdant growth of Eden was mirrored in the contemporary Caribbean world. The projection of Eden as a tropical paradise, whether in the East or the newly encountered West, not only delivered visual symbols that fed the romantic imagination but also stimulated enterprises of various sorts to determine its original site and to preserve its pristine environments.[5]

What made this association of Eden with contemporary landscapes, and the quest for its geographical location, all the more plausible was the widespread belief that the earth was only created four thousand years, or thereabouts, before the birth of Christ.[6] Johannes Kepler, for example, calculated

FIG. 2. Title page of John Parkinson's *Paradisi in Sole Paradisus Terrestris* of 1629. This title is a pun on Parkinson's name, "The Park on Earth of Park-in-Sun."

that the date was 3992 B.C., while Martin Luther thought it exactly 4000 B.C.[7] Many other proposals were made, with estimates ranging between three and six thousand years B.C.[8] Most famously, of course, was the 4004 B.C. date of the Irish archbishop James Ussher, whose calculations were based on extensive research on Hebrew genealogy, ancient Middle Eastern manuscripts, Greek marble inscriptions, and records of astronomical occurrences such as eclipses.[9] The results of his painstaking and extensive research were announced in Latin in one paragraph of his two thousand–page *Annals of World History,* published in 1650. The world's birthday was October 23, 4004 B.C., beginning at sunset on the previous evening, to be precise. That date has continued to decorate the margins of some printings of the Bible right up to the present day.

Regardless of how differently the Garden of Eden may have been conceived from ancient times through the medieval period to more recent days, and no matter the differences in computations of the creation date of the earth, the idea that every member of the human race is descended from the biblical Adam has been a standard doctrine in Islamic, Jewish, and Christian thought. In this respect, if in no other, the catechisms of the seventeenth-century Westminster divines can be taken to speak for them all when they declare that "all mankind" descended from Adam "by ordinary generation." People's sense of themselves, their understanding of their place in the divinely ordered scheme of things, their very identity as human beings created in the image of God, thus rested on a conception of human origins that assumed the literal truth of the biblical narrative and traced the varieties of the human race proximately to the three sons of Noah and ultimately to Adam and Eve.

What lent further confirmation to these convictions were those pictorial representations of the world in map form known as the *mappaemundi,* whose purpose was to educate the faithful in significant events in Christian history, not to provide precise locations or routes to specific places.[10] In particular those tripartite maps known as "T-O" maps delivered to their viewers a comfortable cosmology whose cardinal principles dovetailed with traditional biblical understanding. Take as a very early example the celebrated map of Isidore, an encyclopedist and historian of the early Middle Ages who became bishop of Seville around 600 (fig. 3). Here the three known continents—Asia, Africa, and Europe—are clearly displayed on a symbolic surface orientated to the East and therefore to paradise. But tellingly, superimposed on this iconic geographical plane are the names of Sem, Cham, and Jafeth—the three sons of Noah—who stand as the fathers of the human races. It was an

FIG. 3. Isidorian T-O *mappaemundi*.

altogether tidy arrangement integrating a threefold continental schema with a tripartite racial taxonomy; the geographical shape of the world mirrored the Cross, and its anthropological history followed the contours of the Noachian family. Semites, Hamites, and Japhethites inhabited both the physical and intellectual worlds of Christendom's geographical and historical imagination.

For all its seeming clarity and aesthetic tidiness, however, the assumed connection between these interpretations of the Genesis account and human origins has been disrupted routinely in the Christian West by a beguilingly simple notion that has perennially resurfaced in new guises, namely, the suggestion that Adam was *not* the first man and that human beings existed *before* him. This book is about that alternative story and its periodic reincarnations.

Early Glimpses

Just where the idea first emerged that human beings inhabited the world before Adam is impossible to ascertain. While its presence is plainly discernible in the wake of the voyages of reconnaissance and the revival of classical learning in Europe during the Renaissance, fleeting glimpses of the pre-adamites are detectable among earlier writers. These speculations in large measure surface in esoteric philosophical and metaphysical contexts, and their influence on later ways of conceiving of Adam's ancestors is at best marginal. Yet their lingering, subliminal presence did help to keep the idea of alternative world chronologies flickering on the margins of Western thought, albeit in intellectual and social spaces markedly different from those in which the idea later flourished.

The fourth-century Roman emperor Julian the Apostate, for example, on reverting from Christianity to his earlier pagan ways, suggested that the possibility that the human race was descended from a plurality of original couples was more in keeping both with observed variation in human customs and cultures and with non-Christian traditions of learning.[11] At the same time, Gregory of Nyssa, also in the fourth century, is reputed to have thought that

Adam's physical body was derived from animal forebears, an understandable view perhaps in the light of the outlook of his mentor Origen (ca. 185–ca. 254). Condemned as a heretic by the Western Church, Origen had spoken of the infusion of preexistent souls into human bodies and reckoned that Adam only took physical form as a consequence of the fall from grace.[12] As for Gregory's own thinking on anthropology, it emphasized the continuity *and* discontinuity of human beings with the rest of the natural order. In his view, that distinctively human trait, the *rational* soul, was added to what he called the *vegetative* and *sensitive* (or *animated*) souls—metaphysical features of plant and animal orders, respectively—to produce the rational human species, which was a blend of all three soul types. Because he believed everything existed in spermatic potential from the initial divine impulse of creation, Gregory could and did advance a developmentalist account of the origin of life forms by urging that the human body had been created through the inherent operations of the elements of the earth and thus emerged from animal antecedents.[13] This early evolutionary scheme presumed a natural history of humanity, with Adam emerging from pre-adamic forebears.

Whether any of these accounts actually went so far as to imagine the existence of fully human beings prior to Adam is not certain. But they contributed to a lingering sense that the simple clarity of the Genesis creation narrative might well need complicating in one way or the other. In some other cases—such as the medieval Jewish Midrash literature and in Cabalistic writings—hints of the same class of idea are discernible in commentaries that speculated about the possible existence of other worlds, separate from this one but existing prior to and independently of it. More specifically connected with human prehistory was the suggestion that appeared at the end of the first millennium, in a work entitled *Nabatean Agriculture*—a collection of texts purporting to portray the activities and beliefs of some groups of Arabic stock that developed a hydraulic society during the late Hellenistic period. Whatever their source, these documents sought to defend Babylonian culture against Islam and proposed that Adam had come from India, claiming him as the father not of the human race but of an agricultural civilization.[14] This schema was later recorded in *The Guide of the Perplexed,* the major philosophical work of the celebrated Jewish thinker and medical practitioner of the Middle Ages Moses Maimonides (1135–1204). In this treatise, which gave a philosophical rendering of Jewish doctrine and belief, he argued that the cultural practices of the Sabians (his term for the sect of Hellenistic astrolaters) were deviations from monotheism and that their genesis accounts were steeped in legend and mythology even while drawing inspiration from Jewish sources. Maimonides's apologetics are not my quarry here, of course; what signifies is his reporting of Sabian speculations about Adam that at once served to refute the conjecture yet simultaneously to circulate it: "They deem Adam to have been an individual born of male and female like the other human individuals, but they glorify him and say that he was a prophet, the envoy of the moon, who called people to worship the moon, and there are compilations of his on how to cultivate the soil."[15]

No doubt other traces of pre-adamite inclinations could be excavated from additional ransacking of the classical and medieval records. But in the present context any further rehearsal of such hints would run the risk of constructing a past—and thereby assembling a manufactured history—for a suite of later formulations whose development owed little or nothing to those episodic allusions. What is more important is identifying challenges from several other sources that were harder to dismiss and that progressively raised troubling doubts about the intellectual viability of the traditional annals of world history.

Challenges to Convention

Both from within the boundaries of the Old World and, increasingly, from the New World, misgivings about the conventional adamic narrative were taking deeper root. The sources of these disrupting interventions were diverse, but three in particular conspired to destabilize the comfortable cosmos of tradition. First, the increasing availability of what were called pagan chronicles posed a considerable threat to received wisdom about human origins and history. Second, the genealogy of what were known as the monstrous races raised disturbing queries about how such creatures fit into the adamic story of human descent. Third, the encounter with the New World threw into yet sharper relief the growing tensions between world geography and the Mosaic record. For convenience we can examine these challenges under the labels chronology, monstrosity, and geography.

Chronology

Despite the confidence with which Ussher and other chronologists of world history declared on the creation's date of birth, their calculations did not fit at all well with the time frames enshrined in the cultural memories of other peoples.[16] Their histories, routinely dismissed as pagan chronicles, happily took it for granted that the human story could be traced back *many* thousands of years. In particular, the Egyptians and Chinese were evidently comfortable with a date line that was radically incommensurate with the time available from biblical estimates of the origin of the human race. From the Mesopotamian world, for example, there were stories of Sumerian kings ruling for periods as long as thirty thousand years.[17] Such annals presented Christendom's chronologists with one of the greatest moral problems of the time: how, if at all, to fit these archives into the Judeo-Christian calendar. Already by the fourth century, in an effort to cope with these challenges, Augustine was discrediting the authenticity of such records. Indeed, the continuing dispute over chronology was sufficiently strong that he devoted a whole chapter of *The City of God* to "the falseness of the history which allots many thousand years to the world's past" and another chapter to the "mendacious vanity" and "empty presumption" of the Egyptians in claiming "an antiquity of a hundred thousand years" for their accumulated wisdom. Dismissing these accounts as mythical fables was the standard response of both the rabbis and church fathers. Augustine himself rejected them as "full of fabulous and fictitious antiquities." And yet refuting them served to keep these alternative narratives lurking in the shadowy recesses of Western minds. Considering himself "sustained by divine authority in the history of our religion," Augustine himself harbored "no doubt that whatever is opposed to it is most false."[18]

But it was only a matter of time before the relegation of these annals to the status of mere legend would prove unconvincing. Thus, the Huguenot scholar Joseph Justus Scaliger (1540–1609) came to the point of acknowledging that the Egyptian dynasties predated both the biblical flood and the conventional dating of the creation narrative. His encounter with the chronicles of Eusebius of Caesarea (ca. 260–ca. 339) not only instituted a profound shift in the traditional approach to chronology by bringing nonbiblical sources, philological methods of interrogating classical texts, and astronomical computations within the arc of this historical craft but also began to shake the comfortable calendrical authority on which Renaissance cosmology rested. To be sure, Scaliger still held to a world chronology well within traditional approximations;[19] but his belief that the temporal indications of the Hebrew Bible should be interpreted in tandem with Egyptian, Persian, and Babylonian evidence was revolutionary. And Scaliger's *On the Emendation of Chronology* of 1583 was not the only troublesome intervention. The writings of Martino Martini (1614–31), Athanasius Kircher (1601–80), and Isaac Vossius (1618–89) uncovered Chinese and Egyptian evidence that broke through Western time frontiers and posed the profoundest of challenges to biblical chronology.[20] Thus, Martini, for example, insisted that parts of Asia were well populated before the Mosaic flood. There circulated, too, speculations that the pre-Islamic Iranian world could trace its history back to a time before Adam, to an earlier, androgynous progenitor of humankind— Kayumars.[21] In turn these conjectures, compiled by Azar Kayvan in the late sixteenth century, which challenged the hegemony of both biblical and Koranic orthodoxy about human origins, found their way into the scholarship of the distinguished orientalist Sir William Jones (1746–94), who mobilized it in his writings on the Persians to open up new possibilities for understanding racial and linguistic origins.[22]

The potential implications of this temporal breakthrough were vast. Apart from anything else, it raised grave questions about the status of Hebrew as the original and sacred language. Indeed, at the tail end of the seventeenth century a certain John Webb advanced the novel thesis that after the Flood, Noah and his ark landed not on top of Mount Ararat in Armenia but instead in China. As Umberto Eco remarks, Webb argued that "the Chinese language is the purest version of Adamic Hebrew, and only the Chinese, having lived for millennia without suffering foreign invasions, preserved it in its original purity."[23]

These rather specialized, and in some cases obscure, works were supplemented by writings on universal history more generally that laid before their readers the details of alternative world chronologies from other cultures, even if their authors affected to reject them. Montaigne (1533–92), for example, probably drew on Ludovico Vives's commentary on Augustine's *City of God* to observe that Herodotus had learned from Egyptian priests of their multi-thousand-year history; that according to some the world was eternal; and that "*Cicero* and *Diodorus* said in their daies, that the Chaldeans kept a register of foure hundred thousand and odde yeares. *Aristotle, Plinie,* and others, that Zoroaster lived six thousand yeares before *Plato.* And *Plato* saith, that those of the citty of *Sais,* have memories in writing of eight thousand yeares, and that the towne of Athens, was built a thousand yeares before the citty of *Sais.*"[24] To be sure, such rehearsing of ancient cosmologies with their estimated ages of the world could well be with the hope of refuting them, as in the case of Pierre Charron and Thomas Lanquet.[25] As Paolo Rossi pointed out, two basic tactics were available.[26] Either seek ways of reducing all human histories to sacred history by a range of calculation translations between calendars, almanacs, and date-keeping devices; or deny the validity of nonsacred sources and outlaw them as perverse or mendacious. But whatever the motivation, the tradition of chronological inquiry and its tinkering with antiquarian arithmetic fostered misgivings about the standard six thousand–year–old creation at the heart of traditional Judeo-Christian theology. No doubt that in part accounts for the fact, as Colin Kidd has remarked, that the "study of universal chronology became one of the foremost disciplines of the early modern period. It tackled questions of fundamental importance to the identity of Christendom, and it attracted some of Europe's foremost minds."[27]

Monstrosity

If chronology had the capacity to threaten the foundations on which traditional adamic heredity rested, monstrosity was every bit as troublesome to those who took the time to ponder the matter. Indeed, it was not until the mid-sixteenth century, most conspicuously in Vesalius's 1543 treatise on the fabric of the human body, that anatomical diversities began to be recognized not as monstrous but as variations within the sphere of the natural.[28] Meanwhile, portrayals of what were known as the "monstrous races"—humanlike creatures inhabiting the antipodean fringes of some versions of the medieval *mappaemundi* and cataloged in a range of encyclopedic chronicles—stimulated a whole sequence of questions that rotated around their genesis and genealogy (fig. 4). The reproduction of the anthropological imagination of the first-century Roman encyclopedist Pliny the Elder, in such medieval texts as *Marvels of the East* and *Mandeville's Travels,* served to keep these legendary peoples with all their titillating grotesquerie in the Western consciousness.[29] The list of Pliny's wondrous races—the Plinian races, as they came to be known—was considerable. Among them were Amyctyrae (beings with protruding upper or lower lip), Anthropophagi (cannibals who drank from skulls), Artibatirae (those walking on all fours), Blemmyae (creatures with faces on their chests), Cyclopes (one-eyed beings), Cynocephali (dog-headed tribes), Martikhora (human-headed, four-legged peoples), Sciritae (noseless, flat-faced races), Ethiopians, and Pygmies.[30]

Decorating wonder books devoted to the presentation of the marvelous and the extraordinary, which, according to Mary Campbell, incorporated "the most extreme and exquisite projections of European cultural fantasy," these fabulous beings—traceable back at least to Strabo, Pliny, and Herodotus—proved to be "theological nightmares for Christian writers."[31] In the medieval mind monsters resided in both metaphysical and physical space. Metaphysically, monstrosity occupied its own niche in a complicated taxonomy of the wondrous, which encompassed the entire field of emotions, from

FIG. 4. "Monstrous races," in Sebastian Münster, *Cosmographiae Universalis* of 1544.

admiration and awe to horror and revulsion.[32] Physically, they resided on the southern borders of the mappaemundi, sometimes occupying the extremity of Africa, sometimes dwelling in an obscure antipodean fourth continent beyond Africa's southern tip. Such locations served to reinforce the idea that both nature and culture were at their most hideous at the margins and, conversely, that they were at their most reasonable in the temperate north. The marvelous, the monstrous, and the marginal went together. "The medieval writers on topographical wonders," Lorraine Daston and Katharine Park observe, "depicted the margins of the world as a privileged place of novelty, variety, and exuberant natural transgression."[33] Peter Mason concurs, adding that increasing "geographical remoteness" was "coupled to remoteness in terms of dietary practices, sexual customs and cultural faculties."[34] But wherever the monstrous races were placed, they raised troubling questions. As David Woodward put it: "The monstrous races posed a number of problems for the fathers of the church. If they existed—and there was general agreement that they did—were the human? And if they were human, were they descended from Adam and Noah, possessing souls that could be saved?"[35] David Jeffrey agrees, observing that the "question which continually dogs medieval . . . students of the subject is whether or not such creatures actually participate in the human condition."[36] Geographical imagination thus generated profound dilemmas particularly of a theological nature. Campbell rhetorically captures some of the most pressing anxieties: "How . . . could there be men beneath the burning zone, across which the inhabitants of the *orbis terrarum* could not pass to proselytize? Theological rigor necessitated the construction of a geography . . . conforming at whatever cost to the literal and implicit cosmography of Scripture."[37] By various dogmatic devices global human geography was made to fit with doctrinal prescription.

Coping strategies, of course, were not in short supply, not least since a raft of harmonizing tactics had long been deployed to make sense of what were known as monstrous births. After all, the term *monstrum* was equally applicable to anomalous newborn individuals *and* to unusual races.[38] The jurisprudential and spiritual status of portentous births had long been the subject of canonical concern because the question of their baptism—a burning practical issue for parish priests—routinely raised its head. Sometimes their humanity was taken for granted and the rites of the church administered on what might be called the "better-safe-than-sorry" principle. Sometimes, as was the practice under Roman law, they were put to death. Sometimes *monstra* were interpreted as ominous omens presaging civil disruption or cosmic catastrophe. In many such cases standard human physical features were considered to be a precondition of legal status and thus of essential human nature. Whatever the official judgment, analogies with individuals provided strategies for thinking about human racial difference. And in the case of the monstrous races a whole suite of tactics was available, not least to account for their origin and survival.

Explanations for their presence in the world abounded.[39] Animated in turn by horror, revulsion, antipathy, and anxiety, causes were sought in the idea of degeneration. During the medieval period the Plinian races were regularly thought of as having declined—no doubt on account of sin—from a state of prelapsarian perfection. In particular their deterioration was often attributed to a divine curse pronounced on the guilty crimes of their forebears, Cain (who murdered his brother, Abel) or Noah's son Ham (who mocked his father's undignified nakedness). In such schemas there was little room for ideas of improvement. Monstrosity could emphatically not be considered an early stage in social evolution; racial differences were a consequence of decline, *not* progress. Degenerationist speculations were in plentiful supply. Some supposed that the monstrous races had been exiled by God because of their danger to humanity; some thought that they were children of Adam who corrupted their own offspring by eating forbidden plants or herbs; others conjectured that they had emerged as an accursed consequence of the demonic act of murder carried out by Cain, whose violent nature was matched by physical deformity and anatomical degradation; still others, particularly in the rabbinic tradition, declared they were satanic descendants of Eve, who was infected by the serpent's impurity. Of course, it was not necessary to seek explanations as far back in history as the early days of creation. The narrative of Noah and the curse on his son Ham was another favorite port of call for those looking for ways of making sense of monstrosity. Having been condemned to servitude his name was, as in some T-O maps, assigned to Africa and projected as the forefather of black races. In turn his own son Nimrod was portrayed as red-eyed, black-skinned, and misshapen. Either way, whether drawing on the baneful condition of Cain or of Ham, these narratives became tropes for coming to terms with any group viewed with suspicion or distaste or hostility. Sometimes such peoples were considered degraded human beings. Sometimes their humanity was simply denied. The Blemmyae, Cynocephali, and Pygmies were often believed to lack rational souls and therefore occupied a shadowy existence in which they could display human attributes without actually being human.

And yet these disfigured races were not always relegated to outer darkness. Some more positive assessments did surface from time to time. In the thirteenth-century Ebstorf Map, in which the figure of the world is encompassed by the head, hands, and feet of Christ, the monstrous races are gathered within the embrace of his left hand (fig. 5). This was in keeping with a tradition that considered human difference neither accidental nor a failure nor a consequence of transgression but, rather, an expression of a positive divine plan, a plan in which seeming monstrosities acquired aesthetic value inasmuch as they manifested a kind of anthropological plenitude that celebrated the creator's power.

For all that, the existence of such more or less exoticized species routinely raised disturbing questions about the nature and status of Adam's descendants. Augustine was clear on the matter when he elaborated on the theological centrality of Adamic ancestry. "Whoever is born anywhere as a human being," he insisted, "that is, as a rational moral creature, however strange he may appear to our senses in bodily form or colour or motion or utterance, or in any other faculty, part or quality of his nature whatsoever, let no true believer have any doubt that such an individual is descended from the one man who was first created." For Augustine, plainly, descent from Adam constituted a *definition* of humanness, and this meant acknowledging the possibility that certain Plinian races were beyond the bounds of the human race precisely on account of a non-adamic lineage: "Either the written accounts

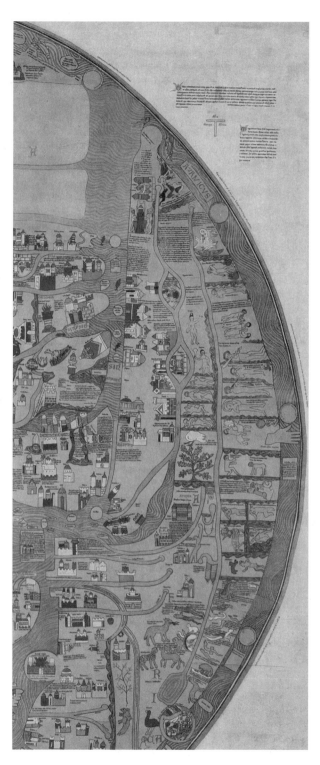

Fig. 5. Detail from the Ebstorf Map showing Christ's hand encompassing the "monstrous races."

of certain races are completely unfounded or, if such races do exist, they are not human; or, if they are human, they are descended from Adam."[40] Because such creatures were not encountered in scripture, anyone who had seen these human races prominently depicted on *mappaemundi* or who had read Pliny on the subject was bound to wonder, as Friedman writes, "if they had descended from Adam, and if so, how they survived the Flood and what should be the attitude of the Christian towards them."[41]

Besides these immediately disconcerting queries, the monstrous races had another role to play. They provided a suite of anthropological templates into which peoples hitherto unknown to Europeans could readily be fitted. And nowhere was this inclination more marked than in Europe's geographical encounter with an entire new continent—America. Ideas of racial monstrosity delivered an imagined geography of the Americas in which ideas of an earthly paradise, for which travelers had long been searching, were conflated with the prodigious exotics laid out in works of wondrous travel. In a profoundly important sense the features of the New World were a projection of the hopes and fears of the Old World. America, in this sense, was imagined before it was discovered. And the result, as Mary Campbell tellingly puts it, was the production of "a Caribbean that belongs as much to the Other World of medieval geographic fantasy as it does to the map Columbus helped realize."[42] It is to the significance of that moment of geographical rendezvous, with all its destabilizing connotations, that we now turn.

Geography

Europe's early understanding of America was as much a projection of the geographical imaginations of the Old World as a topographic description of the New. America, in many ways, was the invention of Europe. Already armed with cultural and physiological taxonomies into which racial "others" and their traditions had long been inserted, it is hardly surprising that European templates of paganism and monstrosity were transferred across the Atlantic. Indeed, it has even been suggested that depicting native Americans as monstrous predated Columbus's arrival in 1492, perhaps deriving from earlier Viking excursions or from South American Indian artwork itself.[43]

Yet whatever the source, and however much the shock of discovery overspilled any conventional categories that were brought to bear on it, the concept of the marvelous (which sometimes, but not always, incorporated the monstrous) provided a lens through which New World peoples could be inspected, one way or another.[44] Columbus, for example, reportedly had in

ADAM'S ANCESTORS

his possession an annotated copy of Pliny's *Natural History* and explicitly informed Luis de Santangel that he had encountered "no human monstrosities, as many had expected," in the islands, though he did consider that there were in existence cannibals—Anthropophagi—as well as men with tails.[45] In his account of *The Discovery of the Large, Rich and Beautiful Empire of Guiana* (1596) Sir Walter Raleigh recorded that he had received reports of "a nation of people, whose heades appeare not aboue their shoulders" and with "mouths in the middle of their breasts."[46] The Blemmyae, as such creatures were known, were illustrated for Raleigh's account by Hondius and simply took their place alongside other representations of New World peoples. The self-same engraving decorated the work of the early-eighteenth-century French Jesuit missionary Joseph-François Lafitau, which compared the customs of "American savages" with the peoples of antiquity (fig. 6).[47]

On various maps and atlases, too, similar depictions surfaced. Among the marginalia of Johann Huttich and Simon Grynaeus's world map of 1537 (which originally appeared in a volume introduced by Sebastian Münster entitled *Novus Orbis*) were portrayals of native Americans engaged in the butchering of human prey, with a second victim arriving slung over a horse. In his famous double-cordiform (heart-shaped) world projection of 1538 Gerard Mercator had "Canibales antropophagi" inscribed across Brazil. And the frontispiece of Abraham Ortelius's *Theatrum Orbis Terrarum* of 1570 used the image of a savage armed with club in one hand and a severed head in the other to symbolize America.[48] In the pictorial representations of sixteenth-century cosmographers such as André Thevet, legendary peoples (like the Amazons) simply occupied the same spaces as newly encountered races, sometimes on the presumption that the American present could reveal much about the European past. As Frank Lestringant puts it: "the American ethnographic treatise could appear . . . to be a manual of European archaeology."[49] Thus, in 1590 Theodore de Bry presented an illustration of a female warrior of the ancient Picts in the first of his multi-volume work, *America,* in order to "demonstrate that the inhabitants of Britain had been no less forest dwellers than these Virginians." Similarly, Jean Bodin thought of primitive Europeans as just that, primitive, and considered that the Bible was only concerned with "the origins of that people whom God alone chose . . . not of the others."[50] In all these cases mapping was an inherently moral project, a rhetorical device to mark out the terrain between savagery and civilization, between otherness and familiarity.

Of course, there were significant differences between the monstrous races from the Old World's margins and those that surfaced in the imagined spaces

FIG. 6. Blemmyae in America, depicted in Joseph-François Lafitau, *Moeurs des sauvages américuains comparées aux moeurs des premiers temps.*

of the New World, and they have been scrutinized by scholars often with an eye to psychoanalytic interpretations. Such analyses, however, are not my concern here. My rehearsal of these modes of depiction is to suggest, rather, why, given the molds into which native American peoples were squeezed, it is not surprising that at least for a time tactics comparable to those used for explaining alternative chronology and medieval monstrosity were deployed to fit America into the European psyche. Resorting to the language of "paganism" had made it possible to dismiss the claims emanating from ancient chronologies and to devise strategies for coping with human monstrosity. And this lexicon provided some with grounds for hope that the apologetic maneuvers that had served the Church fathers in the late Hellenistic world would work just as well for the contemporary heathen that the voyages of geographical reconnaissance were currently exposing. The label "pagan," in other words, could tie together familiar features of Old World historical and anthropological conjecture with the exotic speculative geographies of the New.[51] As Anthony Grafton writes of attempts to make sense of the customs and practices of the peoples encountered by traveling Europeans: "If Indians cherished myths of a great flood or worshipped a single God, for example, they did so for a reason as evident as it was orthodox: the devil, that brilliant dissimulator, had brought them within his spell by teaching them a parodic Black Sacred History, one modelled on the real thing but inverted, like the magicians' Black Mass."[52] But such tactics seemed increasingly shallow, and questions multiplied, as there was a dawning realization that the new America was not joined to Asia. Were the inhabitants of America the descendants of Adam and Noah at all? If so, how did they find their way to the other side of the world? Did they experience a separate fall from grace? Were they encompassed within the scheme of redemption? How did they fit into the racial formula of descent from Shem, Ham, and Japheth? When Pope Julius II decreed in 1512 that the Indians were descended from Adam, it might have seemed to settle the impassioned rhetorical questions posed by the Dominican missionary in Hispaniola Antonio de Montesinos the previous year: "Are these Indians not men? Do they not have rational souls?"[53] Nothing could be further from the truth.

In the half-century or so after Columbus's venture, the debate on the nature and status of the American peoples was engaged and nowhere so conspicuously as in the papal junta at Valladolid, Spain, in 1550, when the doyen of Spanish Aristotelian scholarship, the humanist Juan Ginés de Sepúlveda, and the former Dominican vicar of Guatemala and now bishop of Chiapa, Bartolomé de Las Casas, vigorously disputed the subject of how the American Indians should be treated by Europeans. The papal legate had been dispatched from Rome to Spain to determine once and for all whether the Indians shared the *imago Dei* or were a distinctly other species, whether they were fundamentally bestial and fitted only for slavery or sufficiently advanced that they should not be considered barbarians. The implications were potentially far-reaching, for on its adjudication hung the answer to the question whether waging war on the Indians as a means of civilizing and Christianizing them was justified. So symbolic has this episode become, as a critical moment in the whole history of human rights and the moral economy of imperialism, that it was dramatized for French television in 1992 and staged to critical acclaim in New York's Public Theater in 2005 under the title *The Controversy of Valladolid*. Sepúlveda's proposal was that Aristotle's notion of natural slaves provided the best strategy for dealing with the Indians. Because he considered that they were bereft of rationality, practiced human sacrifice and other barbarous customs, and engaged in degraded idol worship, Sepúlveda felt entirely justified in resorting to the Aristotelian formula that such peoples were destined by nature to be the slaves of superior masters—a proposal that had already been advocated in 1510 by the Scottish professor in Paris John Major. It was both lawful and expedient to bring them forcibly under Spanish rule as a prelude to preaching to them. Las Casas found this whole line of argument utterly obnoxious, and he set about redeeming them from Sepúlveda's bestializing diagnosis (fig. 7). His tactic, articulated in several publications throughout his life, was to dwell on their accomplishments, to insist on their demonstrated rationality, and to compare them and their legal institutions favorably with both ancient and modern "civilizations."[54] He thus insisted that the Creator "has not so despised these peoples of the New World that he willed them to lack reason and made them like brute animals, so that they should be called barbarians, savage, wild men, and brutes as they [i.e., the *Sepúlvedistas*] think or imagine."[55] So, while Sepúlveda considered that Indian cities resembled beehives—the products of mere natural instinct—and their states exhibited extreme simplicity, Las Casas concluded that native Americans "in fact fulfilled every one of Aristotle's requirements for the good life."[56]

At stake in the Valladolid dispute was the moral and judicial status of New World peoples, whether they were truly human, half-human and half-animal, or creatures disguised as humans; whether they were descended from Adam or were the offspring of the devil. Although the tribunal failed to come to any formal verdict, the hearing nonetheless highlighted several things: the lingering influence of Aristotle, whose benediction was sought by interlocutors on both sides of the argument; the political, legal, and moral significance of questions about human origins; and the stresses and strains to which the conventional adamic narrative was progressively subjected.

In the wake of Columbus's transatlantic adventure, genetic theories of native American origins—more or less implausible—abounded. To be sure, the overriding goal was to ensure that the existence of America did not subvert the authority of the Judeo-Christian scriptures. Yet within that "common scaffolding of assumptions," as Grafton terms it, scholars found it possible to "erect wildly different structures."[57] Some speculated that Noah's maritime skills were sufficient to allow for the possibility that America could have been reached by sea in ancient times; some were convinced that the native peoples of America were the ten lost tribes of Israel. Others, notably the Spanish Jesuit José de Acosta in the late sixteenth century, desperately intent on preserving the integrity of scripture, urged that it was possible to cross

ADAM'S ANCESTORS

D. FR. BARTHOLOME DE LAS CASAS
Del Orden de Predicadores, Obispo de Chiapa
Varon apostolico, y el mas zeloso de la felicidad
de los Indios.
Nació en Sevilla el año de 1474, y murio en Ma[...]
el de 1566.

FIG. 7. Bartolomé de Las Casas.

into the New World from the northern wastes of Asia. Still others, in particular Hugo Grotius in 1643, held to the view that the Vikings had colonized America and called upon philological evidence to support his theory, a view that was vigorously challenged by various writers, including Georg Horn.[58]

The combined effects of chronology, monstrosity, and global human geography spawned a frenzy of conjecture, including the heterodox possibility that the earth might be partly inhabited by non-adamic humans, peoples, and cultures who did *not* trace their ancestry to the Edenic couple of the Pentateuchal record. Thus, for all the heroic efforts to encompass the reality of America within the compass of biblical traditionalism, the New World's anthropological and geographical threats to conventional history were sufficiently powerful to encourage some to modify the standard chronicles of creation and, in particular, to speculate that world history predated the adamic narrative. Later, in the mid-seventeenth century, Matthew Hale would concede that the central issue around which debates about human origins rotated was the American Indian. The very fact of their existence, he noted, "hath occasioned some difficulty and dispute touching the Traduction of all Mankind from the two common Parents supposed of all Mankind, namely *Adam* and *Eve*."[59] Nonetheless those flirting with such speculations were soon accused of keeping schools of atheism, peddling skepticism, and harboring heretics. The oriental scholar Jacob Palaeologus, a resident of Prague, was reportedly executed in 1585, for example, for holding to the heresy that because all people were not descended from Adam and Eve, the inheritance of original sin was not universal.[60]

Such was the case with Sir Walter Raleigh (1554–1618) and Thomas Harriot (1560–1621). Raleigh's theorizing in fact was always conducted within the confines of scriptural authority, but his computational strategy for coping with evidence of chronologies of greater antiquity than the biblical record was to seek for the greatest amount of time that the Hebrew text would allow. Indeed, in *Historie of the World,* of 1614, Raleigh presented as the frontispiece a globe displaying Adam and Eve in paradise and later tabulated the Noachian root of every conceivable nation and race. Yet he still admired Egyptian sources and from 1592 onward was often charged with irreligion. As for Harriot, it was his experience of exploration in Virginia that, together with his work on biblical chronology, raised questions about the origin of the Indians. Perhaps it was for these reasons that both Raleigh and Harriot, and indeed Christopher Marlowe, were branded with holding to the heresy that supposed humans existed before the biblical Adam and belonging to a circle of atheists that impiously and impudently persisted in affirming that

American Indian archaeology gave evidence of artifacts that predated Adam by thousands of years.[61] Indeed, for chronologists of world history, not least those amassing data from worldwide geographical exploration, the greatest moral problem according to the Church of England clergyman and fellow of Christ Church Oxford John Dove (1561–1618) was that such annals seemed to confirm the speculations of those infidels who claimed the existence of "genealogies more ancient than Adam."[62]

To be sure, direct evidence of Raleigh's and Harriot's alleged acceptance of the idea that human beings existed before Adam is lacking, but any inklings in that direction would have been reinforced by the writings of Renaissance scholars such as Paracelsus (1493–1541) and Giordano Bruno (1548–1600). Paracelsus, for example, propelled by the sheer presence of newly discovered races, had inclinations toward polygenism, even though he struggled to keep his thoughts within the confines of the biblical account of creation. It was, he confessed, difficult to believe that the inhabitants of the "hidden islands" were descended from Adam, and while he was convinced that they had no souls, and therefore were not fully human, he suggested that "these people are from a different Adam." As he further explained: "the children of Adam did not inhabit the whole world. That is why some hidden countries have not been populated by Adam's children, but through another creature, created like men outside of Adam's creation. For God did not intend to leave them empty, but had populated the miraculously hidden countries with other men." In sum: "It cannot be believed that such newly found people in the islands are of Adam's blood." As for Bruno, he was convinced that the Ethiopians, the American Indians, Pygmies, and various species of giants and troglodytes "cannot be traced to the same descent, nor are they sprung from the generative force of a single progenitor." Rather, their origins could be tracked to one of the three sources that constituted his tripartite cosmological scheme:

> The regions of the heavens are three; three of air; the water
> Is divided into three; the earth is divided into three parts.
> And the three races had three Patriarchs,
> When mother Earth produced animals, first
> Enoch, Leviathan, and the third of which is Adam;
> According to the belief of most of the Jews,
> From whom alone was descended the sacred race.[63]

In Bruno's case such inclinations were all of a piece with his belief in the plurality of worlds, that is, the existence of other planets with inhabitants evidently not traceable to Adam.[64] In his 1584 treatise *De l'Infinito, Universo e Mondi* ("On the Infinite Universe and Worlds") he argued that the universe was infinite and that numerous other worlds existed in which non-adamic humans practiced their own religions. And indeed this heresy was among the key charges brought against him, which resulted in his being burned at the stake in 1600.[65] Others nevertheless shared the same viewpoint. The Dominican philosopher Tommaso Campanella, for example, observed in his 1622 *Defence of Galileo,* "If the inhabitants which may be in other stars are men, they did not originate from Adam and are not infected by his sin."[66] The parallel here advertised between distant space and distant time was one that was to recur frequently in subsequent discussions about Adam's ancestors.

In one way or another, then, the advance of global geographical knowledge rendered troublesome the conventional chronicles on which Christendom had long based its identity. The sheer presence of America, with its own peoples and cultures, the existence of physical artifacts such as the Aztec Calendar Stone, and the uncovering of chronological traditions with time scales hugely incompatible with biblical genealogy, all prompted a number of thinkers to flirt with the suspicion that all races might not be descended from the one biblical Adam. Such encounters cast doubt on Old World assumptions and began to subvert the authority hitherto resident in those ancient texts that had long been the blueprint for Europe's moral architecture.

ℬ And yet, for all these hints at alternative European readings of human origins, they remain largely just that: mere hints, fleeting glimpses, prevenient traces of a monumental heresy still to find full voice. The idea that Adam might not be the progenitor of the entire human race and that there might be non-adamic peoples in existence found expression in print by only a handful of writers. Because of the dangers associated with such speculations, open advocacy was exceptional. Of course, the snippets that I have identified in this chapter could doubtlessly be supplemented by others. William Poole, culling the archival legacy of the early seventeenth century in England, for instance, has succeeded in gathering together a number of scattered comments indicating doubts about the literal truth of the Mosaic record's presentation of the adamic universe. He calls to our attention, for example, the cases of Laurence Clarkson and Gerard Winstanley, members of radical Protestant sects, who gave voice to such views around 1650, though they remained substantially undeveloped.[67] In instances such as these, motivation sprang in large part from the biblical text itself, which gave indications that Adam's family dwelled in a world inhabited by other people. When Cain was banished from Eden, it was noted, he built a city in the land of Nod—an enterprise that presumed the existence of other human beings.

Such speculations were certainly rare given their heterodoxy. And yet the cracks that were appearing in the authority hitherto resident in ancient canonical texts were now opening. Recently disclosed chronologies, the realization of greater and greater human diversity, and new geographical realities all conspired to disrupt the traditional adamic genealogy. The bursting on the scene of a new account of the human story in the mid-seventeenth century, unchained from the restrictions seemingly imposed by descent from Adam, now commands our attention.

2 ❧ HERESY

Isaac La Peyrère and the
Pre-Adamite Scandal

I N FEBRUARY 1656 thirty armed men reportedly burst into the Brussels lodgings of an emissary of the prince of Condé and hauled him off to prison. Four months and multiple interrogations later, the prisoner agreed to be escorted to Rome in order to recant of his heresies and be received into the Catholic Church. On January 17 the following year Isaac La Peyrère—a Calvinist of Portuguese Jewish origins from Bordeaux—arrived at the Vatican to meet Pope Alexander VII, who reportedly welcomed "this man who is before Adam" and roared with laughter while they read together La Peyrère's monumentally heretical treatise, *Prae-Adamitae* (*Men before Adam*).[1] It was on Christmas Day just over a year earlier, 1655, that the book had been denounced by the bishop of Namur, just a month after its condemnation by the president and council of Holland and Zeeland, and since then its author had received nothing but critical censure. Within a year of the book's publication more than a dozen refutations had appeared.[2] Everyone, it seems, hated it.

Born in Bordeaux in 1596, or thereabouts, into a wealthy and influential Protestant family, La Peyrère had already been embroiled in ecclesiastical controversy. At the age of thirty, he was accused of atheism by one of the local synods of the French Reformed Church, though the charges were dropped. Ever since then La Peyrère, a qualified lawyer, had been in the service of the Condé family, enjoying its patronage as their fortunes permitted, and had also been part of a French diplomatic mission to various European countries. Even though La Peyrère had been tinkering with his pre-adamite heresy for fifteen years or more, so long in fact that the circulated version of the manuscript had called forth an extended critique from Hugo Grotius in 1643, he had already established himself as an expert on Greenland (and later Iceland) before *Prae-Adamitae* made its formal appearance. Some reflections on these inquiries into the geography of the northern regions and their significance for La Peyrère's enduring skepticism is a suitable point of departure, not least on account of their connection with the whole issue of the peopling of the Americas.

Northern Geography and the Path to Skepticism

On June 18, 1646, La Peyrère put the finishing touches to a letter to François La Mothe le Vayer, a skeptical French antirationalist and, like La Peyrère himself, a member of a circle of intellectuals enjoying the patronage of the prince of Condé.[3] This was the second such letter La Peyrère had sent to the same individual. The first one, an account of Iceland, had been completed in 1644, but its contents were not published until 1663. The present treatise, on Greenland, appeared in Paris in 1647.[4] Taken together, these two volumes established their author's reputation as a leading authority on the northern regions until well into the nineteenth century.

This account of Greenland, while providing a regional description of the country's features, equally approached a number of the larger issues that absorbed La Peyrère's consciousness. By examining them, we can catch a glimpse of the early complicity of geographical inquiry in the radical project in which he was embroiled and thus in what Richard Popkin has called "the high road to Pyrrhonism"—the sixteenth- and seventeenth-century revival, courtesy of Pierre Bayle, of the ancient skeptical philosophy of Pyrrho of Elis.[5] The map of Greenland that accompanied La Peyrère's text is a useful place to begin because it gave geographical voice to some crucial ingredients in the grand profanity that La Peyrère was sculpting over the years (fig. 8). Having sought the advice of Gabriel Naudé, medical practitioner, librarian, and scholar, and the moderately skeptical anti-Aristotelian philosopher Pierre Gassendi, in preparing his text, he called upon the expertise of the mathematician Gilles Roberval and Nicolas Sanson d'Abbeville, geographer to the king of France, for cartographic assistance.[6] The map revealed La Peyrère's uncertainties about the eastern and western reaches of the subcontinent and depicted the western coastline merging with modern-day Baffin

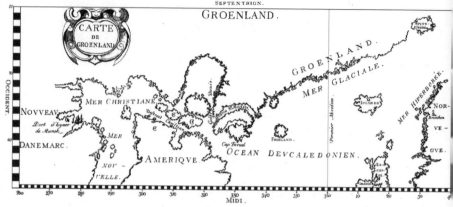

FIG. 8. Map of Greenland from Isaac La Peyrère's *Relation du Groenland.*

Island; Cap Farvel was shown crucially separated by a stretch of ocean from Newfoundland. Thereby he expressed his disagreement with those "who think that Greenland is part of the American continent" and sided with the narrative of "a Danish captain named John Munck, who tried this passage to the East by the north-west of the Gulf of Davis, and according to what he says there is great probability that this land is entirely separated from America."[7] Jens Munck had led an expedition of about thirty men in 1619 in two ships into Hudson Bay. But the mission had to be aborted because, by June of the following year, all but three of the crew had died of scurvy, though the surviving remnant managed to make it back to Denmark.[8]

La Peyrère's cartographic ventures, which showed how Greenland was not geographically joined to the North American continent, were integral to his account of the settlement of the region and disclose something of the connections between geographical investigation and Enlightenment skepticism. After all, as Paul Hazard, reflecting on the meaning of travel at the time, once quipped: geographical exploration meant "comparing manners and customs, rules of life, philosophies, religions; arriving at some notion of the relative; discussing; doubting. Among those who wandered up and down the earth in order to bring the tidings of the great unknown, there was more than one free thinker."[9] Armchair geographers of La Peyrère's stripe were no less subversive. In fact, as he disclosed in correspondence during the mid-1640s with Ole Worm,[10] friend, Danish antiquarian scholar, and gifted physician and polymath at the University of Copenhagen, whom he had met during his time in Scandinavia, La Peyrère's whole project was rooted in his passion to find a persuasive account of the origins of native American peo-

ADAM'S ANCESTORS

ples. And Worm's archaeological data provided La Peyrère with added scholarly authority.[11]

In developing his argument, La Peyrère gathered together information from both ancient and contemporary sources. Two major works in particular provided him with historical data, one of old Icelandic origins, the other modern Danish. From a text described as the "Icelandic Chronicle" (possibly from the early-thirteenth-century survey of an unknown writer named Morkinskinna), La Peyrère learned of the Skreglinguer people who inhabited the western flank of Greenland, and he reported that "Doctor Vormius [the Ole Worm just mentioned], the most learned of all the doctors in northern researches . . . says they were the original savages of Greenland, to whom this name was probably given by the Norwegians." This was a critical intervention: it meant that the native peoples of Greenland were not descended from northern Europeans. What, then, was their origin? Two possibilities suggested themselves. Either they were of American derivation, or they were aboriginal to Greenland itself. La Peyrère believed "there was no need of bringing Americans here at all," but, either way, they predated the advent of the Norwegians, and La Peyrère further suggested that "by the same reasons that Vestrebug had its original inhabitants when the Norwegians arrived there, Ostrebug had them also, and that as the eastern part was nearer the Arctic Sea, was not so fertile, and consequently less inhabited than the west, the Norwegians, who met with less resistance on that side than on the other, took possession more easily of Ostrebug than of Vestrebug." This scenario found corroboration in the *Danish Chronicle* (probably the 1514 work of Christiern Pedersen, who republished Saxo Grammaticus's twelfth-century history of the Danish kings), which, La Peyrère reported, confirmed that Greenland "is inhabited by a variety of races, and that these races are governed by different lords, of whom the Norwegians never knew anything."[12]

In passing, it is worth pausing to note something of the anthropological particulars that La Peyrère conveyed in his regional picture. Indeed, the "richness of detail" that Peter Burke has discerned in this work has suggested to him that "La Peyrère had become interested in this people for their own sake" and not just as a cog in the wheels of his pre-adamite device.[13] Drawing again on the *Danish Chronicle,* he depicted Greenland's indigenous peoples as "savages," "deceitful and ferocious," and incapable of being "tamed, either by present or kindness." Additionally: "They are fat but active, and their skins are of an olive colour; it is believed that there are black among them like Ethiopians . . . The shirts of the men and the chemises of the women are made of the intestines of fish, sewn with very fine sinews. Their clothes are large, and they bind them with straps of prepared skin. They are very dirty and filthy."[14]

Yet for all that, ethnographic portraiture remained secondary to the overriding concern that animated La Peyrère's overall mission: "the question," as Popkin puts it, "of whether the Bible is adequate as an account of how the world developed was challenged both geographically and anthropologically by what was then known about the Americas and about the far north . . . If Eskimos were found in Greenland by the Viking explorers, where did they come from?"[15]

Thus, in the concluding pages of his Greenland excursus, La Peyrère took the opportunity of tackling the views of the recently deceased Dutch jurist and Swedish ambassador Hugo Grotius[16]—as recorded in a treatise by the historian and geographer George Horn, who was himself deeply critical of Grotius's Viking thesis and worked within the temporal economy of the biblical world picture.[17] Grotius, as we have noted, had managed to get his hands on a manuscript version of La Peyrère's as yet unpublished work on the preadamites and issued a scathing refutation back in 1643. Here he claimed that the native peoples of America were of Norwegian descent, and La Peyrère fastened on this assertion to deliver a biting, if belated, retaliation. His rhetorical riposte warrants quoting at length:

> I discover at the same time the errors of the person who has written dissertations upon the origin of the people of America, whom he makes out to have come from Greenland, and makes the first inhabitants of Greenland to have come from Norway. . . . You will judge, sir, by the continuance and the reasoning of my history, that this author errs in every way. First, inasmuch as the Norwegians were not the first inhabitants of Greenland, as it appears from his narrations and the demonstrations I have given you of them; and inasmuch as that M. Vormius, who is very learned in the antiquities of the north, so far from connecting the origin of the people of America with the people of Greenland, thinks that the Skreglingres, the original inhabitants of Vestrebug in Greenland, came from America. Secondly, he is mistaken, inasmuch as there is little or no probability that Greenland was part of the continent of America, and that the passage from the one to the other was not so well known nor so possible as is imagined. Thirdly, he is mistaken in that which I have shown you, that there is no affinity of language or manners between Greenland and Norway; and if, as he says, the Norwegians communicated their language and manners to the Americans, they must have gone elsewhere than by Greenland to get to America. I should here have a good opportunity

of showing up other errors of this dissertation, of making the author eat his own words, and of sending him to the land of visions and dreams; but as he now sleeps his last sleep, we will let him rest in quietness.[18]

La Peyrère's *Relation du Groenland,* then, was a sustained effort to deploy physical, demographic, linguistic, and cultural evidence, to sustain his own suspicions about the adequacy of the traditional story of the development and migration of the human races in favor of one allowing for the possibility of plural origins. The earlier, and very substantially shorter, treatment of Iceland, while much less developed, nonetheless advertised similar concerns. Ethnographically, the Icelanders fared rather better than the inhabitants of Greenland in La Peyrère's geographical imaginary. Even though they inhabited the same frigid reaches, where "it may rationally be supposed, that a Nation living so near the North-Pole, may not be so Refined and Polished as some others," he deemed it "possible the *Iselanders* are not so barbarous as formerly."[19] Indeed, he considered that they were "very strong and courageous" and "had also a sufficient share of Wit, and were so curious in their Annals, that they not only carefully preserved their own History from Oblivion; but also, embellished the same with the most memorable Transactions, that happened in their Neighbouring Kingdoms."[20] These records, alongside the writings of figures such as Dithmarus Blefkenius, who had published in 1607 a work entitled *Islandia,*[21] and Jonasen Arngrim (sometimes referred to, as with La Peyrère, as Angrim Jonas), who had studied with Tycho Brahe and wrote several works of Icelandic history in Latin during the early seventeenth century, drawing largely on Blefkenius, provided La Peyrère with information for his account. Issues of settlement chronology were never far from his thoughts, and he paused to attack those using certain classes of philological data to elucidate the genealogy of human societies. "Many Errors of this Nature," he insisted, "are to be met with in the Writings of most of the best Authors, who have run upon the same Mistake, in looking for the true Origin of Nations among the Interpretation or Etymology of certain *German* or *Hebrew* Words, which to them seem'd to have a near relation to the Language of those Nations they were treating of." Itemizing in particular the efforts of "*Mr. Grotius,* [who] in his Treatise of the Origin of the *Americans,* deduces their Race from the *Germans,* because, says he, many of their Words terminate in *Lan, Land* being a *German* Word," La Peyrère dismissively judged that "nothing can be more fallacious than Conjectures founded upon such-like Etymologies."[22]

The Peyrèrean Formula

If the skeptical thrust of La Peyrère's northern geographies remained to a degree implicit, there was no mistaking its features in the superlative sacrilege he published in 1655, *Prae-Adamitae,* or *Men before Adam.* As we have already noted, a version of this work had been circulating in one form or another at least since 1641, when its banning by Cardinal Richelieu, to whom it had been dedicated, merely stimulated demand for it.[23] Somehow it got into the hands of Marin Mersenne's circle and was picked up by Hugo Grotius, to whose critique we will presently turn. Mersenne himself was initially enthusiastic, judging that the work cast new light on some passages of scripture. Over the following couple of years La Peyrère tinkered with the text in several ways and brought out one section of the work in 1643, *Du rappel des Juifs,* which dwelt on his ideas about a coming Jewish Messiah. Later, during the second half of the 1640s, while he was working on his Greenland and Iceland projects, he corresponded a good deal with Ole Worm, who expressed much interest in the pre-adamite theory. So once again he returned to his pet speculation, revising and reworking it in the light of conversations with a range of interlocutors. And then in 1654, after various travels as part of the

prince of Condé's entourage, he came into contact with the recently abdicated Queen Christina of Sweden, who showed much interest in his ideas. A patron of the arts and literary culture and recent convert to Catholicism, she encouraged him to make the theory public by having it published in Amsterdam. He took her advice, and the work came out anonymously in several editions, three by its Dutch publisher Elzevier and at least one in Basel. But anonymity counted for nothing; there was no doubting who the author was. The work consisted of two parts, both in Latin (52 and 260 pages, respectively), which sometimes appeared together, sometimes separately. The second part came out almost

FIG. 9. Title page of the second part of Isaac La Peyrère's pre-adamite work, originally published anonymously in Latin in 1655.

immediately in an English translation in 1655 under the title *A Theological System upon the Presupposition That Men Were before Adam;* the following year the shorter first part, *Men before Adam,* made its appearance. Sometimes both sections were bound into a single volume, sometimes not (fig. 9). Although efforts were made to suppress its English appearance, the printer Francis Leach went ahead with it nonetheless.[24] Just whose work the translation was remains unknown, though it has been suggested that the identity of "Whitford, gent" was David Whitford, author, clergyman, and sometime Royalist soldier.

Whatever the intricacies of its textual history, the work's central coordinating principle was beguilingly simple: human beings existed before the biblical Adam.[25] This meant that the human species had plural origins; La Peyrère was advocating polygenism. At one level, the whole enterprise was an exercise in biblical hermeneutics. As the work's subtitle made clear—"a Discourse upon the Twelfth, Thirteenth, and Fourteenth Verses of the Fifth Chapter of the Epistle of the Apostle Paul to the Romans. By Which Are Prov'd, That Men Were Created before Adam"—issues rotating around *internal* matters of biblical exegesis were of pressing concern. Here the focal point was how to interpret Saint Paul's words: "Until the law, sin was in the world; but sin was not imputed, when the law was not." La Peyrère found standard interpretations that took this text as referring to the Mosaic law unconvincing, and he seized on the hermeneutic ambiguity of the apostle's declaration to argue that the "law" referred to was not the Mosaic Law but, rather, legislative regulations given to the primeval Adam. La Peyrère thereby voiced his conviction that ceremonial Judaism could be traced back beyond Moses to the Garden of Eden and thus to Adam himself. As he put it: "Long before Moses there were other Ordinances prescribed, and commended to the Jews, other Ceremonies instituted, other Laws of God decreed and confirmed for that holy and elected People. And in this place, I mean the Jews, not onely [*sic*] the Sons of Abraham who are called the Seed of Abraham, but also the fore-fathers of Abraham, the Posterity of Adam." The consequences of this interpretative move were immediate and far-reaching: there must have been human beings on earth before Adam. As he summarized his thesis: "For that law was either to be understood of the Law given to Moses or of the law given to *Adam* . . . if that law were understood of the law given to *Adam,* it must be held that sin was in the world before *Adam* and until *Adam* but that sin was not imputed before *Adam;* Therefore other men were to be allowed before *Adam* who had indeed sinn'd, but without imputation; because before the law sins wer [*sic*] not imputed."[26]

By this neat piece of exegetical reshuffling—a "fresh if rather naïve excursion into biblical exegesis," as Anthony Grafton puts it[27]—a range of irritating inconsistencies in the Genesis record could be conjured out of existence. Here was a ready-made explanation for Cain's fear, after his banishment from the Garden of Eden, that he would encounter hostile individuals seeking to kill him; it delivered a population to inhabit the city he built; it provided a possible answer to the question about where his wife came from. On the standard account there simply were no other people beyond the adamic family to make sense of these details. But now there was a simple answer: preadamites. As La Peyrère himself explained, ever since childhood he had been perplexed by these niggles but had only found resolution when he pondered the fifth chapter of Paul's epistle to the Romans. As he colorfully put it:

> And as he who goes upon Ice, goes warily where he cracks it, being not well frozen, or tender; but where he finds it frozen, and well hardened, walks boldly: So I dreaded first, lest this doubtfull dispute might either cut my soles, or throw me headlong into some deep Heresie, if I should insist upon it; But so soon as I knew by these verses of the Apostle, that sin was in the world before it was imputed; and when I knew, and that certainly, that sin began from *Adam* to be imputed, I took heart, and found all this dispute so solid, that I pass'd through it with lesse fear.[28]

Internal textual issues, then, were the ostensible mainsprings of La Peyrère's intervention, but external historical and geographical data were, if anything, even more decisive. There was, for instance, the pressure exerted by those pagan histories. Using material akin to that for his Greenland dissertation, he culled non-Western chronicles, genealogies, and Renaissance travel books in search of supporting testimony, relying in particular on the recent scholarly work of figures such as Claude Saumaise and Joseph Scaliger. *Prae-Adamitae* was thus replete with details of Egyptian, Greek, Babylonian, and Chinese histories and how they challenged Christian chronology as conventionally understood. La Peyrère's testy impatience with conventional writers who dismissed these records called forth biting commentary: "But as Geographers use [*sic*] to place Seas upon that place of the Globe which they know not: so Chronologers, who are near of kin to them, use to blot out age pasts, which they know not. They drown those Countries which they know not: these with cruel pen kill the times they heard not of, and deny that which they know not."[29]

Fundamental, too, was the evidence he derived from the voyages of reconnaissance and the revelations they delivered on distant cultures and their indigenous traditions. His purpose in amassing such data, of course, was to reveal how elegantly his pre-adamite theory could absorb them and thereby integrate faith and reason. When read in its light, La Peyrère insisted, "the History of *Genesis* appears much clearer and agrees with itself. And it is wonderfully reconciled with all prophane Records whether ancient or new, to wit, those of the *Caldeans, Egyptians, Scythians,* and *Chinensians;* that most ancient Creation which is set down in the first of Genesis is reconciled to those of *Mexico,* not long ago discovered by *Columbus;* It is likewise reconciled to those Northern and Southern nations which are not known, All whom, as likewise those of the first and most ancient creation were, it is probable, created with the Earth it self in all parts thereof, and not propagated from *Adam.*"[30]

Taken together, then, the evidence of old chronological records and new geographical findings, combined with internal biblical exegesis, played a significant role in the move toward bringing the apparatus of textual criticism to bear on the biblical documents. Extrabiblical data just had to be accorded their due role in scriptural hermeneutics, and such considerations led La Peyrère to a range of more or less radical conclusions about the nature of the biblical text. He was convinced, for example, that scripture's documentary history disclosed evidence of fallible human transcription; that Moses was not the sole author of the Pentateuch; and that Noah's flood was a local incident, not a universal event, and that many other cultures had their own flood legends. His reasoning behind the latter suggestions was, in part, a consequence of his conviction that Adam was only the father of the Jews and that it was on account of their forbidden intermarriage with gentiles that they were judged by flood. In such circumstances the deluge was restricted to the vicinity of Palestine. The return of the dove to Noah's ark with an olive branch provided him with further supporting evidence that there were areas that remained untouched by floodwaters. For these reasons La Peyrère has been seen as playing a key role in the foundation of biblical criticism.[31]

Just how subversive La Peyrère intended his proposals to be is not entirely clear. Certainly, there are grounds for seeing his project as radical through and through. Leo Strauss, for example, noting that he was "among the first of those who openly declared their departure from unquestioning acceptance of the Bible," was of the opinion that his "recourse to Scripture, which has caused so much ridicule to be poured on La Peyrère, is no more than an adventitious trimming to a naturalistic and rationalistic theory."[32] Read in this register, La Peyrère played midwife to the inauguration of modern skepticism, an intellectual genealogy that was continued by his friend and biographer, the Oratorian priest Father Richard Simon, and by the renegade Jewish philosopher Baruch Spinoza, who had a copy of *Prae-Adamitae* in his personal library.[33] By the same token, for all the skeptical charges that were laid at La Peyrère's door, there is some reason to suppose that his aim was at least as ecumenical as it was subversive. Adopting his pre-adamite theory, he judged, would aid the conversion to Christianity of various peoples insofar as "*Genesis* and the Gospel, and the astronomy of the Ancients is reconcil'd and the History and Philosophy of the most ancient nations; So that if the *Chaldeans* themselves should come, those most ancient Astronomers, who had calculated the course of the stars, as they say many hundred thousand years ago; or the most ancient *Egyptian* Chronologers, with those most ancient Dynasties of their Kings; . . . They will willingly receive this history of Genesis and more willingly become Christians."[34] As Grafton notes, "by making Genesis a more reasonable text, he would make it more convincing, and thus the heathen would be more receptive to it and would more willingly convert to the true religion, Christianity."[35]

In order to appreciate any proselytizing impetus in La Peyrère's enterprise, the significance of the messianic vision he elaborated in *Du rappel des Juifs* (*The Recall of the Jews*) needs to be registered. Of fundamental importance here was his concern to separate out the Jewish experience from the rest of world history.[36] The theological fulcrum of sacred history was Judaism, and the pre-adamite theory was designed to cut a deep gorge between the Jews and the gentiles, namely, between the story of the adamites and the pre-adamites. In La Peyrère's telling the Bible's prevailing concern was with the Jewish nation, which was cast as principal in the theater of providential history. The gentiles—or, better, pre-adamites—merely look on as Israel's divine drama is played out in the world, although they participate in the benefits of God's election of his own people. Crucial to this vision was La Peyrère's conviction that the Jews would be recalled and that this in-gathering would take place in France prior to their subsequent return to Palestine. This eschatological scenario, of course, had major repercussions for the Christian world: anti-Semitism had to be stamped out and Christianity's debt to its Jewish heritage more fully acknowledged. Integral to the political theology of this salvation history was what might be called La Peyrère's "double messianism," according to which the first-century coming of Jesus was in his role as *Christian* Messiah, his anticipated seventeenth-century return was as political *Jewish* Messiah. Anything that furthered this historical trajectory and fostered closer Jewish-Christian relations was to be welcomed, even if it involved sacrificing much conventional theological orthodoxy.[37] The implications for La Peyrère's France were immediate: France should admit Jews in order to hasten the nation's conversion.[38] All in all La Peyrère's political messianism was a basic ingredient in his pre-adamite formula, which provided a means not only of bridging the gap between Jerusalem and Christendom but also of holding in balance the Bible, scientific knowledge, concern for the Jews, and French national identity.

Whatever its motivations, and whatever the aspirations of its author, La Peyrère's pre-adamite heresy had a significance of epic proportions. That he himself had some inkling of its import may be gleaned from the fact that he compared it with Copernicanism. Neither theory, he noted, changed the world; both changed how the world was viewed.[39] The comparison with the Copernican revolution is not far off the mark. In his exploration of "Isaac La Peyrère and the Old Testament," for example, Grafton has made the observation that in accounting for the intellectual transformation that was effected between the burning of Noël Journet in 1582 for his querying of scripture and Pierre Bayle's *Historical and Critical Dictionary* of 1692, "no one did more to make this revolution happen than the little-remembered French Calvinist Isaac La Peyrère."[40] And indeed, whether the Enlightenment's religious significance is calibrated in what Gay called "the rise of modern paganism" or along the lines of John Locke's "reasonable Christianity," the changed status of the Bible was crucially important.[41] This move, moreover, was not merely theoretical. In the words of Silvia Berti, it "delegitimized the presumed sacredness of the foundations of civil and ecclesiastic authority."[42]

Public Denunciation, Private Admiration

La Peyrère's pre-adamite theory was published to a storm of critical censure. Whatever his own hopes, he found his proposals pilloried by commentators of every conceivable theological hue, not least the hierarchy of the Catholic Church. And having been duly escorted to Rome at the pleasure of Pope Alexander VII in 1657, he embarked on penning an official recantation that was so subtly phrased that he never came to the point of actually denying the existence of pre-adamites. He announced that it was his Calvinist upbringing that had corrupted his thinking yet contended that his opponents had failed to deliver persuasive arguments or evidence to prove his theory false. In such circumstances, he went on, the only way a controversial issue could be resolved was by appeal to an acknowledged ecclesiastical authority—namely, the pope himself. And La Peyrère submitted that he had no option but to accept papal dictate.[43] In Popkin's opinion the whole "apology reeks of hypocrisy,"[44] a judgment supported by the fact that while living out his last days in the seminary of the Oratorian Fathers outside Paris, he continued to accumulate further evidence to support his epic heterodoxy.

In the years that followed, despite the fact that the theory attracted no more than the merest handful of supporters, refutation after refutation was issued in a desperate effort to exorcise the religious mind of the ghost of the pre-adamite. As Colin Kidd has observed, La Peyrère's "ethnic theology ignited one of the largest heresy hunts of the age."[45] And indeed, scholars have enthusiastically joined in the game of tallying the number of rebuttals the work attracted. Kidd himself notes that within eleven years "at least seventeen works had been published with the specific aim of demolishing" the doctrine.[46] More than a century ago, Andrew Dickson White noted that "his book was refuted by seven theologians within a year after its appearance, and within a generation thirty-six elaborate answers to it had appeared."[47] More recently, Grafton has been able to get the count for 1656 alone up to nineteen.[48] Whatever the arithmetic, there can be no doubt that refutation after refutation tumbled from the pens of those alarmed by the specter of the pre-adamites.

The condemnation of Grotius, into whose hands the as yet unpublished document fell, has already been noted. Using the etymology of place names, he insisted that the peoples of America were of Scandinavian descent and that they had settled the New World from Iceland and Greenland. Grotius's concern to save biblical authority prompted him to attack the idea that "before Adam there were other unknown men, as has recently been imagined by some in France"—a conception that brought "great danger to piety."[49] The theologian and librarian to the queen of Sweden, Isaac Voss (Vossius), also confronted La Peyrère's doctrine in his account of the world's origins, in which he sought to provide for a longer chronology through using the Septuagint version (as did Edward Stillingfleeet)—instead of the Masoretic text—of the Old Testament. In his view La Peyrère's reliance on heathen records was both wrongheaded and perverse. Much better, he insisted, to use gentile annals in such a way as to confirm sacred history. His own reworking of dates from the Septuagint rather than the Masoretic Bible, he believed, had shown the right way forward.[50] The greater length of time afforded by using this text—it gained him nearly fifteen hundred years—could even accommodate, he reasoned, Chinese dynasties, which, some had worried, looked as though they predated the Flood. At the same time he expressed his own doubts about the worldwide nature of the Noachian deluge, insisting only on its demographic universality. Apart from anything else, Voss was convinced that there was insufficient water on the earth to submerge all land.[51]

Voss's strategy, however, was not uniformly appealing. George Horn immediately shot back a rebuttal in which he alleged that Voss was a disciple of the "mad" La Peyrère, who had embarked on the task of undermining the authority of scripture. To Horn that enterprise led straight into the arms of atheism and served only to harm the Church. However laudable Voss's intention, it was in the last analysis dangerous and harmful to the Church. He preferred simply to dismiss alternative world genealogies as fabulous.[52] Edward Stillingfleet (1635–99), the orthodox Anglican bishop of Worcester and author of *Origines Sacres* in 1662, was certain that "the peopling of the world from Adam . . . is of great consequence for us to understand, not only for the satisfaction of our curiosity as to the true origin of nations, but also in order to our believing the truth of the scriptures, and of the universal effects of the Fall of man."[53] Indeed, his whole treatise was conducted in the hope that no more would be heard "of men before Adam to salve the authority of the Scriptures," a scheme, he noted, in fact "intended only as a design to undermine them."[54] In a nutshell "the hypothesis of Prae-Adamites is undoubtedly false" and could only be described as a "fiction."[55] His own strategy was to argue for the ambiguity and thus unreliability of those historical records and thereby to reduce them to the status of the merely fabulous. To Stillingfleet the records of the Chaldeans and Egyptians were simply "pretended antiquities."[56] It bothered him, too, that pagan chronology was often the weapon of choice in the arsenal of crusading atheists. As he put it, "The most popular pretences of the Atheists of our age, have been the irreconcilableness of the account of Times in Scripture with that of the learned and ancient Heathen nations."[57] To Stillingfleet a range of fundamental Christian doctrines rested on the assumption of worldwide human descent from Adam, and any tampering with that foundation would result in the collapse of the entire edifice of biblical anthropology. England's lord chief justice, Sir Matthew Hale (1609–76)—jurist, natural philosopher, and religious writer—was similarly disturbed and told the readers of his posthumously published defense of the Mosaic narrative, *The Primitive Origination of Mankind,* in 1677, that the Peyrèrean formula "would necessarily not only weaken but overthrow the Authority and infallibility of the Sacred Scripture."[58]

These rejoinders were typical of a flood of responses dwelling on the theological heterodoxy of La Peyrère's proposals that had been in full flow ever since *Prae-Adamitae* saw the first light of day. Take, for example, three works that appeared in 1656, just the year after the publication of his work.

The French clergyman and Leiden professor of Hebrew Antonius Hulsius (1615–85) brought the full weight of his anti-Semitism to bear on pre-adamism. Because he considered the Jews to be the root cause of all heresy and speculation, he found La Peyrère's project theologically repugnant. "Nothing strange to the Christian truth," he insisted, "has been introduced in the Church, which does not smell of this corrupted Judaism."[59] The controversial, though influential, French Calvinist professor at Groningen Samuel Desmarets (Maresius) (1599–1673), who engaged in critical dialogue with a range of theological systems, also took on the challenge, arguing, in a critique that was later taken up by Diderot, that in essence the heretical pre-adamite doctrine could be traced back to figures such as Maimonides and Paracelsus. He thus sided with the opinion of the Calvinist pastor J. Mestrezat that La Peyrère's treatise merited nothing more than being consigned to the flames because darkness could have nothing to do with light. And in the same year, 1656, the Lutheran theologian Johann Dannhauer (1603–66), who occupied the position of professor of rhetoric at Strasburg, devoted over five hundred pages to a theological critique of the "fable."

These repudiations are indicative of the flurry of consternation that La Peyrère's heresy stirred up during the mid-1600s, and Richard Popkin has been able to identify more than two dozen antidotes appearing in print before the century was out. Indeed, something of the hostile conditions into which the book was launched may be gleaned from the fact that numerous extant copies are to be found bound together with works refuting the doctrine. In these circumstances sympathizers were well advised to keep any admiration they might entertain for the schema to themselves or at most share it with tried and trusted friends. Just how many secret supporters there were is impossible to ascertain, of course. But there is evidence that this strategy was adopted by at least some. Thanks to the researches of William Poole, a couple of English advocates of pre-adamism, who expressed their opinions in private, have come to the surface.[60] In the case of Francis Lodwick (1619–94), a religiously heterodox London merchant and fellow of the Royal Society, who played an early role in bringing philosophical perspectives to bear on language, the motivation sprang largely from linguistic concerns.[61] In print he confined himself to strictly technical matters, in such works as *A Common Writing* (1647) and *The Ground-Work* (1652), and in his proposals for a phonetic alphabet, which he published in the Royal Society's *Philosophical Transactions* in 1686.[62] In private his mind ranged wider. Unpersuaded by the standard narrative of the origin of language as a divinely given, perfect adamic creation, which laid emphasis on the Genesis fall from grace and the

ADAM'S ANCESTORS

Babel episode as the Two Confusions explaining both its corruption and variety, Lodwick opted in his personal memoirs for a polygenist account that stressed the arbitrary nature of speech and its diverse sources. His own work on the Irish language confirmed him in these judgments. He simply found it impossible to see in "the Confusion of Babel" a persuasive explanation for a language without affinity to other Continental tongues. Had Irish resulted from the Babel judgment, traces of its dispersal would have been left by migrants who eventually found their way to Ireland. An entirely independent origin of Irish seemed much more plausible.[63] It was the same, too, with Welsh, Slavonic, and Dutch, which he deemed incompatible languages. Besides all this, Lodwick could see other benefits in La Peyrère's offering, not least of which was that it allowed him to retain his preconceptions about the fundamental difference between black and white races. And yet, however central La Peyrère–style pre-adamism was to Lodwick's philosophical linguistics, it remained for him a secret, heretical allegiance at best only hinted at in public discourse.

Unlike Lodwick, whose proposals were in substantial measure grounded in La Peyrère's speculations—he reportedly owned both Latin and English versions of *Prae-Adamitae*—the second, anonymous supporter seems to have cultivated his own version of the pre-adamite theory independently of the Continental source. The existence of worlds before the biblical Adam is certainly used to make sense of some conventional conundrums that had long bothered Bible readers—the behavior of Cain after the slaying of his brother and the identity of the "sons of gods" who took wives from the "sons of men." But the inspiration for positing a pre-adamite universe was much more metaphysical than textual and showed no interest in questions of chronology or geography. It was derived, rather, from a Neoplatonic conception of divine necessity that supposed that God necessarily and eternally engaged in works of creation as a consequence of the nature of His being. The eternity of the world, and the existence of pre-adamites, were nothing less than the expressions of an inescapable divine impulse to create.[64]

Yet for all the condemnation heaped on the theory, and for all the dangers associated with it, a number of more open advocates of pre-adamism, of one stripe or another, are discernible. Typically, however, when the pre-adamites did surface, they were to be found in the dubious company of radicals, skeptics, or esoterics. In certain fringe religious circles in England, for example, the idea showed up from time to time, in some cases even predating the publication of La Peyrère's system. Indeed, Philip Almond maintains that "Pre-adamitism was probably a not uncommon belief during the revolutionary period," as much for its fostering of an increasing desire to contest "dominant readings of the Bible with more personal and individual ones" as for the need to account for stories about Indians or challenges to chronology.[65] Laurence Clarkson, a Muggletonian—a radical millenarian group that held the soul to be mortal and found no need for formal religious ceremonies—reflected on his youthful, wilder days in the 1650s, when he had denied that "*Adam* was the first Creature, but that there was a Creation before him . . . judging that land of *Nod* where *Cain* took his wife, was inhabited a long time before *Cain.*" And Gerard Winstanley, a member of the agrarian reformist group the Diggers, warned against literalistic readings of the Genesis account, highlighting the contradictions and absurdities such hermeneutics would deliver. To him the idea that Cain was "the third man in the world" could not be taken literally because a few verses later Bible readers were told that he needed protection from others who would seek to take his life. To Winstanley it was obvious that "there were men in the world before that time"— namely, the time of Adam.[66] The English freethinker, libertine, and disciple of Hobbes Charles Blount (1654–93), who was accused by Josiah King of adopting pre-adamism,[67] plainly told the readers of his *Oracles of Reason,* published just before his death in 1693, that "there were two creations both of Man and Woman, and that *Adam* was not the first Man, nor *Eve* the first Woman, only the first of the Holy race."[68] It was simply one of the "great Errors committed in the manner of reading Scripture" to imagine that Adam was the father of all humankind. Any sensible hermeneutic would recognize that "*Moses* made [Adam] only to be the first Father of the *Iews,* whilst others Hyperbolically make him to be the first Father of all Men," and that Noah's flood only covered the "Land of the *Iews.*"[69]

In a later generation versions of pre-adamism, following this latter trajectory, were deployed for increasingly skeptical purposes. Voltaire (1694–1778), for example, doubted that the peoples of the Americas were of Noachic descent and urged that Chinese and Indian genealogies massively predated biblical antiquity. In Voltaire's case an anti-Semitic spirit was mobilized to undercut Christian claims about the origins of the world; by rocking the foundations on which Judaism was erected, Christianity would be destabilized. Moreover, as Colin Kidd puts it, in "Voltaire's deistic brand of anti-Semitism . . . the Hebrews were parvenus, their Abraham a corruption of the Hindu Brahma and Adam an obvious derivation from Adimo, the first Indian."[70]

If Voltaire's interventions may be taken to symbolize the skeptical ends to which pre-adamic theorizing could be put, its association with esoterics is nowhere more conspicuous than in the case of the Irish Parliamentarian and barrister-at-law Francis Dobbs (1750–1811). In a note dealing with La Peyrère's proposals in his account of the rise of rationalism in Europe, W.E.H. Lecky insisted that "existence of a race of men not descended from Adam was very strenuously maintained . . . by an eccentric member of the Irish Parliament named Dobbs."[71] While Popkin queries the claim that Dobbs endorsed the idea of non-adamic humans descended from an intrigue of Eve and the Devil, contending that this interpretation rests on a misreading of his interpretation of Genesis in his *Universal History*,[72] Dobbs did in fact elaborate on the idea at length in the published 1800 version of a set of twenty-five letters addressed to his eldest son. Here, in the twenty-first installment of *A Concise View, from History and Prophecy, of the Great Predictions in the Sacred Writings*, Dobbs spelled out his belief "that there are two distinct Races of Men, the one from God through Adam, the other through a Creation of the Devil."[73] Central to Dobbs's scheme was the proposal that the world was peopled by two quite separate races. The first, descended from Adam, was destined for immortality, but there was "another race upon the earth, that had not their origin from God."[74] This other race was the offspring of the devil, and it was among them that Cain (who himself had Satan's nature) settled and took a wife when he was banished from the Garden of Eden. The descendants of this race were corrupt from inception and fated for "annihilation."[75] Intriguingly, Dobbs also hinted at the possibility of a hybrid race springing from a union of these two creations, though he did not further develop the idea. As he reflected on the early chapters of Genesis: "Now here are, in the plainest terms, two distinct races pointed out. The one sprung from Adam, who was the son of God, and of course his offspring were the sons of God also; and another race, distinguished by the name of men only; and from the blenditure of the two, a spurious race, but which had only, in regard to life, what belonged to that race that had its origin from Satan."[76]

There Dobbs left matters. But the esoteric Irish millennialism, according to which the Messiah would return first to Ireland (the battle of Armageddon being appropriately staged in "Armagh") before going on to Palestine,[77] into which his polygenetic account was inserted, discloses something of the arcane atmosphere in which advocates of non-adamic humans sometimes crystallized. For Lecky this Hibernicized eschatology, which drew on Isaac Newton's commentary on the books of Daniel and Revelation,[78] and which he delivered in a speech to the Irish House of Commons in June 1800, had immediate political ramifications. The special place of Ireland in the divine economy underwrote the nation's independence; Ireland's freedom from British colonial dominance was "written in the immutable records of Heaven."[79] Biblical prophecy, evidently, confirmed Dobbs's radical views on Ireland's constitutional right to formulate its own laws. Regardless of its eccentricities, the speech was a runaway success, and thirty thousand copies were immediately sold.[80]

Whatever its taint of heresy, and despite the unsavory company in which the idea frequently surfaced, there were nonetheless those who felt themselves attracted to pre-adamism, even if they could not bring themselves to embrace it fully. In some cases half-hearted rejection was the outcome, not least because they could see the theory's potential to resolve mounting tensions that conventional readings of sacred texts increasingly faced. Others set about modifying the basic Peyrèrean formula to suit their own purposes. Nathaniel Lardner (1684–1768), a nonconformist scholar whose theological convictions developed from Baxterian Calvinism through Arminianism to a modified Arianism, for example, flirted with a version of the theory but could not quite come to accept it, whatever its attractions. His greatest achievements were his *Credibility of the Gospel History,* published between 1727 and 1757, and *A Large Collection of Ancient Jewish and Heathen Testimonies to the Truth of the Christian Religion,* a four-volume treatise issued between 1764 and 1767; in these works he strove to defend the authenticity of the Bible against deistic critics.[81] Certainly, Lardner's own reading of the Genesis narrative was self-confessedly literalistic, but in an essay he penned in 1753 on the subject of the Creation and the Fall, he admitted that there were "not a few difficulties in the account, which Moses has given of the world, and of the formation, and temptation, and fall of our first parents." The greatest of these rotated around the question of human origins, and Lardner conceded that the belief that "all mankind have proceeded from one pair"—the sheet anchor of the Christian conception of Creation—could not be established with absolute certainty from contemporary empirical evidence without the bolstering support of scripture. Indeed, he did come to the point of confessing that "many pairs, resembling each other, might have been formed by God, the Creator, at once, in several, and remote countries, that the earth might be peopled thereby." This suggestion—that there may have been co-adamites, if not pre-adamites—was one that had already begun to receive an airing during the mid-eighteenth century. The idea here was that God had created several humans in different locations at the same time as Adam. Lardner himself felt constrained to reject this possibility but diffidently added that "the account of Moses, I suppose, may be relied upon." This left him with the need to find some means of explaining the source of racial differentiation, and he was forced to resort to the "difference of climates, with the varieties of air, earth, water," as the means of making "sensible alterations and differences in one and the same species." Moreover, Lardner did not hesitate to point to what he saw as the moral implications of the Mosaic monogenetic account—namely, that common descent from the same parents should "abate exorbitant pride" and reaffirm the universal brotherhood of the human race.[82] Yet Lardner's rejection of polygenism was anything but vigorous, and his denial of it was such that its attractions were plainly displayed for all to see.

Whereas Lardner could not bring himself to endorse co-adamism, others less unnerved by charges of heresy found in this option a means of retaining pre-adamism's explanatory power while deflecting its heterodoxy. To them the co-adamite rendition presented real possibilities because it removed the stigma attached to the notion that the world was inhabited *before* Adam. This version was anonymously promulgated in 1732 in a fifty-nine-page tract entitled *Co-Adamitae, or an Essay to Prove the Two Following Paradoxes, viz. I. That There Were Other Men Created at the Same Time with Adam, and II. That the Angels Did Not Fall.* The author's main objective was to relieve internal tensions in the biblical narrative by answering exegetical questions such as where Cain got his wife, why he had cause to fear when he fled Eden, and who peopled his city and supplied building materials. To this anonymous writer the conventional interpretation did violence both to divine economy and to common sense. For the former it was plain that "*Adam* singly, neither could use, nor attain the Benefit of all the other Creatures, therefore there were more Men than *Adam,* some to attend his particular Service, and others which God distributed, to reap the Increase of more remote Parts of the Earth."[83] On the latter score it was incredible to conjecture that Adam could have attended to a globe whose "Circuit . . . according to the most modest Compute of all Geographers, is 226000 *Italian* miles."[84] Besides, there were places where the name Adam was not "a Denomination solely appropriated to one,"—that is, to an individual—but, rather, was a label that "signifies Man indefinitely," namely, generically.[85] As for the doctrine of original sin, the author proposed that Adam's action was representative of all humanity in something like the way "Princes, when they make a League with any Neighbour King, do not only oblige themselves faithfully to observe such Articles as are then agreed to, but their Subjects also, at least inclusively, though they are not at all made privy to his Proceedings."[86] Evidently, humanity was implicated in Adam's transgression representatively, not genealogically. Taken overall, then, this account was essentially theological and exegetical in its intent and scope, but subsequently, as we shall see, this self-same system was advanced with rather more scientific objectives in mind.

It is time to pause and take stock of pre-adamism's fortunes thus far. While the context within which the pre-adamite was talked about was almost invariably theological, it would clearly be mistaken to think of that vocabulary as hermetically sealed off from other fields of discourse. Indeed, it was not so much that there were a range of demarcated conversations—geographical, anthropological, theological, and so on—between which there was cross-communication. Rather, geographical and anthropological considerations were simply part of the theological frame of reference, and vice versa. Of course, this does not mean that there was universal agreement among the conversationalists. And as we have seen, advocates of pre-adamism, either by self appointment or public accusation, frequently found themselves marginalized on the skeptical fringes of the interchange. All this, however, would soon change as pre-adamite vocabulary, in one form or another, became more common in the emerging language of the human sciences.

Time, Texts, and Talk

Whatever was to be made of La Peyrère's intervention, things could never be the same afterward. The monumental heresy to which he gave voice was at once remarkably simple—there were humans on earth before Adam—and yet extraordinarily far-reaching. However superficial the scholarship of La Peyrère himself may have been—Richard Simon, his friend and correspondent, thought his learning extremely limited and complained of his ignorance of both Greek and Hebrew[87]—such was the theory's conceptual versatility that it swept into its orbit a wide range of matters occupying the minds of contemporary thinkers. Thus, while Grafton observes that La Peyrère's "erudition disappears on inspection," he at once adds that his intelligence was correspondingly "all the more impressive." La Peyrère "fused for a poignant moment the popular scepticism that could call any text into question and a few powerful fragments of learned scholarship. The combination was potent enough to stimulate, to split, and eventually transform the world of learning to which he never really belonged."[88]

The pre-adamite theory forced open new possibilities in at least three scholarly arenas, which can be grouped under these labels: time, texts, and talk—just some of the more conspicuous intellectual regions into which the pre-adamite theory was catapulted. Pre-adamism had a direct bearing on matters of time; it challenged standard accounts of the age of the world and the understanding of humanity's deep past and its foreseeable future. It also raised fundamental questions about how ancient texts, not least sacred texts, were composed and how they should be read. And it provoked troubling questions about the origins and nature of human language and of human beings themselves.

Time

La Peyrère's postulation of a pre-adamic cosmos had profound implications both for time past and for time future. It was, in part, to meet the challenges arising from world chronology that his theory was brought into the orbit of those seeking to come to grips with competing conceptions of the human past. Le Peyrère's program, moreover, was highly imaginative in that it retained the standard dates for biblical chronology even while opening the gateway to infinite times past. His own conception of post-adamic time was entirely in keeping with the conventional conception of roughly four thousand years that the Latin Vulgate had assigned to the epoch between the creation of Adam and the birth of Christ. But by drawing back the curtain on an endlessly extended pre-adamic cosmos, he uncovered a mechanism by which the annals of the Chaldeans, Chinese, Egyptians, Phoenicians, Scythians, and what he described as the yet undiscovered antipodean worlds could be embraced. Thousands, hundreds of thousands, even millions, of years could be accommodated without disturbing the customary chronicles of the adamic narrative. It delivered a remarkable concordance. The idea of infinite time, moreover, impelled La Peyrère to ponder the possibilities of infinite space and cosmic geography.[89] And indeed, later generations would likewise link questions about how to think about primeval time with issues rotating around the plurality of worlds.

Massively extended past time, however, was only part of the temporal dimensions of La Peyrère's pre-adamite project. As we have already seen, his pre-adamite theory was intimately intertwined with an eschatological future that granted pride of place to the Jews. Their anticipated restoration—that is, the reuniting of the adamic stock—courtesy of a massively retracted Christianity would presage their return to the land of Israel and the ushering in of the millennial kingdom. His idiosyncratic biblical hermeneutics were, at least in part, designed to preserve the Old Testament as an exclusively Jewish domain.

La Peyrère's pre-adamite scheme, however heretical its opponents deemed it to be, thus opened up debate on both past and future time. Pagan chronology and the Judeo-Christian eschaton alike were folded into the fabric of his eccentric speculation.

Texts

Another suite of intellectual concerns congregating around how to read ancient texts, particularly the Bible, and how such canonical authorities had been subject to compositional editing, was dramatically engaged in the wake of La Peyrère's interventions. Now it became increasingly difficult to maintain that the Bible was the sole authoritative account of human origins, and this suspicion, taken with La Peyrère's own doubts about the Mosaic authorship of the Pentateuch and his observations on the unreliability of the *textus receptus,* served to keep public consciousness of the pre-adamite theory firmly tied to skeptical moorings. Given that he considered scripture "a heap of Copie confusedly taken,"[90] it is hardly surprising that La Peyrère's pre-adamism should be stage-cast by W. E. H. Lecky as a kind of halfway house toward free thought and that Andrew Dickson White, in his pugilistic recounting of the warfare between science and religion, should not hesitate to marshal La Peyrère in the cause of reason over dogma.[91] For in his wake it became harder to accept the Mosaic authorship of the Pentateuch uncritically; it became harder to approach ancient sacred texts with unalloyed reverence; and it became harder to ignore extrabiblical data in scriptural hermeneutics.

In the interim between his vilification as a heretic and his canonization as "a heroic martyr, the Galileo of the exegetes,"[92] the stream of textual criticism that La Peyrère opened up widened into a flood. "Within a generation," Anthony Grafton remarks, "far deeper thinkers—who normally denied that they owed him any debts—would stand beside him in the moat, using far more powerful engines to besiege the castle of biblical authority." The Jewish thinker Benedict Spinoza denied the absolute authority of the Old Testament, insisting that its message was addressed to primitive Jews in categories meaningful to them, not to modern society. The Catholic writer Richard Simon resorted to the most recent Near Eastern scholarship "to prove beyond reasonable doubt that the Pentateuch was a mosaic of lost earlier texts." And the Protestant Isaac Vossius—like La Peyrère an admirer of ancient Egypt and modern China—would influentially insist that one must follow the Greek text of the Old Testament, which gave a longer chronology than the Hebrew, in order to resolve the problems La Peyrère had brought to light.[93] Whatever the direct genealogy of these latter proposals, and whatever their indebtedness—or lack of it—to La Peyrère, there is no doubt that his theory raised fundamental hermeneutical questions. In bringing together pagan sources and biblical interpretation, La Peyrère believed he was combing out exegetical snarls that had long perturbed conservative exegetes. For this reason La Peyrère plausibly takes his place as a prime mover in the inauguration of modern biblical criticism.[94]

Talk

The pre-adamite theory, as we have seen, was also profoundly implicated in speculations about the origins and development of language and whether human speech had degenerated from a pristine past or progressed from primitive antecedents. By proposing the existence of humans, and therefore language, before Adam, it was profoundly troubling to those for whom Adam's speech was the foundation of all human speech. Thus, George Dalgarno (ca. 1620–87), an Aberdonian schoolmaster and preacher, whose work on the genesis of language was influential, found it necessary to contest pre-adamite speculations. His own careful philosophical-theological analysis of the relationship between Adam's perfect (but not divinely determined) speech and its imperfect state in the wake of both the Fall and Babel amounted to nothing more than a belated, contingent set of events, if indeed human language was spoken by pre-adamic beings created separately across the face of the earth. The pre-adamites deeply disturbed the fundamental architecture of Dalgarno's theory of adamic naming, in which he sought to explain Adam's linguistics as both the natural consequence of divine creation and an arbitrary system of signification.[95]

The "search for the perfect language," as Umberto Eco styles it, was a long obsession with language thinkers. The passion to recover the lost perfection of speech in paradise, where there was what Eco calls a "primordial affinity between words and objects,"[96] took many forms. Saint Augustine and most of the early Church fathers were convinced that primeval Hebrew was the primary and perfect language; in the fourteenth century Dante sought less for a specific language than for a more generic form of speech (forma locutionis), which had been lost after the Babel episode; a little earlier Boethius of Dacia thought it possible to distill a universal grammar from existing languages and thus to catch a glimpse of the lost language of Adam; in the mid-sixteenth century Conrad Gessner, sharing the Hebrew hypothesis, claimed that all existing languages contained traces of corrupted Hebrew vocabulary.[97] Others, such as John Wilkins and Dalgarno during the seventeenth century, sought in different ways to create an entirely rational, universal language, based on a rigorous system of classification, that recovered the essence of Eden's lost speech. La Peyrère's advocacy of entire pre-adamic worlds was not so much a challenge to these particular proposals as a massive undermining of the entire enterprise of searching for any perfect adamic language. For that project presupposed linguistic monogenism, the idea that all languages had developed from a single Edenic mother tongue; by contrast La Peyrère's system was resolutely polygenetic—and therefore polylingual. As Paolo Rossi summarizes it: "Lapeyrère proposes no theories concerning the rise and development of language among the pre-adamites. He limits himself to showing that traditional solutions to the problem were untenable."[98] Eco concurs, commenting on the proposals of the heretic La Peyrère: "Quite apart from the obvious theological implications of such an assumption (and the works of La Peyrère were condemned to be burnt), it was clear that, by now, Hebrew civilization—along with its holy language—was falling from its throne. If one accepted that species had developed differentially in differing conditions, and that their linguistic capacity reflected their degree of evolution and of adaptation to environment, it was easy to accept the polygenetic hypothesis."[99]

Whether or not contemporaries found linguistic polygenism an easy hypothesis to accept is open to debate. But the fact is that La Peyrère had opened the door to a more general anthropological polygenism, which was taken up in one way or another in generations to come. It earned for him an acknowledged place in the early history of the human sciences.[100] In La Peyrère's scheme polygenism was bereft of its later associations with certain forms of racial prejudice. Indeed, its thrust was profoundly humanitarian inasmuch as it sought to integrate all peoples within the divine economy and rejected any notion of racial superiority, even of the Jews, through whom the redemption of the world was secured. It is important to register these sentiments because it is frequently assumed that polygenism was necessarily implicated in a racist ideology. Indeed, one recent set of commentators has egregiously blamed La Peyrère for establishing "the polygenist basis for racism" and that his pre-adamites "were descended from a different 'Adam' and hence sundered from 'true men,' which means primarily Europeans."[101] Such accusations are entirely without foundation.

Besides its reformulation of human ancestry, La Peyrère's theory had more general biogeographical implications. Some of them were seized upon by a Dutch naturalist, Abraham van der Myl, who used it to argue that the entire fauna of the New World, animal as well as human, had been separately created.[102] Then there was Peyrère's suggestion that the Mosaic flood was not universal. Even critics of pre-adamism could find this idea attractive, as in the case of Bishop Stillingfleet, who, while fiercely repudiating the suggestion of the existence of men and women before Adam, nevertheless found the idea of a local flood distinctly appealing. For him the deluge was demographically, not geographically, universal. Nor did he consider that it wiped out all other plant and animal life, for "what reason can there be to extend the flood beyond the occasion of it, which was the corruption of mankind?"[103] Indeed, even while defending the monogenetic origin of the human race with all vigor, he was entirely sanguine about what might be called animal polygenism as an explanation for worldwide biogeographical distributions. But because he believed that all humankind at the time dwelled in the Near East, no larger geographical scope was required for the Flood.

⌘ According to Richard Popkin, La Peyrère "was regarded as perhaps the greatest heretic of the age, even worse than Spinoza, who took over some of his most challenging ideas."[104] His name remains connected with what Noel Malcolm calls "an unholy trinity of writers: Spinoza, La Peyrère, and Hobbes."[105] And from the mid-seventeenth century to the present day his work has been constantly refuted—from Grotius's assault on the theory, even before the published version had seen the light of day, to twenty-first-century online Catholic encyclopedias. Yet its ramifications spiraled out into a wide range of different domains. If La Peyrère's heresy was of monumental proportions, so too was its legacy. Anthony Grafton thus nominates *Prae-Adamitae,* alongside Blaeu's *Great Atlas,* as the two key texts that widened into "frightening crevasses" the long-standing "cracks in the canon." Both, he observes, "reveal in complementary ways the tremors of what would finally become an intellectual earthquake."[106]

3 ✌ POLITY

The Cultural Politics of
the Adamic Narrative

OR ALL THE SCORN POURED ON La Peyrère and the idea of non-adamic races, the political significance of the narrative of human origins and diversity reasserted itself with ever greater force during the eighteenth century. Certainly, refutations continued to surface. Itemizing the strategies that several of his critics mounted in their effort to preserve received wisdom about Adam as the father of all humankind will help to convey something of the disquiet La Peyrère's heresy continued to induce. These reflections will serve as background to a consideration of debates congregating around the role that climate was thought to play in the processes of racial differentiation. The politicization of human origins that preoccupied pre-adamism's proponents and detractors throughout the eighteenth century will thereafter command attention as we examine something of the ways in which scientific inquiry, linguistic diversification, moral philosophy, race relations, and cultural politics were intimately interwoven.

Criticism and Conjecture

In 1718 the French Catholic apologist and orientalist Eusebius Renaudot (1648–1720) used his translation of two ancient Arabic travelers to snipe at La Peyrère's theory. Numbered among those chronologists who had "put Weapons into the hands of Libertins and Free-thinkers," he told the readers of his appended dissertation on Chinese learning, was "the Author of the *Preadamite System.*" Although La Peyrère's acquaintances themselves reported that "he was so ignorant that he scarce understood *Latin,* yet having formed a System by wresting some Passages of Scripture to his own mind . . . he laid hold on it not only as a very valid Proof of his own Whims, but also of the infinit [*sic*] number of Years the *Assyrians, Babylonians,* and *Egyptians* reckoned, which the very Heathens themselves rejected as fabulous." Many had been "insnared thereby," Renaudot judged, "not to become *Preadimites* [*sic*] indeed, but to harbor other Notions equally subversive of Religion."[1]

Still, those who sought, even with seeming casualness, to refute the speculation found themselves having to multiply ever more extreme conjectures in the effort to preserve literal readings of the Genesis text. Illustrative here was the doctrinal conspectus by the minister of the independent church at Three Cranes in London, Thomas Ridgley (1667–1734). A proponent of strict Calvinism, Ridgley rejected what he saw as various liberalizing tendencies in the Anglican Church and produced a number of publications on creed making and other religious subjects. In his two-volume *Body of Divinity* of 1731, Ridgley turned his readers' attention to the adamic origins of the human race and paused to attack the "bold writer" who "about the middle of the last *Century,* published a book, in which he advanced a new and fabulous notion; that there was a world of men, who lived before *Adam* was created; and that these were all heathen; and that *Moses* speaks of their creation, as what was many ages before *Adam,* in *Gen.* i. and of *Adam's* in *Chap.* ii. whom he supposes to have been created in some part of the world, which was then uninhabited." Ridgley found this theory offensive and subversive of the entire doctrinal structure of Christianity, and it was therefore no surprise to him that, even though "Peirerius'" book was "not much known in the world, yet the notion is propagated and defended by many *Atheists* and *Deists.*"[2] Accordingly, he felt compelled to attack the idea in several catechetical responses to questions about the adamic world. And here the need for increasingly speculative extravagance to save biblical appearances manifested itself.

In seeking to explain global demography, for example, and to account for the populating of Cain's city from Adam's own offspring, Ridgley resorted to what might best be described as Edenic hyper-fecundity. In primeval days, he surmised, procreation took place "in an uncommon degree, the necessity of things requiring it." Indeed, he did not consider it "absurd to suppose, that, at least, as many children were generally born at a birth, and in as early an age of the mother's life, as have been or are, in any uncommon instances in later ages." In addition to this natal prodigality, Ridgley added that "the time of child-bearing continued many years longer than it now doth . . . and, if the age of man was extended to *eight* or *nine hundred* years, we may conclude that there were but few that died young." Ridgley was certain that by such means, in a relatively short space of time, Adam's offspring "might spread themselves through many countries, far distant from the place where *Adam* dwelt; and therefore there is no need to suppose, that those, with whom *Cain* dwelt in the *land* of Nod, were persons that lived before *Adam* was created."[3]

The anonymous authors of *An Universal History,* which began to appear in London in 1736, likewise resorted, in part, to what they called "primitive fecundity" in order to account for the growth of population in the adamic world. And this explanation, together with the power of climate to induce organic modifications, was called upon to preserve intact the standard biblical story in the face of "those who think mankind were in being before *Adam.*"[4] Indeed, these writers, specifically attacking the idea that "*Moses* intended only to give an account of the origin of the *Jews,*" as put forward by La Peyrère and Blount, expressed their disquiet at the insistence "of the *Pre-adamites*" that both white and black races could not have descended from Adam and Eve.[5] Arguing that climate was sufficiently powerful to bring about changes in complexion, they went on to advance an account of racial differentiation in terms of what, perhaps a little anachronistically, might be called sexual selection and psychic influence. After the initial change in skin tone that any group might experience in moving to a "very hot country," the authors conjectured that "in a generation or two, that high degree of tawniness might become natural and at length the pride of the natives. The men might begin to value themselves upon this complexion, and the women to affect them the better for it; so that their love for their husbands, and daily conversation with them, might have a considerable influence upon the fruit of their wombs, and make each child grown blacker and blacker, according to the fancy and imagination of the mother."[6] This explanation, the authors were convinced, was infinitely superior to those accounts explaining that "blackness was at first supernatural, and a judgment inflicted upon *Ham.*"[7]

Evidently, as the eighteenth century wore on, questions about human origins and the status of Adam as the father of all humanity spread well beyond matters of theological heterodoxy and scientific explanation and further into the realms of national politics, social relations, and moral philosophy. The reason for this shift, of course, is that the adamic narrative not only laid out some of foundations of civil society—marriage, family, agriculture, ritual, urban life—but it also delivered a universal anthropology that knit together, in one way or another, every human being regardless of racial, religious, ethnic, or national identity. Tampering with that received story was itself a political intervention, the implications of which were the subject of concern to various interlocutors. Here we focus on the period somewhat casually referred to as the Enlightenment in hopes of disentangling some of the differing threads woven into what I call "adamic politics" on both sides of the Atlantic and to disclose something of the continuing political investment in the chronicle of Adam and his descendants. For challenging the traditional story was taken to have monumental social implications that were every bit as far-reaching as its theological connotations.

Climate's Imperial Power

It was in the fourteenth book of his celebrated 1748 work on comparative governmental institutions that the French political theorist Montesquieu elaborated on what might best be called the imperialism of climate. To him everything from human physiology and national character to social customs and moral standards bore the indelible stamp of regional climate. It was Montesquieu's familiarity with the physiological theories of figures such as Boerhaave that encouraged him to conceive of the nervous system as consisting of tiny tubules that carried what he referred to as animal spirits, or nerve fluid, around the body. This understanding of the human constitution allowed him to speculate that, for example, inhabitants of cold climates were more vigorous because the fibers in their cardiovascular systems contracted in cold air and stimulated faster flowing of the blood.[8] Warm air, by contrast, relaxed and lengthened "the extremes of the fibres."[9] The implications were far-reaching. It meant that "people are therefore more vigorous in cold climates. Here the action of the heart and the reaction of the extremities of the fibres are better performed, the temperature of the humours is greater, the blood moves freer towards the heart, and reciprocally the heart has more power. This superiority of strength must produce a great many effects, for instance, a greater boldness, that is more courage; a greater sense of superiority ... more frankness, less suspicion, policy, and cunning."[10] By contrast, the inhabitants of warm climates were, "like old men, timorous."[11] Evidently, the force of climate was as irresistible as the might of an omnipotent imperial power. As Montesquieu famously put it in a later section of his celebrated treatise, "The empire of the climate is the first, the most powerful of all empires."[12]

In one sphere after another Montesquieu elaborated on the imperatives of climate—for education, law, religion, customs, diet, agriculture. In so doing, he compared the national character of India with China, much to the detriment of the former; he proposed as a natural law that, in moving from the equator to the poles, "drunkenness increased[ed] together with the degree of latitude,"[13] and he found differences in attitudes to suicide between the Romans and the English. In climate, then, lay the sources of human variation; what looked like inherent racial differences were simply the outworkings of climate's imperial agency.

This view caught on. During the middle decades of the eighteenth century Georges Buffon added his voice in support of the omnipotent agency of climate in producing human differences. "The influence of climate," "food, which has a great dependence on climate," and "manners, on which climate has, perhaps, a still greater influence" were, to Buffon, the three fundamental causes of racial variety. Indeed, the human constitution was remarkably pliant before the iron hand of climate. Human migration, Buffon believed, provided ample testimony. Under different regimes human "nature was subject to various alterations."[14] In the New World, he famously insisted, much to the irritation of Thomas Jefferson, an inferior climate inevitably produced a degenerate anthropology, a mediocre zoology, and a substandard botany.[15] But these differences were not inherent; they were derived. So, while on the surface it might look as though "the Negro, the Laplander, and the White were really different species," in fact "those marks which distinguish men who inhabit different regions of the earth, are not original, but purely superficial. It is the same identical being who is varnished with black under the Torrid Zone, and tawned and contracted by extreme cold under the Polar Circle."[16]

Later in the century Johann Friedrich Blumenbach (1752–1840) added his voice to the environmentalist chorus, arguing in his medical doctoral thesis (completed in 1775 and going into its third edition in 1795) that racial divergence was due to the influence of climate and habitat. What he called "the continuous action, carried on for several series of generations, of some peculiar stimuli in organic bodies"—notably, climate—was such that firm lines of racial difference could never be drawn. Although "there seems to be so great a difference between widely separate nations, that you might easily take the inhabitants of the Cape of Good Hope, the Greenlanders, and the Circassians for so many different species of man," he wrote, "yet when the matter is thoroughly considered, you see that all do so run into one another, and that one variety of mankind does so sensibly pass into the other, that you cannot mark out limits between them."[17] For a rather more popular audience the celebrated Anglo-Irish man of letters and sometime physician Oliver Goldsmith turned the attention of the readers of his *History of the Earth*, which first came out the year he died, 1774, to the subject "Of the Varieties

ADAM'S ANCESTORS

of the Human Race." Goldsmith was convinced that "there is nothing in the shape, nothing in the faculties, that shews their coming from different originals; and the varieties of climate, of nourishment, and custom, arte sufficient to produce every change."[18] In holding this view, he sided with the seventeenth-century merchant and oriental traveler Sir John Chardin, whom he quoted as confirming that all "nations are derived from the same original, however different either their complexion or their manners may appear: for as to the complexion, that proceeds entirely from the climate and the food."[19] He thus was unpersuaded by those who sought to attribute black skins to causes other than climate. Such "ungrounded conjecture" must be rejected in favor of "the ancient opinion, that the deepness of the colour proceeds from the excessive heat of the climate."[20] Both "reason and religion," Goldsmith happily concluded, conspired to confirm that "we have all sprung from one common parent" and that "a kinder climate" would eventually obliterate the "accidental deformities" that he believed separated African, Asian, and American peoples from "more civilized" races.[21] By environmentalist strategy, then, the traditional Christian belief in human monogenesis could be sustained; a single human species had simply diversified under the evolutionary influence of differing climatic conditions. The empire of climate preserved the unity of the human race; evolution, as it were, saved scripture. And in so doing, as we will presently note, it also went some way toward giving sustenance to those who opposed slavery, by combating writers who erected racial classifications on polygenist foundations.

Races for Places

Henry Home, Lord Kames (1696–1782), the Scottish lawyer and philosopher, did not see things the same way at all (fig. 10).[22] Indeed, some of the commentators noted earlier had Kames in their crosshairs when they issued robust defenses of biblical monogenism. With a distinguished reputation as a leading legal authority already secured and a long-standing involvement in agricultural and commercial affairs behind him, he turned his attention in his late seventies to the question of human history. In 1774, at the height of the Scottish Enlightenment, his famous *Sketches of the History of Man* first appeared, with variant editions surfacing in the United States.[23] The treatise's "Preliminary Discourse" constituted a lengthy attack on the power of climate to explain human variation and a robust denial of the organic unity of the human race. And in declaring war on the empire of climate, Kames considered his "most formidable antagonist" to be "the celebrated Montesquieu"—that "great champion for the climate"—though he also spent time challenging those such as Vitruvius, Mallet, and Buffon, who likewise ascribed "supreme efficacy to the climate."[24] In a nutshell they reversed reality: climate did not make human varieties; rather, human varieties were made for different climates. As he put it, "there are different races of men fitted by nature for different climates."[25] And this was the arrangement of divine Providence: "If we have any belief in Providence, it ought to be so. Plants were created of different kinds, to fit them for different climates, and so were brute animals. Certain it is, that all men are not fitted equally for every climate. Is there not reason to conclude, that as there are different climates, so there are different species of men fitted for those different climates?"[26] Cultural mores, social customs, skin color, skeletal structure, differences in the very "fibres" on which Montesquieu's entire edifice was erected—these physical and moral traits were not to be thought of as the products of environment; they were "from original nature."[27] All of this pointed toward a seeming inevitability: human diversity was primitive, not derived, and different races had their proper places.[28]

Like other key figures in the Scottish Enlightenment, notably Thomas Reid, Dugald Stewart, and Francis Hutcheson, Kames sought for a coherent theory of the human constitution. But whereas Scottish Enlightenment philosophy generally lodged its conception of the universality of human nature in an assumed monogenism, the data accumulated by geographers and anthropologists brought Kames to the brink of polygenism. According to George W. Stocking Jr., Kames "did not see any contradiction between arguing on the one hand that men were not all members of the same species, and on the other, that they were all equal and shared a common human nature";[29] Paul B. Wood, by contrast, is of the opinion that Kames's contemporaries among the Scottish men of letters rejected his polygenetic inclinations "as subverting the foundations of religion and morality."[30] Whatever the implications for human ontology and wherever Kames should be located within the orbit of a Scottish Common Sense tradition erected on the foundations of a universal human constitution, it seemed to Kames that God simply had created in different climates many human pairs with appropriately regulated physiologies. Indeed, reversing the very argument that Buffon had so confidently championed, Kames was sure that the history of colonization bore ample witness to the accuracy of *his* assertions. He could identify numerous instances in which colonists found it impossible to acclimatize to alien climatic regimes. "Several European colonies," he reported, "have subsisted in the torrid zone of America more than two centuries; and yet even that

ADAM'S ANCESTORS

Fig. 10. Henry Home, Lord Kames, from an engraving after a portrait by David Martin.

length of time has not familiarized them to the climate: they cannot bear heat like the original inhabitants."[31] Moreover, Kames could not square Buffon's belief that Africans, Asians, native Americans, and Europeans were the same human species, tinged only with the colors of climate, with the fact that diverse racial groups were to be found in places with the same climatic conditions. North America's regional climates, for example, had not produced corresponding racial types. Looking to climate was thus entirely wrongheaded. The American Indians must be a separate post-adamic creation. As he explained in book 2 of his treatise, "America has not been peopled from any part of the old world"; rather, he supposed "the human race to have been planted in America by the hand of God later than the days of Moses" and that "Adam and Eve might have been the first parents of mankind, *i.e.,* of all who at that time existed, without being the first parents of the Americans."[32] They were constitutionally different, and this enabled Kames to account for Americans' presumed demographic stagnation as a consequence of their inherent lack of sexual vigor.[33]

For all this unquestionable clarity, however, Kames ultimately felt the need to stop short of explicitly endorsing the idea of non-adamic humanity. Perhaps it was because he had earlier been embroiled in theological controversy on the freedom of the will; perhaps it was because the idea was just too shocking. Whichever, Kames felt compelled to concede that the idea that "God created many pairs of the human race, differing from each other externally and internally; [and] that he fitted these pairs for different climates, and placed each pair in its proper climate," was heterodox. For "this opinion," he went on, "however plausible, we are not permitted to adopt; being taught a different lesson by revelation, namely, That God created but a single pair of the human species."[34] But while he affected to draw back from the skeptical heresy of pre- and co-adamism and, in a desperate effort to avoid the charge of infidelity, proposed that God had impressed upon the human race an immediate change of constitution at the time of the Tower of Babel to fit it for its diaspora, his concession to orthodoxy was nothing but halfhearted: "Though we cannot doubt of the authority of Moses, yet his account of the creation of man is not a little puzzling, as it seems to contradict every one of the facts mentioned above."[35] It would be hard to conceive of a more hesitant endorsement of the standard adamic story nor one more shot through with insinuations about original human diversity. Certainly, Kames's critics could read between the lines. The Scottish Common Sense thinker James Beattie (1735–1803), for instance, vigorously opposed the polygenism he detected in Kames—and indeed in David Hume—not least because of the racial bigotry he could discern in the very fabric of the theory.[36]

Kames was not a lone voice in raising skeptical doubts about the efficacy of climate's power. Griffith Hughes (1707–ca. 1758), an Anglican clergyman-naturalist in Barbados and a fellow of the Royal Society, was certain that climate's imperial rule was not without limits. In book 1 of his *Natural History of Barbados*—a work that was not well received by his fellow natural philosophers—Hughes elaborated on a Montesquieu-like thesis to explain climate's influence on human temperament and in particular why it was that inhabitants "of hot Countries are of a more volatile and lively Disposition, and more Irascible in general, than the Inhabitants of the Northern Part of the World."[37] His account centered on the impact of heat on the "Velocity of the Fluids" or "Animal Spirits" in the circulation of the blood. In climates where cold conditions restricted the motion of "Globules of Blood" through the capillary vessels, temperaments lacked the cheerfulness and levity of those in hotter climes. Yet for all that, the influence of climate was circumscribed.[38] Crucially, it was impotent to explain differences in skin complexion and hair character. To be sure, some claimed that these features were the consequence of "the intense Heat of the Sun." But Hughes was certain that this was "far from being the truth." Instead, he considered that "the *Whites,* the

Indians, and the *Blacks,* differ not accidentally, but originally and really."[39] Like Kames, Hughes thought racial constitution was impervious to climate's power.

Others, too, were prepared to risk heterodoxy in opting for originally different sources of racial derivation. The Royal Navy surgeon John Atkins (1685?–1757), for example, who became something of an expert in tropical diseases, appended some observations on the climatic and anthropological conditions of the Guinea coast to his 1732 work *The Navy-Surgeon: Or, a Practical System of Surgery,* in which he sought to account for differences in skin color. It was a phenomenon, he protested, that was "extremely difficult to account for," and he confessed himself drawn to the conclusion that "White and Black must have descended of different Protoplasts; and that there is no other Way of accounting for it."[40] He reiterated this judgment a couple of years later in his account of *A Voyage to Guinea, Brasil, and the West-Indies,* in which he announced that "tho' it be a little Heterodox, I am persuaded the black and white Race have, *ab origine,* sprung from different-coloured first Parents."[41] Plainly, the idea of multiple Adams was now receiving a more widespread airing.

It surfaced, too, in the work of the English surgeon Charles White (1728–1813), who came to the conclusion that "the various species of men were originally created and separated, by marks sufficiently discriminative."[42] White had devised his multiple creation hypothesis after poring over the cabinet of human skulls in the possession of John Hunter and announced the opinion in his *Account of the Regular Gradation in Man* (1799) as a means of explaining the apparent graduated orders of human and animal species. A fellow of the Royal Society renowned for his influential *Treatise on the Management of Pregnant and Lying-in Women* (1773) and a founding member of the Manchester Literary and Philosophical Society, White drew on his own knowledge of human and animal anatomy, along with the reports of numerous natural philosophers and travelers (among them Edward Tyson, Carolus Linnaeus, Georges-Louis Leclerc de Buffon, Johann Kaspar Lavater, Petrus Camper, William Hunter, and in particular Samuel Thomas von Sömmering, some of whose findings he translated in an appendix), in his argument for a general gradation in physiological form from one species to another. It was emphatically not, however, a proto-evolutionary portrait; rather, it was domesticated in the tradition of the great chain of being that connected all livings things together in a seriated hierarchy from the lowest to the highest. Indeed, he argued explicitly against any evolutionary-style arrangement, insisting that those who preserved traditional monogenism—and thereby acquiesced to a too-pliable interpretation of racial features—could find no way of drawing a stable boundary between humans and apes. As he put it, if "we admit that such great varieties can be produced in the same species as we find to exist in man, it would be easy to maintain the probability that several species of *simiae* are but varieties of the species Man . . . And if the argument be still further extended, almost all the animal kingdom might be deduced from one pair, and be considered as one family; than which a more degrading notion certainly cannot be entertained."[43] Contrary to common assumption, monogenesis did not preserve human dignity; it subverted it. White's presentation of material on everything from skeletal structure and brain size to perspiration and varieties of hair thus led him to the conclusion that climate was powerless to explain the differences and that the only option was to concede "that species were originally so created and constituted, as to be kept apart from each other."[44] As for the biblical narrative, White was of the opinion that the "Mosaic account of the creation is believed, by most rational Christians, to be allegorical." But even if it were literally true, there was internal textual evidence for "another kind of mankind besides that descended from Adam . . . for we nowhere read of Adam and Eve having any daughters . . . Who then was Cain's wife, and whence did she come?"[45] In these judgments he consciously opposed Camper (1722–89), who thought "the whole human race descended from a single pair" and sided with the opinions of Kames and Edward Long (1734–1813).[46]

The lawyer and antiquary Edward King (1734/5–1807), who was elected to the Royal Society in 1767, pressed the internal biblical evidence for non-adamic humans even more strongly in his "Dissertation concerning the Creation of Man," which appeared in 1800 as part of a set of additions to the second edition of his *Morsels of Criticism*. References to Cain's wife, the inhabitants of his city, the identity of the giants recorded in the Pentateuch, and such like, rehearsed many of the niggles that had long troubled attentive Bible readers, and King presented the pre-adamites as the solution par excellence. While this treatise did not do much to enhance his reputation as a natural philosopher, it was popularly successful and delivered to readers a range of exegetical devices to support the existence of non-adamic races. The central plank of his argument was that "*Man* . . . was at first created *of one* GENUS indeed,—and all of *one blood,* and *in the image of* GOD;—but *of different species;*—with different capacities,—and powers,—and dispositions,—for very wise purposes."[47] It was, he confessed, "almost impossible to rest satisfied with believing . . . that the *White* European,—or Asiatic,—and the *Black long-haired South American,*—the Black *curled-haired African*

Negro,—the *Cossack Tartar,*—the *Eskimaux,*—and the *Malayan,*—were all descended from one common ancestor, and mother."[48] However "ingenious" and "plausible" Montesquieu's climatic thesis appeared to be, King remained unconvinced of its power to mold such dramatic racial differences.[49]

King's project embraced several dimensions, some of which invited political engagements. His advocacy of different species or classes or what the "Brahmins, in India, would call . . . *Different Casts*" of human, albeit within a single genus, for example, soon led him into talk about hybrid marriages, adulterated heritage, and racial mixing;[50] and his designation of the adamic family as "*That Transcending Class,* supereminent in all qualifications, and usefulness,—who were distinguished by the appellation of *The Sons of God,*" reinforced that sense of racial hierarchy that a subsequent generation would press into the harshest of supremacist theologies.[51] "Well-formed language," "true natural science," "sound philosophy," architecture, and other "blessings of high cultivation" were the progeny of Adam's instrumentality.[52] And, unsurprisingly, it was into this "unmixed line of clear descent" that the Messiah was born—a lineage markedly different from Cain, who, having "*debased his descent from Adam . . .* had already married into an inferior *cast,* or species of mankind."[53] Only the line from Adam's son Seth, "this *one* family of the highest *cast* continued *unmixed*" down to the time of Noah.[54] Besides all this, King claimed that even if the fall from grace had never taken place, the hierarchy of racial excellence that existed on account of "different original species" would have been advantageous for every racial type. Even "the lower species," he conjectured, "might, in consequence of that very difference, have received *more* enjoyment than they could have done in any state of excellence, where there was no such distinction."[55] Evidently, hierarchical difference was built into the very fabric of the universe, and it had eternal ramifications. An eschatological component was thus intrinsic to his speculations about extra-adamic humanity, who, "to the utmost possibility that their respective faculties afford," would participate in the salvific accomplishments of the adamic Messiah.[56] But even if King concluded his excursus by expanding on the universal benefits of the true adamic family, the potential for exploiting the narrative for different purposes was always there. King might be certain that "the ruins of Paradise, the *Garden of delights,* are the riches of the world;—and the feint resemblance of its arrangement, and improved order, the best blessing left by our first parents to the whole race of man," but others would readily read a less benign politics out of such a primeval anthropology.[57]

Adamic Politics

Whatever their different accounts of racial character and human origins, politics of one form or another were folded into the genesis narratives that monogenists *and* polygenists, supporters of the traditional Mosaic account *and* advocates of pre-adamism, alike advanced.

Moral Cartography

As much as anything else, Montesquieu was engaged in the construction of a global moral geography on climatically controlled physiological principles. The distribution of "the nervous glands" in the human body in different parts of the world was such that "different degrees of latitude" were characterized by "degrees of sensibility."[58] The consequences for cultural politics were massive. Medically, susceptibility to pain was regionally differentiated. Morally, inhabitants of northern climates had "few vices, many virtues," while in southerners "the strongest passions multiply all manner of crimes."[59] Mentally, hot climates produced people of indolence, with "no curiosity, no noble enterprise, no generous sentiment."[60] And there were military implications, even if undeclared, in such casual remarks as "You must flay a Muscovite alive to make him feel pain."[61] Muscovite Russia, we might recall, was regarded by Montesquieu as an instance of oriental despotism, a regime in which the lives and property of servile subjects were imperiously ruled over, it was supposed, by a single master.[62] Goldsmith, too, elaborated a world-scale moral climatology, judging the Laplanders as "rude, superstitious, and stupid" and exhibiting "all the virtues of simplicity, and all the vices of ignorance"; the Tartars as composed "chiefly of robbers"; the southern Asiatics as "a feeble race of sensualists, too dull to find rapture in any pleasures, and too indolent to turn their gravity into wisdom"; and the "Negroes of Africa" as a "gloomy race" whose climate had tended to "relax their mental powers" and render them "stupid, indolent, and mischievous."[63] The Europeans, of course, surpassed all these other groups and brought to "perfection" all those "arts which might have had their invention among the other races of mankind."[64]

From his polygenist stance Kames was no less willing to indulge in moral cartographics. Just as Montesquieu and Goldsmith elaborated a moral climatology of ethnic character, so Kames, working within a different explanatory framework, was every bit as happy to lay out a universal moral geography, with all the political ramifications that this view might imply. To him the "savages of Guiana" were "indolent," the Ladrone islanders "great cowards," the Laplanders "most timid," and the Macassars people of "active courage."[65] Montesquieu and Kames might differ on whether such characteristics were reducible to climate or were manifestations of "original disposition,"[66] but the moral maps of humanity that they each constructed bore an uncanny likeness to each other.

Human Origins and the Politics of Slavery

How Adam was understood had implications, too, for questions congregating around the slave trade. Certainly, there was no simple correlation between beliefs about adamic anthropology and positions adopted on slavery, though Ann Thomson has recently argued for significant links between materialism, irreligion, polygenism, and pro-slavery sentiments, on the one hand, and monogenesis, religious belief, and abolitionism, on the other, during the eighteenth century.[67] But even if that boundary cannot be quite so confidently drawn, it is still the case that claims about human origins were often intertwined with debates about the morality of slavery.

As early as 1680, the Church of England clergyman and missionary, first to Virginia and later Barbados, Morgan Godwyn, wrote at length in support of the right of African slaves and native Americans to be admitted to church membership in a tract for the times addressed to the archbishop of Canterbury. Rather sanguine about the practice of slavery itself, he vigorously argued their case in his lengthy 1680 plea on *The Negro's and Indians Advocate, Suing for Their Admission into the Church.* Godwyn was fully aware that what he called the "*Pre-adamites* whimsey" was being deployed first to "derive our *Negro's* from a stock *different* from *Adam's*" and then to "*Brutifie*" them.[68] His intention, by contrast, was "to prove the Negro's Humanity."[69] It was a strategy diametrically opposed to those Spaniards—he seems to have had Sepúlveda in mind—who had concluded that certain races were not human in order to "justifie *their murthering the Americans.*"[70] For all that, he acknowledged that fantastic, "false," "empty and silly," though the "foul *Heresie*" of pre-adamism was,[71] its original author himself had never used it to dehumanize any racial group but, rather, had acknowledged the full humanity of the pre-adamites:

We are to take notice that their objected *Pre-Adamitism,* doth not at all prejudice, nor even relate to this Question, which is touching the *Homoniety* of our *Negro's,* not their *Origination* or *Descent.* That Creation which they pretend to be *Antecedent* to *Adam's,* being according to the Principle of its first

Author [La Peyrère?], no less of *Real* Men than this latter; their Posterity therefore must needs be such . . . so this *Pre-Adamitism* doth utterly evacuate and overthrow whatever they from thence might otherwise *infer*.[72]

Yet his exemption of La Peyrère from the charge of bestializing any ethnic group only served to underscore how well aware Godwyn was of the theory's potential to shore up the politics of racial and ecclesiastical prejudice and so, advertising the refutation that Matthew Hale had issued, he took pains to detail what he considered to be the incoherences of this "irrational *Novelty*."[73] The political investment in the adamic narrative was plainly of epic proportions and not least in the realm of extending human rights to all racial groups.

A century or so later, in his prize-winning Cambridge essay of 1785 on the topic "Is it lawful to enslave the unconsenting?" the abolitionist and Quaker sympathizer Thomas Clarkson called upon the "hypothesis of climate" to account for the emergence of the different skin tones of all races from an original "dark olive" color, a shade that was both "beautiful" and "a just medium between black and white." He, too, was quick to draw out political consequences. This original skin tone, he reckoned, afforded a "valuable lesson to the Europeans, to be cautious how they deride those of the opposite complexion, as there is great reason to presume, *that the purest white is as far removed from the primitive colour as the deepest black*."[74] Such considerations, Clarkson was certain, refuted the claims of those who urged that the black races "were of a different species" and were "an inferior link in the 'chain of nature.'"[75] And here the politics of climate plainly surfaced. The foundation was laid for his resolute rejection of the view that "Africans [had] been *made for slavery*."[76] On every front slavery was abhorrent. As he concluded: "It is contrary to *reason, justice, nature, the principles of law and government, the whole doctrine, in short, of natural religion, and the revealed voice of God*."[77] The published text of his essay, which appeared in 1786 under the title *An Essay on the Slavery and Commerce of the Human Species, Particularly the African,* had the effect of recruiting William Wilberforce to the abolitionist cause.[78] It was the first of nearly two dozen works dealing with slavery by one of Britain's most enthusiastic abolitionist pamphleteers.

Around the same time, Reverend James Ramsay (1733–89), surgeon, abolitionist, and sometime student of the Scottish philosopher Thomas Reid, was adamant in his discourse on African slaves that climate and other local conditions explained different cultural traits, and he thereby sought to rebut the infidel tendencies of those speculating about separate human creations. He had Kames explicitly in his sights as he rejected the speculation that the different human families had "distinct progenitors" in favor of the view that "men had one common ancestor." With the backing of scripture, this latter teaching was so conspicuous a feature of divine revelation that Ramsay was certain that "we are not at liberty to pursue every wild conjecture."[79] Much better, in his opinion, to recall that "climate, diet and various modes of life have great power over the features, form and stature of man."[80] Again, all of this was part and parcel of an antislavery political project stretching back to the 1760s that aroused the antagonism of influential planters in the Caribbean, where, as an ordained Anglican, he worked among slaves. Renewed vituperation from that quarter also followed the publication of the *Essay on the Treatment and Conversion of African Slaves in the British Sugar Colonies* when it appeared in 1784, as readers were told that "Vice never appeared in Africa, in a more barbarous and shocking garb, than she is seen every day in the most polished parts of Europe . . . [C]limate, mode of living, and accidental prevalence of particular customs, will account for many national characteristics. But the soul is a simple substance, not to be distinguished by squat or tall, black, brown, or fair."[81]

In different guises similar alignments are discernible elsewhere. In 1775 James Adair, an Indian trader in the southern Appalachians, possibly of Irish birth, urged that the American Indians were of Jewish origin and claimed thereby to have overthrown Kames's entire polygenist system, and more particularly the "wild notion which some have espoused of the North American Indians being Prae-Adamites, or a separate race of men, created for that continent." Despite their isolation and absence "of the use of letters," they had retained "the ancient standard of speech," and this feature, to Adair, pointed to their incorporation within "the common laws of God in the creation of Adam."[82] His conclusion was plain: "the Indians have lineally descended from Adam, the first, and the great parent of all the human species."[83] Adair's purpose in advancing this theory was in part to work toward a plea, in the final pages of his treatise, for a renewed society in which native Americans would enjoy freedom and equality with their colonial neighbors. Although "his voice and message were drowned out by war," Adair believed that, by delivering to them a Jewish heritage that antedated even European traditions, he was enhancing the Indian cause and providing them with high status.[84]

The politics of human genesis were no less a feature of those undermining in various ways the monogenism of adamic anthropology. For all the apparent tentativeness that Kames advertised in his endorsement of a multiplicity of post-Edenic Adams, and however suspect the doctrine among the orthodox, such views were happily promulgated by those less unnerved by its taint of profanity, and not least for the theory's political possibilities. And yet the flow of the political tide did not always run in predictable channels. Charles White, who was attracted to the idea of multiple Adams, for example, declared himself an opponent of slavery. And an advertising flyleaf of the *Account of the Regular Gradation in Man* expressed its author's "hopes that nothing advanced will be construed so as to give the smallest countenance to the pernicious practice of enslaving mankind, which he wishes to see abolished throughout the world."[85] Again in the body of the text he declared himself "persuaded [that] the Slave trade is indefensible on any hypothesis, and . . . would rejoice at its abolition."[86] And yet for all that, the project's political bias lay exposed on virtually every page. The "African," he mused, "seems to approach nearer to the brute creation than any other of the human species," and a few pages later he asserted, "In whatever respect the African differs from the European, the particularity brings him nearer to the ape."[87] On page after page European superiority in anatomy, intelligence, and beauty found expression, and carried away with a rhetorical cadence as he reached toward the finale of his text, he waxed Solomonic on the charms of the European female—the allure of a "blush . . . that emblem of modesty" and the graces of a "bosom . . . tipt with vermillion."[88] To cap things he provided a classification table of race mixtures, elaborating on the differences between a "mulatto," a "quadroon," a "samboe," a "mestize," and so on, in terms of the proportions of white and black in the race mixture, down to detail of $15/16$ black and $1/16$ white,[89] and further added a statistical theorem to predict the colonial "effects of free and indiscriminate marriages of white and blacks."[90] His reading of polygenism into Genesis was nothing if not a project in cultural apologetics.

White's contemporary, the Jamaican slave owner and planter Edward Long, was at pains to establish that the black races constituted a different human species and in so doing sought to undermine any suggestion that racial difference was simply an epiphenomenon of climate. It was in his three-volume *History of Jamaica,* published in 1774, a work aptly described as a combination of polemics and propaganda,[91] that Long challenged Buffon's climatic account of racial difference. The transplantation of Africans into colder climates had failed to induce any hereditary modification, and this fact, Long maintained, subverted "the whole fabric of Mr. Buffon's hypothesis."[92] To him it was much better to attribute dark skins to the direct work of the Creator. This polygenist apologia, moreover, was part of a strategy to justify the enslaving of Africans, whom he considered a subhuman, inferior species. They were "void of genius, and seem almost incapable of making any progress in civility or science"; they had "no plan or system of morality," no taste for anything but women, "gormandizing, and drinking to excess."[93] With adjectival exuberance Long stockpiled foul adjective upon foul adjective to bestialize Africans: in a single paragraph they were excoriated as "brutish, ignorant, idle, crafty, treacherous, bloody, thievish, mistrustful, and superstitious" and a sentence later as "proud, lazy, deceitful, thievish, addicted to all kinds of lust . . . incestuous, savage, cruel, and vindictive, devourers of human flesh, and quaffers of human blood, inconstant, base, and cowardly."[94] All this was marshaled in favor of Long's rhetorical question, "must we not conclude, that they are a different species of the same *genus?*" The politics of plural origins were plain for all to see.

The Orang Connection

To drive home his conclusion, Long had exploited human diversity to its fullest possible extent. Thus, while Oliver Goldsmith minimized differences between human races—in all the animal kingdom "the differences between mankind are the smallest," he reported[95]—Long did everything in his power to maximize them. And in so doing he worked hard to shorten the space between human races and the orangutan—indeed, to incorporate the orang within the human family and present it as "savage man." Strenuously opposing Buffon's views on this matter, Long traced in some detail anatomical and behavioral correspondences between orang and human. Even on the issue of that distinctively human trait, speech, Long speculated that "human organs were not given him [the orangutan] for nothing" and that orangs had "some language by which their meaning is communicated."[96] To be sure, the orangutan was far from "our idea of a *perfect* human being"[97]—but it did come exceedingly close to certain human races.

Orangs and blacks provided decisive links in the chain of being that connected "brute creation" with civilized humanity. And in this connection some of his most scandalous opinions saw the light of day. Not only did Long consider that both apes and slaves displayed comparable lecherous behavior, but he even went so far as to contemplate sexual relations between orangutans and certain African tribes. In language that abolitionists found utterly repulsive, he suggested that "amorous intercourse between them may be frequent; the Negroes themselves bear testimony that such intercourses actually happen; and it is certain, that both races agree perfectly well in lasciviousness of disposition."[98] This of course confirmed the judgment he had issued just a few pages earlier that, "ludicrous as the opinion may seem, I do not think that an orang-outang husband would be any dishonour to an Hottentot female; for what are these Hottentots? They are, say the most credible writers, a people certainly very stupid, and very brutal. In many respects they are more like beasts than men."[99] Shocking it might seem, but this, in Long's mind, was precisely what had been "ordained by the Divine Fabricator" when he created beings "ascending" through the various "gradations of the intellectual faculty," from its rudimentary presence in apes to "the utmost limit of perfection in the pure White."[100] Such a polygenist doctrine fit perfectly the needs of a trader in Guinea slaves whose "natural baseness of mind," he contended, "seems to afford least hope of their being (except by miraculous interposition of the divine Providence) so far refined as to *think*, as well as act like *perfect men*."[101] All this was in keeping with his claims in *Candid Reflections* (1772) that African biology equipped black slaves for plantation labor.[102]

It was at this point of conjunction between human and orangutan that Long most conspicuously approached the writings of such other students of human nature as the seventeenth-century physician, anatomist, and fellow of the Royal Society Edward Tyson (1651–1708), and his own Scottish contemporary, the legal philosopher James Burnett, Lord Monboddo (1714–99). Tyson, who is credited with having conducted the first anatomical dissection of a chimpanzee, reported his findings in a 1699 discourse on what he called the "orang-utan," or "pygmy."[103] In a treatise that incorporated detailed illustrations of the creature's skeletal structure and other physiological features (fig. 11), he argued that the orang-pygmy did not belong to the human species, as earlier writers (whose opinions he surveyed at some length) had supposed. In the past, he noted, "the *Antients* were fond of making *Brutes* to be *Men:* on the contrary now, most unphilosophically, the *Humour* is, to make *Men* but mere *Brutes* and *Matter.* Whereas in truth *Man* is part of *Brute*, part an *Angel;* and is that link in the *Creation*, that joyneth them both together."[104] Yet he did acknowledge the creature's remarkable proximity to the human species and elaborated in detail how the "Orang-Outang or Pygmie more resembled a Man, than Apes and Monkeys do" and, correspondingly, how it differed from the human and approximated the ape.[105] The specimen evidently occupied a transitional location in the chain of being (something like the way he considered the porpoise to be intermediate between fish and land animals),[106] albeit firmly on the side of the animal kingdom. Nevertheless, he conceded that the connection was so strong that when he was "de-

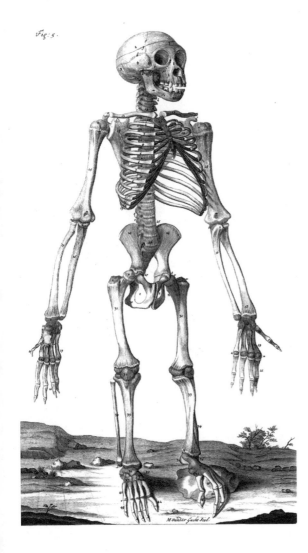

Fig. 11. Pygmy skeleton, from Edward Tyson's 1699 investigations into pygmy anatomy in *Orang-Outang*.

scribing the *Brain* of our *Pygmie*," readers could well "justly suspect" that he was "describing that of a Man."[107]

Long, who was certainly familiar with Tyson's conspectus,[108] was evidently willing to narrow the gap further between orang and human—no doubt at least in part for the purposes of slave apologetics—and to that degree approximated yet more closely the views of James Burnett–Lord Monboddo, the brilliant Scottish controversialist and associate of men such as Joseph Banks, William Jones, David Hume, Adam Smith, James Hutton,

Dugald Stewart, and James Boswell. The contemporary debates surrounding whether or not humans shared with apes the upper intermaxillary bone (the bone that in human infants fuses with others to form the upper jaw but remains separate in apes) certainly kept the question open. Goethe's 1784 challenge to Petrus Camper, who, having dissected an orangutan between 1778 and 1782, insisted that humans did not possess the *os intermaxillare,* not only revealed his agreement with Gall, that all species enjoyed a unity of structure based on homologous forms,[109] but left the relationship between human and orang tantalizingly ambiguous.

In Monboddo's case it was a concern with the character and origins of speech that led him to his controversial claims about the orangutan and his conviction that the creature was human. With a characteristic Scottish Enlightenment interest in the nature of human nature,[110] Monboddo focused his concerns on the question of language, whether it was a natural attribute or an acquired achievement. His own position was crystal clear. Siding with Condillac, he was certain that language was a human invention, not a natural endowment, and therefore not essential to human nature. To be sure, he considered that the *capacity* to acquire language was a human trait, but this emphatically did not mean that its *acquisition* was a necessary condition for human status. Bolstering his account with the testimony of ancient philosophers, "none of whom ever dreamed, that any thing else was essential to men, except reason, and intelligence," he confidently asserted that "articulation is not natural to man."[111] While his point of departure was an elucidation of the progress of the human mind, Monboddo came to acknowledge that the existence of society was a prior necessary circumstance for the evolution of language to take place. Having thus uncoupled the faculty of speech from intrinsic human constitution, Monboddo was certain he had opened the door to identifying creatures without articulate speech as human. This included what he believed to be certain African tribal groups bereft of language and, controversially, the orangutan.[112]

It was in the second book of his unfinished inquiry into the origin and progress of language that Monboddo turned his attention specifically to the account of the orangutan that Buffon and Linnaeus had delivered. By this stage of the argument he had already established to his own satisfaction that language had been constructed, that "there must have been society before language was invented,"[113] and that, as he put it, while "man in his natural state, is the work of God . . . as we now see him, he may be said, properly enough, to be *the work of man.*"[114] As such, Monboddo could confidently identify speechless human races, notably those "wild races of men" inhabiting "the woods of Angola, and other parts of Africa."[115] This sentiment, he acknowledged, had evidently given offense to some Christian readers, and so he went out of his way to protest the theological propriety of his speculations, maintaining that in the prelapsarian world the first human enjoyed "many faculties belonging to his nature, of which he is now no longer possessed; and, among others, he may have had the faculty of communicating his thoughts by articulate sounds, which were understood, as soon as uttered, by those who heard them. But this natural faculty, as well as others, we may suppose that he lost upon his fall." The need to acquire language in a fallen world was, in fact, the consequence of "the curse then pronounced upon him."[116] Perfection-degeneration-progress: that was the linguistic sequence through which speech had passed. And with this schema firmly in place, Monboddo now turned attention to the orang and to his "hypothesis" that here was "a barbarous nation, which has not yet learned the use of speech."[117] Drawing on the earlier investigations of Tyson, Buffon, Purchas, Gassendi, and numerous other travelers and natural philosophers, he did everything in his power to humanize the orangutan and to bring the creature within the orbit of the human family. As he concluded: "The substance of all these different relations is, that the Orang Outang is an animal of the human form, inside as well as outside: That he has the human intelligence, as much as can be expected in an animal living without civility or arts: That he has a disposition of mind, mild, docile, and humane: That he has the sentiments and affections peculiar to our species, such as those of modesty, of honour, and of justice . . . That they live in society, and have some arts of life."[118]

While Monboddo's account, taken in the round, was comparatively bereft of the crudely racialist, pro-slavery language of Long, he nonetheless associated the orangutan with certain peoples whom he considered to be in a savage stage of human development, notably the Hottentots, the Iroquois, and various African peoples "without the use of Speech."[119] His exposition, moreover, revealed him as sympathetic to a polygenetic reading of the origins of post-Fall human language, when he declared that there was "no reason to believe, that it [language] was invented only by one nation, and in one part of the earth; and that all the many different languages . . . are derived all from this common parent." There just was no "one primitive language."[120]

Stanhope Smith and the Foundations of the American Republic

Monboddo's proto-evolutionary account of the emergence of articulate speech from more primitive sounds and signs and his stadial portrayal of the progress of humanity rising from an "originally a wild savage animal, till he was tamed, and . . . *humanized,* by civility and arts," have latterly earned him much "appreciation" from those seeking for anticipations of present-day evolutionary orthodoxy.[121] In his own time it was a different story. Indeed, he anticipated that his account would "give offence to some,"[122] and in this judgment he was certainly not mistaken. Nowhere was this more conspicuously expressed than in the writings of the American clergyman-philosopher Samuel Stanhope Smith (1751–1819), professor of moral philosophy and subsequently president of the College of New Jersey (Princeton), who divined profoundly sinister political implications in Monboddo's orangutan philosophy, in Kames's toying with non-adamic human races, and in White's polygenist anatomy. There was "nothing more contemptible than philosophers with solemn faces, retailing like maids and nurses, the stories of giants—of tailed men—of a people without teeth—and of some absolutely without necks. It is a shame for philosophy at this day," he went on, "to be swallowing the falsehoods and accounting for the absurdities of sailors." The reason for this state of affairs, he reckoned, was at least in part an American one. "We in America," he pondered, "receive such tales with more contempt than other nations; because we perceive in such a strong light, the falsehood of similar wonder, with regard to this continent, that were a few years ago reported, believed, and philosophized on in Europe."[123] In some ways Smith's critique, and his resort to environmentalism to preserve human unity, paralleled the reprimands that others had already issued to those who departed from humanity's traditional adamic genealogy. But Kames's most conspicuous critic had domestic concerns in mind, too, and the particular setting he occupied at Princeton during the early years of the new republic added a distinctively American inflection to the debate.

It was in his *Essay on the Causes of the Variety of Complexion and Figure in the Human Species,* which first appeared in 1787 and then in an expanded form in 1810, that Smith determined to show that the unity of the human species was not only good science; it was also good theology and, even more important, good moral philosophy. Presented initially as an oration before the American Philosophical Society, Smith's analysis was "the first American treatise devoted to the causes of racial variation in the human species,"[124] and it secured for its author an acknowledged place in the annals of American anthropology.[125] William Stanton, for example, recorded that his book was "the first ambitious American treatise on ethnology and long a standard work in the United States."[126] The work was, in fact, an extended exercise in climatic determinism, so much so that Marvin Harris believed that here the "ultimate pitch of environmentalism" had been reached.[127] In a nutshell Smith insisted that climatic factors predominantly, but also social conditions, could fully explain racial variation. Polygenism was thus to be rejected. According to Winthrop Jordan, "Smith's *Essay* stands as a monument to American faith in environment. It was a special faith. Environmentalism had been an underlying animus in the Revolutionary era and it towered as Americans stood poised and triumphant overlooking their magnificent continent."[128]

The salient features of Smith's assessment can readily be surveyed. Climate, migration, and attendant social circumstance were more than adequate to account for human variability. The distinguishing features of human societies were nothing more than responses to "the influences of the climate, of the sterility or richness of the soil, of the elevation or depression of the face of their country, of the vicinity of seas or desarts [*sic*], of their insular, or continental situation: or the modifications of all these, resulting from their occupation, and their habits of living."[129] By reference to these geographical factors a single human constitution was preserved—flexible, to be sure, but common nonetheless. One implication of this commitment, of course, was that the human species had undergone some evolutionary modification, and Smith plainly embraced the doctrine of the inheritance of acquired characteristics as the explanatory mechanism. Another was human cosmopolitanism: humankind was evidently capable of acclimatizing to the new climatic regimes, which typified zones of colonization.[130] Smith, like his opponents, went on to expand on the regional geography of the human complexion. "From the Baltic to the Mediterranean," he confirmed, "the different latitudes of Europe are marked by different shades of skin colour."[131] Human hair, skeletal stature, forehead shape, physiognomic features, even mental powers and moral inclinations, were all correlated with "climates . . . some states of society, and modes of living."[132] Along the way he relegated to the realm of mythology the exotic tales of travelers who were only too willing to generalize from any "little deviation from . . . [the] ordinary standard" in order to indulge in "the most hyperbolic relations."[133] Supremely, Smith's mechanistic environmental determinism—thoroughly characteristic as it was of Newtonian science—seemed to be confirmed in America's "ethnological laboratory."[134]

A common origin and a shared nature, of course, did not entail ethnic or cultural uniformity, and Smith, evidently as accepting of cultural relativism as Montesquieu, accounted for "savage life" in terms of degeneration. Nevertheless, he remained convinced that differences between the races were literally only skin-deep. He was certain, for example, that "if the Anglo-

American, and the indian [*sic*] were placed from infancy in the same state of society, in this climate which is common to them both, the principal differences which now subsist between the two races, would, in a great measure, be removed when they should arrive at the period of puberty."[135] To be sure, this did not amount to modern egalitarianism, but it did confirm the unity of the human species.

Whatever its scientific dimensions, adamic rhetoric lay at the foundations of Smith's refutation of polygenism. He paraded Kames, for example, as engaged in "laudable attempts to disprove the truth of the Mosaic story" and considered him an "excellent specimen of the easy faith of infidelity!"[136] In the "Strictures on Kaims [*sic*]" he quoted at length the Scotsman's seeming loyalty to the Genesis narrative, adding: "When ignorance or profligacy pretends to sneer at revelation and at opinions held sacred by mankind, it is too humble to provoke resentment. But when a philosopher affects the dishonest task, he renders himself equally the object of indignation and contempt."[137] It was much same with his riposte to Charles White. "I cannot close these observations," he snapped

> without reprobating in the strongest manner, that disingenuous mode of assailing the holy scriptures which has become fashionable with a certain class of writers, and which this gentleman affects to imitate. They speak of them with oblique and ambiguous respect . . . while at the same time, it is suggested that if we do believe them, it must be in spite of nature, and of the most certain physical facts. Thus do these authors study to undermine revealed religion by hinting that its friends require only implicit faith in opposition to all the truth of science. This mode of attack I cannot regard as either fair, or manly.[138]

The ironies in these exchanges, of course, are manifold. For in Smith, the self-appointed defender of orthodoxy, we find one who rejected special providences and thereby naturalized explanations by conceiving of God as building certain intrinsic qualities into the very stuff of reality. By contrast Kames, the supposed infidel, resorted to miracle to explain the diversification of languages at Babel and clearly was far more supernaturalistic in his belief in separate human creations. Smith's achievement in branding Kames as impious required, as Mark Noll has put it, "considerable rhetorical finesse."[139] And yet above and beyond matters of scientific evidence *and* theological decorum was politics. At base Smith's fundamental concern was the management of the moral economy. Science and theology alike were directly implicated in matters of social polity. For Smith what was at stake in researches into historical anthropology was nothing less than the constitution of society and the maintenance of the social order.

Certainly, Smith was delighted that his own Baconian investigations served "to confirm the facts vouched to us by the authority of holy writ,"[140] and he did not hesitate to elaborate in detail on what he considered to be the empirical errors of Charles White, whose claims were variously challenged as incoherent, inconsistent, and shy of evidence. For it was crucial to Smith that his anthropology did not rest solely on theological foundations. As he had made clear on the very first page of the treatise: "The unity of the human race, notwithstanding the diversity of colour and form under which it appears in different portions of the globe, is a doctrine, independently of the authority of divine revelation, much more consistent with the principles of sound philosophy, than any of those numerous hypotheses which have referred its varieties to a radical and original diversity of species."[141] What *was* critical were the political implications of non-adamic humanity. Were there different species of humans, they would "be subject to different laws both in the physical and moral constitution of their nature."[142] The empire of climate at once preserved human nature from pre-adamite subversion, secured the intellectual integrity of moral philosophy, and delivered the promise of political stability:

I must repeat here an observation which I made in the beginning of this essay, and which I trust I am now entitled to make with more confidence, that the denial of the unity of the human species tends to impair, if not entirely destroy, the foundations of duty and morals, and, in a word, of the whole science of human nature. No general principles of conduct, or religion, or even of civil policy could be derived from natures originally and essentially different from one another, and, afterwards, in the perpetual changes of the world, infinitely mixed and compounded. The principles and rules which a philosopher might derive from a study of his own nature, could not be applied with certainty to regulate the conduct of other men, and other nations, who might be of totally different species. The terms which one man would frame to express the ideas and emotions of his own mind must convey to another a meaning as different as the organization of their respective natures. But when the whole human race is known to compose only one species, this confusion and uncertainty is removed, and the science of human nature, in all its relations, becomes susceptible of system. The principles of morals rest on sure and immutable foundations.[143]

At bottom what Smith found so offensive about the speculations of Kames, Monboddo, and White about non-adamic humans was that they were unsuited to the American project, in which it was vital, Smith believed, to have a common constitution to underwrite human morality. In the early days of the new American Republic a confidence in a common human constitution was precisely the philosophy that was needed if public virtue were to be retained in a society "that was busily repudiating the props upon which virtue had traditionally rested—tradition itself, divine revelation, history, social hierarchy, an inherited government, and the authority of religious denominations."[144] And here the Scottish Common Sense philosophy that Smith had imbibed during his student years from his soon-to-be father-in-law, John Witherspoon, was vitally important, for it insisted fundamentally on a universal human sensibility that netted together every single member of the human race. Without that unifying constitution, as he put it in the first edition of the work, any "science of morals would be absurd; the law of nature and nations would be annihilated."[145] The Scottish philosophy was thus, as Norman Fiering observes, "uniquely suited to the needs of an era still strongly committed to traditional religious values and yet searching for alternative modes of justification for those values."[146] After all, James Madison (himself a former student of Witherspoon at Princeton) and the other architects of the Constitution required a ubiquitous human nature in order to undergird public moral order.[147] Otherwise, the "science of politics" that Madison adumbrated in the famous *Tenth Federalist*, showing his whole-hearted endorsement of the Enlightenment style, would dissolve.

From Witherspoon, Stanhope Smith knew how the Scottish philosophy, articulated by the Irish-born Francis Hutcheson, had developed a scientific and optimistic social theory that did not need to resort to the dictates of tradition and cultivated a moral Newtonianism that rendered ethics a species of physical science by appealing to a moral sense intrinsic to all humanity.[148] It was only because there *was* a common human nature that it could be the object of empirical scrutiny. And it was precisely this idea that Smith's assault on non-adamic humanity was designed to preserve in the face of seeming anthropological disruption. If, as Witherspoon was certain, a republic was right to break both with its monarchical past and religious establishment; and if republics, dependent as they were on the performance of public virtue, were the fragile thing Montesquieu had shown them to be, then morality could only be preserved in the public square if a universal ethical sense could be extracted from human nature by the methods of empirical science. Talk of non-adamic races, different human species, multiple origins, and the like

ADAM'S ANCESTORS

struck a blow at the very heart of the new America's moral economy. Small wonder that Smith and other moral philosophers in the fledgling republic did everything in their power to convey to their students the universality of the human constitution. Only that way could the nascent nation endure the stresses and strains that must inevitably come its way and prevent the fissiparousness that otherwise would ensue. Adamic politics was the foundation on which the edifice of public order must be erected.

⅋ The debate over the status of the biblical Adam, whether he had ancestors and whether he shared the earth with other human beings independently generated, flourished throughout the period of the Enlightenment. To the earlier speculations of La Peyrère and his contemporaries were added the reflections of a new century's suite of natural philosophers, political theorists, overseas travelers, medical practitioners, and moral theologians. Claims about comparative anatomy, cultural anthropology, natural history, regional climatology, and colonial history were all folded into the competing narratives of humanity's genesis that both monogenists and polygenists offered. Yet underlying these scientific preoccupations were fundamentally political fixations: the elaboration of moral maps of the globe, the ethics of plantation economies, and the management of domestic polities. Whether or not Adam had ancestors, whether or not too much had been made of the empire of climate, whether or not there were originally different varieties of the human constitution—these questions had the profoundest implications for everything from the cultural politics of human anatomy to biblical hermeneutics, from the justification of the slave system to the regulation of republics.

4 ❀ APOLOGETICS

Pre-Adamism and the Harmony
of Science and Religion

A S THE NINETEENTH CENTURY DAWNED, the sense that strategies needed to be developed to keep scientific findings and religious faith in tandem gripped more and more minds. Challenges from geology, ethnology, and philology, supplemented
later by the yet more dramatic gauntlet thrown down by the Darwinians,
encouraged the cultivation of harmonizing tactics. Among these the postulation of a pre-adamite world—sometimes inhabited by human or humanlike creatures, sometimes not—began to attract an increasing coterie of advocates. Pre-adamite vocabulary thus obtruded with increasing frequency
into the dialogue between scientific theory and religious conviction. Some
of these occurrences connected the pre-adamite device to matters of racial
difference; others only touched on racial implications, though a later generation, as we will see, would muster their conjectures in the cause of supremacist politics. Either way, the suspicion spread that there was an ever-growing
need to find ways of reading the Bible so as to accommodate the findings and
speculations of science, and pre-adamite talk was one way of dealing with the
problem. The very idea that had originated in skepticism and had provided
sustenance to those radicals who sought to exploit the cracks beginning to
appear in the canon was now beginning to be harnessed for the purpose of
protecting the Bible from the results of infidel science.

As Edward King announced at the very birth of the new century, inter-
preters of the Bible should "carefully avoid *rashness*" and "examine most in-
quisitively, what *really is, or is not declared therein;*—using exceeding great

caution that we may not be misled by any prejudices, and mistakes, however riveted, by long prescription, and timid allowance."[1] This declaration appeared in an addendum to the second edition of King's *Morsels of Criticism,* which had first appeared in 1788, which was specifically dedicated to the task of showing, as its title page advertised, "the most perfect Consistency of Philosophical Discoveries, and of Historical Facts, with the Holy Scriptures." In particular he was aware that the traditional story of universal human descent from Adam had "caused a great many stumbling-blocks to be laid in the way of those who wish to understand the Sacred Writings," and he insisted that the received doctrine was "far from being really founded on Scripture."[2] His own course of action was to point out that the Genesis narrative housed two quite different creation stories, the first dealing with humankind in general, the second with an individual, Adam. This realization provided one of "many proofs and arguments that may be derived from the Holy Scriptures themselves, which tend to show . . . that *the commonly received opinion, that all mankind are the sons of Adam* . . . is directly contrary to what is contained there."[3]

Such hermeneutic maneuvers opened up the possibility of there having been other inhabited regions of the earth at the time of the Mosaic story and thus removed any need to believe "that the *Kongouro* (for instance), or certain other strange animals of Africa, and of America, were in the ark with Noah." It was the same with "*very noxious animals* . . . such as serpents, and toads, and most venomous snakes, of *such* kinds as have never been met with in Armenia, Mesopotamia, Syria,—or indeed in any part of Asia."[4] The point of the whole exercise was to "take away at once all grounds of those taunting scoffs, and mockings of blasphemers, that have been the disgrace of all ages" and at the same time to "remove many of the difficulties of the natural historian."[5] In one way or another, then, pre-adamite speculations were increasingly evident in works that sought to retain harmony between science and scripture.

Pre-Adamite Worlds

During the early decades of the nineteenth century traditional understandings of the Mosaic cosmogony had come under increasing strain from the encounter with new geological data. An ever-expanding time frame put more and more pressure on standard chronologies of the earth, and a growing fascination with deep time fostered by visual representations of primitive life forms raised the public's consciousness of a distant primeval world.[6]

Engravings and illustrations, together with the famed models of dinosaurs erected in Crystal Palace, which were opened to the public in 1854, raised in some minds the question of why so much history had elapsed before the appearance of the crowning glory of creation—the human race. Commenting on this "Pre-Adamite World," one observer gave voice to "the strangeness of the fact, that myriads of creatures should have existed, and that generation after generation should have lived and died and passed away, ere yet man saw the light."[7] The need to find some rationale for this state of affairs gripped the pens of theological writers.

The Primeval Earth

Among those to take up this precise challenge was John Harris (1802–56), a Congregationalist clergyman and, from 1838, president of Cheshunt College in Hertfordshire. Subsequently, after the amalgamation of several independent colleges, he assumed the principalship in 1851 of New College, London, another Congregationalist institution. As the title of his 1846 volume makes clear, his *Pre-Adamite Earth* mobilized ideas about pre-adamic geology as a set of "Contributions to Theological Science." A popular author, he published on both metaphysical theology and the importance of clerical education. *The Pre-Adamite Earth* was a substantial philosophical essay in which he sought to establish a workable relationship between theology and the emerging sciences of the earth. The whole treatise was erected on teleological foundations and showed Harris's familiarity with a wide range of scientific writers, including John Herschel, William Whewell, Jean-Baptiste de Lamarck, Pierre-Simon Laplace, Hugh Miller, and Alexander von Humboldt. His treatment of the subject revealed him to be profoundly antipathetic to the new evolutionary speculations emanating both from France and from the recently published, anonymous *Vestiges of the Natural History of Creation.* Certainly, he defended the notion of development in earth history, but it was a creationist progression and emphatically not the "wild speculation, reckless alike of inductive facts and of moral consequences," being promulgated by those advocating radical evolutionary continuities.[8] That "fanciful hypothesis," he quipped, was "contradicted at every step."[9] But he was persuaded nonetheless of a lengthy geological history and considered it his task to elucidate how the development of the earth constituted an extended preparation for its later human occupants. A good deal of time was spent on demonstrating to his own satisfaction the intellectual coherence of a pre-adamite world essentially anticipating the subsequent adamic drama and thus playing out what he called the law of progression. In analyzing the geological strata, observers were "looking on the successive steps of a scene preparatory for a new stage of the Divine Plan . . . The inorganic world was designed by the Divine Creator to become the scene of organic forms—of life."[10] Packing all of geological history into the first verse of the book of Genesis enabled Harris to leave the subsequent six-day Creation story literally intact. To him a pre-adamite cosmos predated the Creation narrative, and it was a world restricted to prehuman inorganic and organic nature.

In pursuing this strategy, of course, Harris was simply revisiting a scheme that had been in favor among a number of geological writers for quite some time. The Scottish evangelical intellectual Thomas Chalmers, for example, had been telling his followers at least since 1814 that a gap of as yet unspecified time—into which global geological history could be squeezed—existed between the initial act of Creation and the point where the Genesis account takes up the story.[11] The vast temporal possibilities that his gap theory opened up were matched by the limitless spatial extension that he meditated on in a series of astronomical discourses that showed his theological enthusiasm for a plurality of worlds inhabited by divinely created beings.[12] By the end of that same decade the Oxford geologist William Buckland had adopted the same harmonizing system and later, in his 1836 *Bridgewater Treatise,* supplemented the argument with linguistic evidence from Hebrew that he had gleaned from the biblical scholar E. B. Pusey.[13] Where Harris went beyond these standard concordist schemes was in his attempt to craft a metaphysics capable of mounting a robust defense of the idea that pre-adamite earth history was orchestrated by the Creator to be the theater for human occupancy.

Three years later, in 1849, when Harris brought out *Man Primeval,* he reconfirmed these convictions and defended the traditional adamic six-day Creation as part of another metaphysically inclined excursus, this time concentrated on the essential nature of the human constitution. "Man was not made for the earth," he proclaimed in his opening sentence; "the earth, from the first, had been preparing for man, and we are to suppose that now, at length, the hour of his creation had arrived."[14] His self-appointed duty in this text was to elucidate the essence of human nature. Understanding, reason, imagination, emotion, conscience, language—these were the lenses through which the human subject should be investigated. Here he paused to reflect on the whole question of the unity of the human species and reviewed the polygenetic proposals that were on offer, making use of the critique advanced by physical anthropologists such as Johann Friedrich Blumenbach and James

Cowles Prichard and philologists such as Bunsen, all of whom defended the unity of the human race. And while he dispassionately reviewed the polygenist case and acknowledged that "the identity of humanity cannot be regarded as dependent on, or necessitated by, an identity in the original means of production," he was of the opinion that the doctrine that "all the families of mankind have actually descended from a single pair, appears . . . to be taught in the account of the Adamic creation."[15] Experience confirmed the verdict. For even the "most dissimilar races," he judged, "are found . . . to be Psychologically identical. Tribes rashly proscribed as on a level with the brute, have in our own day vindicated their claim to a common humanity. The metropolis of civilization is not without its degraded Bushmen . . . As far as we know, no race of men stands in intellectual or moral isolation."[16] His conclusion was thus crystal clear: "All the branches of evidence appropriate to the inquiry support each other, and unite in authenticating the conclusion that the human species is one, and that all the differences which it exhibits are to be regarded merely as varieties."[17] To Harris, in sum, the designation *pre-adamite* was no more than an adjectival description of the primeval earth prior to the advent of the human species, whose genesis was recorded in the Mosaic story of Adam. Certainly, he emphasized the significance of progression in creation, but he resisted any recasting of such a trajectory in terms of evolutionary transformism. "The development of which we speak," he insisted, "is not of one thing from another, but of the Divine plan of creation, and of our conceptions of that plan."[18]

Reviewers agreed. One, writing from Columbia, South Carolina, waxed lyrical with enthusiasm for Harris's "sound logic, lucid reasoning, great research, systematized facts, and well-digested science" even while abominating "the development hypothesis—the most dangerous form of modern infidelity."[19] To this commentator Harris had done a superlative job in showing that Christian culture had nothing to fear from the science of geology. And indeed, the writer did not miss the opportunity that Harris's *Pre-Adamite Earth* afforded for pressing for the teaching of natural history in his own state and bemoaning its absence in particular from the South Carolina College. Another anonymous writer concurred and, after delivering a thumbnail sketch of the various strata that constituted the pre-adamite world, noted that because geology had contributed "incalculably to the comforts of life . . . the Government of every civilized country, has been employing Geologists to explore its territory."[20] Plainly, writers supportive of geological revelations about deep time found talk of a pre-adamic earth history attractive not only because it allowed them the opportunity to tell religious believers that they had nothing to fear from this branch of science but also because it rendered the subject safe for inclusion in college curricula.

The idea of an uninhabited pre-adamite earth continued to show up amongst religious writers wanting to retain a traditional interpretation of the days of creation and yet allow for a lengthy global history. This was the case, for example, with Herbert William Morris—described as professor of mathematics at Newington Collegiate Institute, London—who addressed the issue in the final decades of the nineteenth century. Morris adhered to a literal reading of the days of creation and put forth his views in an 1871 volume on *Science and the Bible; or, the Mosaic Creation and Modern Discoveries,* subsequently reprinted in Britain under the title *Work Days of Creation,* and then in an 1885 work, *Present Conflict of Science with the Christian Religion.*[21] In 1890 he brought out a second edition of the first of these works, noting in the new preface that he had "added a large amount of new and interesting matter, particularly respecting *the pre-Adamite history of the Earth*—its numerous and surprising revolutions of continents and oceans; the deposition of iron, coal, and salt rocks . . . its mysterious changes of climate; the cold and submergence and desolation of the Glacial Period, and the beneficial results these were designed to work out."[22] All of the geological periods— the Laurentian, Cambrian, Silurian, Carboniferous, Permian, Triassic, Jurassic, and so on—were identified and relegated to the pre-adamite earth, and Morris provided his readers with luscious engravings to shape their imaginations of a lost primeval cosmos that had sunk into chaos between the first two verses of the Bible (fig. 12). "Here, then, is a *Hiatus*—a vast gap—in the Mosaic narrative," Morris told his readers, "which it is important to observe. Between the *creation* of the earth, as stated in the first verse, and the *condition* in which it is found and described, in the second verse, there must have elapsed a long and indefinite period of time."[23] Creation then ruin then restoration and re-creation—this was the sequence through which the globe had passed.[24] As with Harris, Morris's pre-adamite world was devoid of human beings and prefatory to the literal six-day set of creative transformations that was recorded in the Mosaic account.[25]

Hints of a Primordial Race

Other writers concerned to keep Genesis and geology in harmony, however, were less sure that the pre-adamite earth was free of humanlike inhabitants, and the idea of pre-adamite races, of one sort or another, began to attract supporters. John Thomas (1805–71), for example, one of the founders of

FIG. 12. "Scenes in the Pre-Adamite World," in H. W. Morris, *Work Days of God.*

the religious sect known as the Christadelphians—an anti-trinitarian, millenarian, Protestant group without an ordained clergy—used the idea of pre-adamites specifically as a means of accommodating a long earth history. In his *Elpis Israel* (The Hope of Israel), one of the founding documents of the denomination, written in 1849, he advanced the view that there "are hints, casually dropped in the Scriptures, which would seem to indicate, that our planet was inhabited by a race of beings anterior to the formation of man" and that these "pre-Adameral inhabitants" were to be understood as fallen angels.[26] These beings were subject to the "total wreck of their abode, and their entire submergence, with all the mammoths of their estate, under the waters of an overwhelming flood" reduced the earth to the formless emptiness that readers encountered in the first verses of the Bible. The possibilities of this conjectural prehistory for understanding recent geological revelations was clear:

> Fragments . . . of the wreck of this pre-Adameral world have been brought to light by geological research, to the records of which we refer the reader, for a detailed account of its discoveries, with this remark, that its organic remains, coal fields, and strata, belong to the ages before the formation of man, rather than to the era of the creation, or the Noachic flood. This view of the matter will remove a host of difficulties, which have hitherto disturbed the harmony between the conclusions of geologists and the Mosaic account of the physical constitution of our globe.

ADAM'S ANCESTORS

Geologists have endeavoured to extend the six days into six thousand years. But this, with the Scriptural data we have adduced is quite unnecessary. Instead of six thousand, they can avail themselves of sixty thousand; for the Scriptures reveal no length of time during which the terrene angels dwelt on our globe.[27]

Again, the numismatist and Anglican clergyman Henry Christmas (1811–68) suggested in 1850 that there might well have been pre-adamic people and that acknowledgment of their existence provided a suitable way of coping with geological time. Editor of several Anglican periodicals and author of works on theological and historical-geographical subjects, he contributed numerous pieces to the *Numismatic Chronicle* and enjoyed fellowship of the Royal Society.[28] Christmas's *Echoes of the Universe,* which began as a series of lectures to young Anglicans, was intended to buttress the "connection between human science and divine truth."[29] In the fourth of these excursions Christmas turned to the task of seeking to "make the discoveries of modern geologists tally with the account given in the book of Holy Writ," supportively quoting the contributions made in Harris's *Pre-Adamite Earth.*[30] His approach was anticipative. It sketched out a possible theory of early times that could be called upon to meet the challenge forthcoming from recent scientific advances. Evidence of geological time, together with the old challenge from Chinese chronology, pressed him to consider the possibility that the early Genesis narrative incorporated two accounts and that the former referred to a "previous creation."[31] This pre-adamic world, in turn, was subject to some great "convulsion" that cleared away "the debris of a former race," thereby "preparing the globe for that which we now behold."[32] In earlier times "among the antediluvians themselves" there were traditions "concerning races of men in existence before Adam," and this in itself, Christmas observed, provided grounds for "that theory which we have laid before the reader."[33] To be sure, there were difficulties in the way of such a reading, not least over the whole question of death as a consequence of Adam's Fall, and Christmas devoted some energies to establishing that mortality was actually a "condition of creation."[34] But there he left the matter, as a potential means of keeping science and religion in harmony.

Isabelle Duncan's Conjecture

For all the rather fleeting suggestions that a pre-adamite world could rescue the Bible from the assaults of the geological column, the most sustained prosecution of such a scheme appeared in 1860, in an idiosyncratic theological oddity sharing diagnostic features with the version tersely sketched by John Thomas in *Elpis Israel*. Initially anonymous, *Pre-Adamite Man; or, the Story of Our Old Planet and Its Inhabitants* was the work of Isabelle (sometimes Isabella) Duncan. Within the previous year or two, evidence of human artifacts in strata with long-extinct animals had been excavated at Brixham Cave, and these findings brought into sharper focus the archaeological questions that had earlier been raised with Jacques Boucher de Perthes's *Antiquités celtiques et antédiluviennes* of 1847.[35] Developments of this stripe posed new questions for Bible readers, and Duncan's book was one of the earliest, from an evangelical standpoint, to try to address the issue. Throughout it she demonstrated considerable familiarity with the works of natural scientists, including David Thomas Ansted, William Buckland, Jacques Boucher de Perthes, Alcide d'Orbigny, Charles Darwin, William Herschel, Gideon Mantell, Roderick Murchison, Adam Sedgwick, and William Whewell, and showed evidence of visits to a number of key geological and archaeological sites.[36] Her strategy was to discriminate vigorously between the two Creation stories in Genesis, not as two versions of the same event but as records of two separate creations. Distinguishing between these two narratives and identifying their discrepancies, of course, was already part of the stock-in-trade of higher critics, particularly in Germany, who attributed them to different ancient sources brought together by a redactor. But Duncan had no critical purpose in mind. She wanted to preserve the accuracy of scripture and to bring forward a proposal for reconciling the two accounts so as to achieve, as she put it, "harmony between the Bible and the rocks" and, therefore, "a better and truer understanding of the Inspired Record."[37] Her answer? Two creations—one pre-adamite, the other adamite.

This line of attack, she conceded, bore some resemblance to that of "Isaac de Peyrère," but the similarities were so superficial, she insisted, as to bear not "the remotest analogy to the views which the writer has ventured to propound."[38] And indeed, there were crucial differences: for Duncan there was no continuity between the pre-adamite and adamite worlds; matters of racial variation were of no importance in her account; and she vigorously opposed the polygenism of the Peyrèrean arrangement in an effort to develop a more theologically conservative rendering. In other respects the work followed the concordist tactics of other apologists not least in backing the idea that the Genesis days (of the first Creation account) correlated with geological epochs. As she herself put it, "The six days of creation were in fact six ages, or cycles of ages," and numerous exegetical problems evaporated "when

FIG. 13. Engraving by W. R. Woods of the modern adamic world separated from its pre-adamite predecessor by a lifeless glacial age, in Isabelle Duncan's *Pre-Adamite Man.*

we adopt the idea suggested by several modern geologists, and acquiesced in by Dr. Chalmers, Pye Smith, and Hugh Miller, that the events of Creation must have passed in six successive visions before the mind of Moses."[39] But the radical chasm that Duncan proposed between the pre- and post-adamic worlds when she illustrated the differences between these two phases of earth history has been interpreted as one of the very first recognitions of geological sequencing in the pictorial genre of depicting extinct fauna. Her use of a foldout panorama of the different worlds of the pre-adamite and the adamite etched by W. R. Woods, as Martin Rudwick observes, broke the "standard mold by suggesting a sequence in deep time."[40] Her visualization was a four-part composition, with the lowest layer showing reptiles; the platform above amalgamating mammals from various stages of the Tertiary, including the Great Irish Elk; the third tier disclosing the lifeless icy world inspired by Agassiz's glacial theory; and finally, on top, the adamic world of modern creatures, with the pyramids rising in the far distance (fig. 13). The icy desolation that Duncan interjected between Adam's world and that of its earlier inhabitants enabled her to accept both the existence of a pre-adamite humanity and to adopt a monogenetic account of the descent of modern humans. In so doing, she was certain that the doctrinal orthodoxy of her own proposals was well and truly secured. Nevertheless, it was clear that for Duncan the earth had been host to a human race before Adam, with a history that had run its course prior to the appearance of the adamic family and which had experienced its own fall from perfection—a fall correlated with worldwide geological disruption. It was into this chaotic wasteland that the new Eden was inserted, "a little spot" of perfection that had been "retrieved by special providence from the ruin that still pervaded the world."[41] Perfect, yes—but narrowly circumscribed. Adam enjoyed nothing of the worldwide empire that his pre-adamite forebears had for their pleasure and delight.

Pre-Adamite Man's novelty with respect to its graphic portrayal of discrete phases in earth history, however, was nothing compared to the eccentricity it advanced in its later pages. As Stephen Jay Gould pointed out in an examination (composed in his own inimitable style) of Duncan's work, unambiguous evidence of human remains would not be located until 1891, when Eugène Dubois unearthed the remains of *Homo erectus* in Java, and in an attempt to account for the presence of remnants of human stone tools, but the absence of human bones in the fossil record, she proposed what he called "a stunning solution to her greatest conundrum."[42] Her question: "We have the bones of the lower animals in abundance in the rocks of their respective eras, where are those of the Pre-Adamites?"[43] Her answer: the pre-adamites had been bodily resurrected to become "the Angel Host, whose mysterious visits to our world are so often recorded in the Bible—whose origin is so obscure—whose relations to Adam's family are so close, yet so unexplained."[44] From that point Duncan's book turned into a discourse on angelology—their pre-adamite status, their relationship to the departed, and their connection to human history. Not surprisingly, in Gould's view Duncan's geological representations emerge not as a contribution to scientific innovation "but as a theological scenario for Earth's history in the pre-Adamite tradition of textual analysis. The key white strip of Agassiz's ice-covered world may be validated by geological science, but this catastrophe represents, for Duncan, the agency of God's wrath after the Satanic group of pre-Adamites fell from grace."[45] Nevertheless, the geological detail in her reconstruction provided readers with a pretty good review of the scientific evidence for an ancient world.

Thanks to the painstaking researches of Stephen Snobolen, the hitherto elusive circumstances surrounding the publication of Duncan's *Pre-Adamite Man;* the potential ambiguities in its authorship, given its initial anonymity; the editorial involvement of Isabelle's husband, Reverend George Duncan; and its reception in the popular press have all been uncovered. Published at the very time—during the winter of 1859–60—when Darwin's *Origin of Species* and *Essays and Reviews* made their appearance, and when human artifacts coexistent with extinct animals were being uncovered, *Pre-Adamite Man* enjoyed considerable popularity, quickly going through several editions. Its reception was mixed, ranging from enthusiasm to ridicule. Thus, while the *Morning Advertiser* found "much ingenuity" in the book, the *Athenaeum* mercilessly satirized it; and whereas the *Illustrated News of the World* judged that it would create as much of a stir as *Vestiges,* not least on account of its "frank modesty which immediately wins the reader's favour," the *Morning Post* was scathing in accusing its author of engaging in nothing less than the perversion of scripture.[46] Snobolen's account, moreover, tellingly locates Duncan's evangelical apologetics, however fanciful it might seem to modern eyes, in the context of biblicist attempts to come to terms with recent archaeological revelations. He identifies, in its presentation of an "Eden before Eden, a fertile paradise unencumbered with sin, civilization or war," a kind of "myth-making" eschatological romance in reverse—a future paradise projected into the past.[47] Crucially, too, he perceives that Duncan's treatise "should be seen as an inherently conservative theological work—the radical early history of preadamism notwithstanding," and that her "genius lies in her thorough conversion and baptism of a theory with sceptical origins."[48] In Duncan the migration of pre-adamism from heresy to orthodoxy, from skepticism to apologetics, was now under way.[49]

Gall's Projections

Whereas Duncan divined angelic agents behind pre-adamite artifacts, Reverend James Gall (1808–94) assigned them a very different identity. Gall, a Church of Scotland minister and founder of an interdenominational mission in Edinburgh, has achieved fame for his cartographic endeavors, celestial and terrestrial. In an attempt to provide a graphic portrayal of the heavenly bodies, he devised a number of map projections, which he also applied to the earth's surface. One of these, the Gall orthographic projection, originally put forward at the 1855 meeting of the British Association for the Advancement of Science, was an equal-area projection that reduced the polar distortion inherent in the standard Mercator map.[50] Reinvented by Arno Peters, who publicized it in the 1960s, it became stunningly popular and nowadays is often used to dramatize the extensive land area of the third world.[51] Gall's significance in our story, however, springs from a somewhat different set of preoccupations—a concern with harmonizing science and religion. And here, too, the idea of adamic forebears came to the rescue.

In *Primeval Man Unveiled* (1871), a work that further expanded suggestions he had put forward in *The Stars and the Angels* (1858), Gall sought ways of accommodating traditional faith to both geology and Darwinian evolution in such a way as to keep intact the long-standing compatibility of the book of nature and the book of scripture. The key, he believed, lay in rethinking the authorship of the Genesis narrative, and he argued in some detail for the view that Moses had put together documents from earlier witnesses—Noah, Seth, and Adam himself. This led him to distinguish, as others were doing, between the two Creation narratives of Genesis chapters 1 and 2, to describe the first of these as an apocalypse, and to identify the second as history. The purpose of the latter was to describe the creation of an individual Adam, "the first of the Messianic race"; it was the story of the inauguration of the line through which the Messiah would come.[52] The former account, or what he called "the natural history of the universe," was an apocalyptic rendering of the creation, through the agency of natural law, of a world that had fallen into ruin through some global catastrophe.[53] This primitive world was an inhabited one, and the remains that archaeologists were uncovering were those of "a previous population of our globe."[54] The existence of these pre-adamites, who had been swept from the planet prior to the coming of Adam, provided Gall with a via media between the polygenists who spoke of "several Adams—a white Adam, and a black Adam, a red Adam and a yellow Adam," and those monogenists who, while maintaining the unity of the human race through their common descent from Adam, had to push his creation so far back in time to accommodate the dating of fossil remains being excavated in Denmark, Egypt, and America as to subvert the integrity of biblical chronology.[55] The pre-adamites allowed him to keep to a traditional chronology *and* retain the unity of the race.

Thus far Gall's account paralleled Isabelle Duncan's in several crucial respects, not least in postulating that the pre-adamites had become "extinct thousands of years before the time of Adam," but he went out of his way in the preface to insist that there was in fact "very little similarity" between his views and hers.[56] The reason, it seems, was to do with the identity of the pre-adamites. The savage state of society that he believed was indicated by the archaeological record, and indeed was in keeping with "the theory of gradual development," indicated a degraded race much different both from Adam's original perfection and from Duncan's angelic residents.[57] To Gall it simply pointed to a botched humanity under the influence of satanic forces. Their lingering spiritual malevolence and the remains of their "material bodies . . . which have been found in Denmark, England, France &c," together attested "to the degeneracy at which they had arrived before they became altogether extinct."[58] From that point in the narrative Gall strayed off into a variety of spiritualist themes about the character of satanic forces, magical practices, and the identity of the Antichrist. These preoccupations disclose the context in which his postulation of pre-adamites was domiciled—namely, as a biblicist device to retain primitive supernaturalism and yet keep faith with the revelations of geological and archaeological science.

Snatches of similar pre-adamites were also fleetingly glimpsed by other writers. George Hawkins Pember (1837–1910), for example, a champion of the gap theory, told the readers of his *Earth's Earliest Ages* in 1876 of pre-adamite beings whose absence from the fossil record might be on account of their bodies being "resolved into primal elements" or that "death did not touch these primeval men until the final destruction" prior to the creation of Adam.[59] The parallel with Gall's proposals are evident. To Pember the pre-adamites were called upon both to "remove some of the Geological and other difficulties usually associated with the commencing verses of Genesis"[60] and, perhaps even more so, to provide an explanation for the origin of supernatural agents. Demons, he speculated, might well be "the spirits of those who trod this earth in the flesh before the ruin described in the second verse of Genesis."[61] For Pember, it seems, pre-adamite beings occupied a necessary niche in the ontology of a former world, the remnants of which were detectable in the geological strata.

The racial fixations that frequently dominated pre-adamite schemes, it is clear, are conspicuously absent from these advocates of pre-adamite humanity. Their concern was apologetic; it was designed to uphold harmony between science and theology. Indeed, twentieth-century fundamentalists have continued to resort explicitly to the speculations of both Duncan and Gall in their attempts to reconcile the Genesis story with scientific developments.[62] Nevertheless, a substantial body of pre-adamite writing was used, as we will continue to observe, in the service of racial politics and as an explanation for human variability.[63]

Human Diversity and Pre-Adamite Apologetics

Just as the pre-adamites were resorted to as a means of keeping faith with both earth science and the Old Testament scriptures, they were no less called upon to explain those racial differences and distributions that students of anthropology and biogeography fastened upon. Certainly, critical voices could be heard from those intent on retaining traditional Mosaic monogenism. The physician and linguist John Mason Good (1764–1827), for example, shielded the biblical account from the critique of "the celebrated Isaac Peyrere" in his lecture, later published as "On the Varieties of the Human Race," which was delivered as part of a three-course series at Surrey Institution in 1811–12.[64] A fellow of the Royal Society and man of letters, he rejected the Peyrèrean speculation that the Mosaic narrative contained accounts of two separate human creations and provided, to his own satisfaction at least, answers to the sorts of conundrum that La Peyrère had presented to conventional interpretations. In response to the old chestnut about who peopled Cain's world, for instance, Good considered that because the "first fall of man . . . did not take place till one hundred and twenty-nine years after the creation of Adam," there was ample time, "especially if we take into consideration the peculiar fecundity of both animals and vegetables in their primaeval state, for a multiplication of the race of man, to the extent of many thousand souls."[65] Besides this, he was sure that students of human variety had made much too much of anatomical differences between the races; they were just far too inclined to generalize from rare occurrences. Merely expounding their claims, he felt, was "sufficient to show the absurdity of the argument for a plurality of human stocks or species." Their "altogether superficial" arguments were "not more than *skin-deep*."[66] With Montesquieu he attributed diversity to the effects of climate, diet, and manners, to which he added "morbid and hereditary affection."[67]

Natural Histories of the Human Species and Centers of Creation

Notwithstanding such censure, pre-adamism became increasingly evident among writers who directed their attention to ethnological subjects. William Frederick Van Amringe (1791–1873), for example, protested the doctrinal propriety of polygenetic pre-adamism in his 1848 *Investigation of the Theories of the Natural History of Man*.[68] In this work he was cautious, however, not to commit himself explicitly to the pre-adamite scheme but was at pains all the while to insist that if the conjectures of the most recent physiological anthropologists were to be verified, any theological shock waves could easily be absorbed. This was a line of attack that became increasingly popular among orthodox Christian apologists. Indeed, Van Amringe had the suspicion that the Mosaic history sat rather well with the presumption of multiple human species of different creative origins. The testimony of the Genesis narrative was just too fragmentary, he felt, and all sorts of long-standing questions about Cain and his city were left unanswered. All of these problems evaporated with the coming of La Peyrère's pre-adamite. Here the concern was to find ways of keeping the most recent physiological and anthropological findings consistent with scripture.

Emphatically more partisan, and indeed decidedly more racist, were the articles published a couple of years later by the eminent Swiss naturalist at Harvard Louis Agassiz. In two papers published in the *Christian Examiner* for 1850, Agassiz took up the issue of animal and human geographical distributions.[69] His argument was simple: there were distinct zoological zones or provinces—an arctic, a European temperate, an African, a tropical Asiatic, and so on—in which the Creator had placed discrete species. Agassiz's idealist *Naturphilosophie* encouraged him to consider that each race had a separate point of origin and that any blurring of its transcendental individuality was both biologically and socially repugnant. Not only had animals and plants originated in centers of creation across the globe, but they remained "within fixed bounds in their geographical distribution."[70] Simply put, races were made for places.[71] But Agassiz's arguments did not stop with his interpretation of the natural history of the globe's inhabitants; he was no less concerned to demonstrate the doctrinal propriety of his polygenist message. He thus began his two-part series for that Boston Unitarian journal by arguing that the traditional monogenetic account of origins was a "very modern invention" not to be found in the book of Genesis.[72] To the contrary. "That Adam and Eve were neither the only nor the first human beings created is intimated in the statement of Moses himself, where Cain is represented to us as wandering among foreign nations after he was cursed . . . Thus we maintain that the view of mankind as originating from a single pair, Adam and Eve . . . is neither a Biblical view nor a correct view, nor one agreeing with the results of science."[73] Moreover, the view that the Creator only produced one pair of each animal species ran into the knotty problem that the "primitive pair of lions" would have had to "abstain from food until the gazelles and other antelopes had sufficiently multiplied to preserve their races from the persecution of these ferocious beasts."[74] One lion breakfast would have wiped out a whole species. Presumably, Adam's daily diet, too, would have rapidly resulted in mass extinction. As for humanity, Agassiz insisted, "mankind cannot have originated in single individuals, but must have been created in that numeric harmony which is characteristic of each species; men must have originated as nations, as the bees have originated in swarms."[75]

Despite his self-conscious disavowal of any interest in political matters or in justifying slavery, the racial bias that Agassiz happily exhibited in his portrayal of the different racial types—he found in African races a "peculiar apathy, a peculiar indifference to the advantages afforded by civilized society"[76]—reinforced the earlier pluralist arguments of figures such as Charles Caldwell and Samuel George Morton, who vigorously promoted the multiple origins of humanity.[77] Soon talk of separate beginnings would be aggressively taken up both by writers with little concern to reconcile science and religion as well as by those who cultivated a white supremacist theology. For the meantime our focus remains on those who, even if their work exhibited racial overtones, showed some apologetic concern with retaining concord between anthropology and the Bible. Thus, Agassiz devoted his energies to arguing that the "unity of mankind does not imply a community of origin for men; we believe, on the contrary, that a higher view of mankind can be taken than that which is derived from a mere sensual connection,— that we need not search for the highest bond of humanity in a mere animal function, whereby we are most closely related to the brutes."[78] Spiritualizing all allowed for bestializing some.

Needless to say, Agassiz's proposals were not universally welcomed by theological commentators. W. H. Moore, a Presbyterian minister from Richmond, Virginia, excoriated both Agassiz and his polygenist predecessors. Anthropological theorists who were forever going on about the parallels between apes and races, he noted wryly, had been inclined, with all due "magnanimity and abnegation," to "generously disclaim the honour of this simial relationship, and benevolently bestow it upon poor Quashee"![79] To Moore the "difference between the fairest Caucasian and the sootiest African, is not nearly so great as that between the little, shaggy Shetland pony, and the gigantic dray-horse of London," and Agassiz's attempt to enlist the Bible in his cause of different human species was odious.[80] "We would be glad to know," Moore sniped, "how he has discovered that Adam and Noah belonged to the white race at all."[81] In all there was to Moore something utterly "cold blooded" in Agassiz's "haughty assignment of more than half the human race to a doom of hopeless, irreversible degradation."[82]

Others, by contrast, protested the doctrinal appropriateness of Agassiz's polygenism. Both it and the pre-adamism of La Peyrère, for example, were defended by the Boston Unitarian minister N. L. Frothingham, also writing in the *Christian Examiner* the year following the appearance of Agassiz's pair of articles. Even while affecting agnosticism on the whole subject, Frothingham exploited Agassiz's recent intervention in three conspicuous ways. First, he used the occasion of Agassiz's multiple-adamite proposals to castigate those whose theological attachments stood in the way of intellectual freedom. Yet he was glad to report that there had been remarkably little acrimonious comment on Agassiz's essays, a condition he attributed to an increased and welcome "freedom of philosophical inquiry."[83] Nevertheless, he warned of the dangers of those who would "cast reproach or suspicion upon an investigation in the department of zoology, or ethnology" in the name of "divine truth" and urged that "in no case whatever, and in no degree whatever, should the student of physical science be checked or limited in his inquiries by the supposed authority of any ancient writing, however sacred."[84] Calling readers' attention to Galileo's fate, he cautioned against efforts to repress anthropological inquiry in the name of orthodoxy. Second, he took the opportunity that Agassiz had afforded to attack the idea of "the plenary inspiration" of scripture and to liberalize exegesis of doctrines such as that of the Fall in ways that facilitated pre-adamite humanity.[85] Indeed, it was clear to him that adopting pre-adamism necessarily rubbed Augustinian theology the wrong way, not least in its understanding of Adam's headship of all humanity. But that caused him no concern. For Frothingham hermeneutics should be practiced "in the light and height" of biblical freedom, "and not by the smoky lamp of a scholastic dogmatism."[86] Third, he happily exploited Agassiz's exposition to rehearse in considerable detail, both biographical and theological, the original pre-adamite scheme of La Peyrère and to record his admiration of this "learned and ingenious man"—whatever his shortcomings—as an exegete. Frothingham found La Peyrère "frequently original, far-sighted, modestly bold. His book abounds with wise, sober views of Scripture story, that surprise us with their superiority to the current religious notions of even our own times."[87]

Also in that same year, 1851, Samuel Kneeland, a Boston medical naturalist and polygenist, found much to commend in the arguments of both Van Amringe and Agassiz in an eighty-four-page introduction to the American edition of *The Natural History of the Human Species,* a work first published in 1848 by Lieutenant Colonel Charles Hamilton Smith (1776–1859). An amateur naturalist and solider, Smith produced numerous works on natural history, which he illustrated with his own watercolors, and this particular volume allowed him considerable scope to demonstrate his qualities as a draftsman through his depictions of "typical" human head forms, cranial structure, and eye forms (fig. 14). Here he cautiously kept open the monogenist-polygenist matter; it remained, he said, "a question in systematic zoology, whether mankind is wholly derived from a single species ... or produced by the hand of nature at different epochs."[88] But the whole tenor of the work was polygenist, as he specified the "centres of existence of the three typical forms of man"—"the intertropical region of Africa, for the woolly haired—the open elevated regions of north-eastern Asia for the beardless—and the mountain ranges towards the south and west for the bearded Caucasian."[89] In three different geographical locations the three typical human stocks had

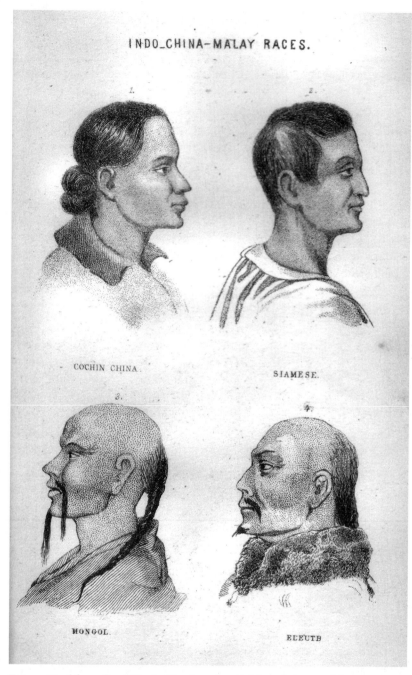

FIG. 14. Head forms from Charles Hamilton Smith's *Natural History of the Human Species.*

come into being, at different points in time. The most ancient was what he called the tropical type, the latest, the Caucasian. Indeed, he speculated that it was only when the other racial types fused with the Caucasian that they achieved the intellectual advancement that enabled them to attain "the ultimate conditions for which Man is created" and thereby to display humanity's "essential unity."[90] What is notable is that Smith's entire inventory of the human species was charted without recourse either to biblical exegesis or to the significance of Adam, an account demonstrating that caution must be taken not to equate pre-adamism as a theological construct with polygenism as an anthropological thesis.

Nevertheless, Kneeland's lengthy introduction—it amounted to nearly a quarter the length of the whole book—did bring the American edition of Smith's compendium into the realm of theological dialogue. His declared intention was to present an even-handed overview of scientific works on race, but his judgments were ever biased toward polygenism and Agassiz-style creationism.[91] And while he was quick to deplore as unscientific the way in which monogenists sought scriptural self-justification, he himself did not hesitate to assure his readers that the Bible was entirely compatible with multiple Adams. Thus, he complained that the language of Reverend Thomas Smyth's *Unity of the Human Race* (1850), which defended the idea that humans were descended from Adam, was more that of "the polemic theologian, and the advocate of southern institutions, than the scientific naturalist, and ethnologist; and, however appropriate in other places, is quite irrelevant on the subject of the origin of mankind."[92] And yet he was at pains to insist in his attack on Prichard's monogenism that there "is no evidence that the sacred writers considered the colored races as descended from the same stock as themselves. This is a modern and human invention for political or other purposes."[93] Such interventions routinely asserted themselves as Kneeland expounded the views of Agassiz and Van Amringe. Were not "many local centres of animal and vegetable creation . . . most consistent with the wisdom of God"? Was attributing racial differences to the accidents of climate not "equivalent to making physical influences more powerful than the Creator"? Did not the "incompleteness and obscurity of the Mosaic account of man" point unequivocally "to a race of men independent of Adam"?[94] Whether as an exercise in cultural apologetics or animated by a sense that if polygenism were to succeed, its doctrinal orthodoxy needed securing, by rhetorical gesture and partisan review, Kneeland did his best to domesticate human plurality to the religious ethos of his mid-nineteenth-century American audience.

The Apologetics of Lane, McCausland, and Nemo

An even more concerted effort to render pre-adamic polygenism congenial to scripture appeared a few years later, in 1856, and in an enlarged edition in 1860, by the Arabic scholar and lexicographer Edward William Lane (1801–76).[95] Edited and introduced by his nephew, the distinguished British archaeologist and orientalist Reginald Stuart Poole, head of the Coin Department at the British Museum and later professor of archaeology at University College, London, Lane's *Genesis of the Earth and of Man* aimed to do for anthropology and archaeology what Hugh Miller and Edward Hitchcock had already done for geology—namely, to show its profound compatibility with Christianity. Fundamentally, Lane was convinced that a much more sophisticated hermeneutic arsenal was needed to cope with the new threats to biblical orthodoxy, and he used his considerable expertise in Semitic languages to construct his edifice. This project, his nephew was not slow to point out, was exceptionally timely in the light of two prevailing circumstances. First, there were those who high-handedly dismissed scientific inquiry as false because it "disagrees with their theological opinions." That irresponsible attitude was, Poole implied, characteristic of the Catholic Church. He hoped for better from "our Protestant clergy," whose "full acknowledgement of the sole authority of the Scriptures and the right of private judgment" affirmed the "two great principles upon which hinged the Reformation." Lane's preadamism was thus staged as an essentially Protestant intervention. The second state of affairs had more clearly emerged from the shadows as the second edition of the book was going through the press. This was the reemergence of the "old battle" between science and religion that had manifested itself dramatically at the 1860 meeting of the British Association for the Advancement of Science, the very meeting at which Huxley's infamous confrontation with Bishop Samuel Wilberforce reportedly took place.[96] Given this situation, Lane's analysis was needed more than ever.

Blending some fairly labyrinthine exegesis of various obscure passages in Genesis and displaying an awareness of recent trends in textual criticism and developments in the elucidation of ancient hieroglyphs, Lane urged his case for two separate human creations—one pre-adamic, the other adamic. The flint arrowheads currently being excavated in cave deposits from the Pliocene epoch, alongside teeth and bones of extinct species, were positive evidence that "the origin of man is very far more ancient than it is commonly believed to be," and the only way of preserving biblical chronology was to attribute them to a pre-adamite race.[97] But crucially here—in contrast to the pre-adamism of Duncan and Gall—Lane believed that pre-adamite stock contin-ued in existence during and after the time of Adam. Its members' blood ran in the veins of later post-adamic generations. A certain amount of doctrinal architecture needed rearranging to cope with this proposal—standard inter-pretations of the scope of the fall from grace, the inheritance of original sin, the theological ontology of death, the consanguinity of the human species in the absence of a common origin, and the demographic scope of the Flood all required adjustment. And having addressed these to his own satisfaction, Lane pursued his task. In essence he considered that the first creation of hu-mans had taken place in the valley of the upper Nile, from which location they diffused through Africa and Asia and branched out into a range of ra-cial types. Much later the adamite races—which he took to be Caucasian—were divinely created, and they intermarried with various pre-adamite strains to produce the diversity of racial types that are extant today. This perspective not only provided an explanation of human racial differentiation, but it fa-cilitated an adamic chronology that was closely in keeping with a conserva-tive genealogical time scale. For only by viewing Adam as a recent newcomer on the stage of human history could the authority of science and scripture remain unimpeached. As Lane put it:

> Some men of science and learning, holding the common belief that Adam was the first of our species, have confessed themselves to be compelled, by geological and other considerations, to adopt [the] opinion . . . that he was created twenty thousand years, or more, before the Christian era; and the early Biblical history has been enormously distorted to accommodate it to this belief. But our own opinion, that the Bible itself indicates the existence of Pre-Adamites, relieves us from the necessity of requiring a more extended Biblical chronology than that which appears to be advocated by most of the best judges in the present day.[98]

There is no doubt that a sense of racial superiority pervaded Lane's analy-sis. His supposition that "the Caucasians alone are descended from Adam," his talk of the deterioration of "Adamite blood," and his representation of adamic degeneration through marriage "with descendants of the primi-tive stock" are indicative.[99] It is therefore not surprising that the American polygenist and racial apologist George Gliddon soon approvingly and ex-tensively quoted from Lane's work, observing that it "augurs well for eth-nological progress in Great Britain" and pointing out that it continued the argument earlier put forward by La Peyrère.[100] Yet taken overall, the thrust of Lane's account was decidedly more exegetical than racial in tone and more concerned with amalgamating data on a wide scale to support his conjecture. In this latter regard his use of philological evidence to bolster his version of double-adamism is particularly conspicuous and was picked up in Poole's introductory editorial. Here Poole lambasted the monogenetic supposition underpinning Bunsen's linguistic archaeology as grounded on nothing more than mere sentimentality. "Had it not been that Bunsen had started with [this] fixed persuasion," Poole quipped, his philology would have been entirely different.[101]

For his part Lane launched his own attack on the Bunsen-Müller model of language development and, siding with Ernest Renan, advanced his own scheme, based on the discrimination of monosyllabic, agglutinate, and amalgamate forms. This approach ran counter to Bunsen and Müller's taxonomies, first, in "representing the Semitic stock as in no way derived from a primeval language," and, second, "in representing the Egyptian, which he terms 'Khamitism,' or 'Chamitism,' as being, from the first, collateral to the Semitic." The implications of this polygenetic taxonomy were, of course, obvious: "there have existed Pre-Adamites of our species" (fig. 15).[102] These beings spoke a primitive monosyllabic language that developed into early forms of agglutinate speech—perhaps akin to ancient Chinese—which in turn branched out along two separate axes, Nigritian and Turanian. Both monosyllabic and agglutinate forms had "originated with artless, uncivilized, illiterate races."[103] The Adamites, by contrast, were speakers of an amalgamate tongue, and their descendants were predominantly Caucasian.

This contrast between earlier primitive and later complex languages had prompted Bunsen to discern an evolutionary descent of the latter from the former and a presumed monogenetic origin for language. Lane strenuously disagreed, perceiving no genealogical links between these linguistic systems and postulating a separate point of origin for each. A sequence of "successive creations" was being misread as development from a common origin.[104] Bunsen and Müller's accounts, moreover, put unbearable pressure on the biblical documents, requiring a "very-extended chronology of the Post-Adamic period" and, Lane inferred, the reduction of "a great part of the history of the book of Genesis to the category of faulty and vague traditions."[105] Both empirically and exegetically, the postulation of pre-adamite languages seemed irresistible. Philologically, adamic monogenism undermined biblical authority; pre-adamite polygenism rescued it. As he explained, Bunsen and Müller's "belief in the descent of all mankind from Adam . . . demands concessions enormously at variance with the Scripture-history of the time anterior to the Dispersion, unless we admit the evidences in favour of the existence of

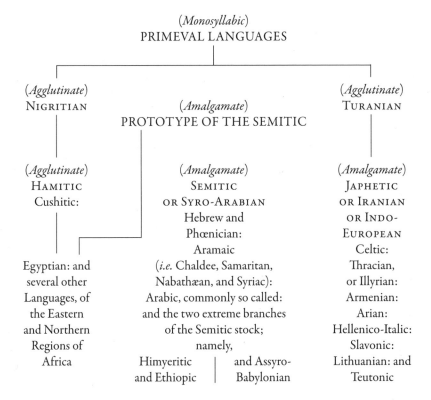

(*Monosyllabic*)
PRIMEVAL LANGUAGES

| (*Agglutinate*) NIGRITIAN | (*Amalgamate*) PROTOTYPE OF THE SEMITIC | (*Agglutinate*) TURANIAN |

| (*Agglutinate*) HAMITIC Cushitic: | (*Amalgamate*) SEMITIC OR SYRO-ARABIAN Hebrew and Phœnician: Aramaic | (*Amalgamate*) JAPHETIC OR IRANIAN OR INDO-EUROPEAN Celtic: |

Egyptian: and several other Languages, of the Eastern and Northern Regions of Africa

(*i.e.* Chaldee, Samaritan, Nabathæan, and Syriac): Arabic, commonly so called: and the two extreme branches of the Semitic stock; namely,

Himyeritic and Ethiopic | and Assyro-Babylonian

Thracian, or Illyrian: Armenian: Arian: Hellenico-Italic: Slavonic: Lithuanian: and Teutonic

FIG. 15. Edward Lane's dual model of linguistic development.

Pre-Adamites presented by the Bible itself, and by physical and historical, as well as linguistic, facts."[106]

The question of language history also played a critical role in another work of pre-adamite apologetic, *Adam and the Adamite,* which appeared in 1864 shortly after the second edition of Lane's volume. Its author, the Irish judge Dominick McCausland (1806–73), was by now Queen's counsel and recently appointed Crown prosecutor under the second administration of Lord Derby.[107] McCausland was the author of a number of successful religious works, and the tenor of this particular volume can be inferred from its subtitle: "The Harmony of Scripture and Ethnology." The aim was simple: to uphold the detailed accuracy of scripture in the face both of challenges from the human sciences and of the critical spirit of Bishop John William Colenso and *Essays and Reviews.* From the outset McCausland freely admitted that the biblical text "was not written to instruct us in the knowledge of the physical sciences."[108] But this did not for a moment mean that science and scripture dealt with different spheres and remained cognitively dissonant; it was, he insisted, a "mischievous error to seek to divide Scripture and science, and separate the deep mysteries and important truths of revealed religion from the rich treasures of philosophy which God has provided for our instruction."[109] And so, to preserve the accuracy of Mosaic primeval history in the face of revelations from ethnology, McCausland turned to the pre-adamite.

The findings of the geologists, historians, archaeologists, philologists, and ethnologists all came within this purview and pressed him to concede that if Adam was to be taken as "the progenitor of *all mankind*," the chronology of the Old Testament would have to be abandoned, "and all that is written in the Book of Genesis of antediluvian members of his family must be treated as the fanciful speculations of some visionary mythologist."[110] Happily, such a judgment would be too precipitate, for the biblical Adam was only the last of a series of human types that God had created. Superior to his forebears, Adam's appearance ex nihilo in the image of God was recorded for all posterity, although in a short time the pure adamic line would be sullied through intermarriage with pre-adamite stock. In McCausland's case, as earlier with Lane, the pre-adamite theory was engaged to safeguard the integrity of scripture from the assaults of higher critics—a quite remarkable reversal of its earlier deployment as a source of skepticism.

McCausland's rhetorical strategy was to review the scientific state of affairs in a number of scientific specialisms—geology, ethnology, anthropometry, archaeology—reserving the lion's share for developments in philology. Only then did he present his pre-adamite scheme as the sole way of keeping the Bible in harmony with these advances. The text is laced throughout with racist interpolations, all boiling down to the superiority of the adamic Caucasians and the inferiority of pre-adamic stocks. But the major concentration was on linguistic evidence for a separate and superior adamic language inserted into preexistent uncivilized linguistic families. Using evidence from comparative vocabulary and grammatical forms, and distinguishing monosyllabic from inflectional systems, he urged a fundamental chasm between primitive Turanian languages and the later Iranian (or Indo-European or Aryan) family. The Japhetic component of this latter cluster branched out into European languages, while another stream developed Sanskrit—the "most complete and polished of all the languages of the earth."[111] The Semitic element incorporated Aramaic, Arabic, and Hebrew, whose speakers occupied a more circumscribed territory than that of their Japhetic counterparts. In McCausland's telling the former family, the Japhetic, developed more advanced civilizations, while the Semitic branch took the "most prominent part in the religious development of mankind."[112] Taken together, these two communities of inflectional language speakers dominated the history of civilization and exhibited their superiority over those nations using agglutinate Turanian tongues, whose structure could be reduced to a mechanistic "gluing or affixing words to the original roots."[113] The superiority of the former linguistic systems reflected the perfection of adamite language, "the parent of *the inflectional languages*"; it displayed "no monosyllabic infancy, and never passed through an agglutinative stage" because it was the result of divine creation.[114] And this was the secret of the success of the Noachic dispersion. As superior language speakers, they easily broke up pre-adamite social groups and came to dominate vast stretches of the globe. As McCausland explained, this was because "the language which originated in Paradise attained a richness and luxuriance in vocabulary and grammatical structure, that is not to be found in the mechanically-framed languages of the Turanian pre-Adamite."[115]

Some of McCausland's readers certainly found his views persuasive. An anonymous contributor to *Scribner's* magazine, for example, juxtaposed McCausland's thesis to the Duke of Argyll's monogenetic *Primeval Man,* much to the former's advantage. To this commentator there was no other way of protecting the Mosaic chronology from the challenge of deep anthropological time. Just as the Copernican system had eventually escaped the fangs of religious dogma, so, too, must the pre-adamite. Bolstering his stance by exegetical attendance to the Genesis text, and by rehearsing criticisms of the climatic thesis, this writer insisted that the pre-adamite theory provided good grounds "for adhering strictly to the Bible narrative, while at the same time [accepting] the theory that Adam was the progenitor of the Caucasian race only, whose history is related in that narrative."[116] It also had the advantage of resolving long-standing difficulties—such as the differences between inflectional languages and the monosyllabic families—in language history, without resorting to the tactics of Bunsen, who, "assuming the unity of mankind in Adam," was compelled to allocate much too early a date to the deluge.[117] Another observer, Alice Wayne, writing for the *Ladies' Repository* on McCausland's later *The Builders of Babel* (1871), was just as supportive and recommended his works to those "inclined to waver in their religious belief."[118] Not everyone, however, was convinced by such philological arguments. Writing from Montreal in 1880, John Campbell used his expertise in native American languages to attack linguistic polygenism and the "doctrine of a multiplicity of human protoplasts, or of many Adams, which the Frenchman, Isaac de la Peyrère first brought prominently before the Christian world . . . [and which] found great acceptance in America some years ago."[119] Campbell painstakingly scrutinized the languages, customs, and religions of American Indians to disclose their proximity to Old World peoples, to undermine perceptions of them as "incorrigible savages," and to expose the way in which "American traders plunder them, and the American troops slaughter their defenceless old men and women."[120]

Another text originating in Ireland and marshaling pre-adamite races to defend Genesis from the attacks of both natural science and biblical criticism appeared in 1876 under the pseudonym Nemo. Its purpose was simple: to establish that the existence of pre-adamite human races was revealed in scripture itself. Drawing on a wide range of writers such as the Duke of Argyll, William Herschel, William Thomson (Lord Kelvin), and the Dublin clergyman-geologist Samuel Haughton, Nemo labored long and hard to confirm a Miller-style reading of geological history and indeed wove a version of the astronomical nebular hypothesis into his reading of the Mosaic days of Creation—six visionary stages that "passed before the mental eye of Moses."[121] The second "day" of Creation, for example, spanned the Old Red Sandstone and Devonian, the Silurian, and the Cambrian epochs. But Nemo went well beyond Miller by integrating various species of early humans into the scenario. "Palaeolithic man," the anonymous writer noted, had come into being 350,000 years ago in a tropical climate with such creatures as the lion, tiger, and hyena; 50,000 years ago "Neolithic man," remnants of whose tools were increasingly coming to light through recently excavated kitchen middens, was created alongside the reindeer and woolly rhinoceros; 6,000 years ago, alongside Neolithic humans, "our Adam" made his appearance, as did modern fauna and flora. Indeed, Nemo was even willing to countenance the possibility of human origins being traced back to the Miocene epoch and thus to "pre-glacial times, and to thousands, or it may be a million of years, before the days of our Adam—to a time at which the earth may have been fully inhabited with human beings, and with races afterwards swept away or destroyed during the rapidity of the drift of the glacial epoch."[122] This staging of Paleolithic, Neolithic, and adamic humanity, however, was not to be understood as an evolutionary development. Nemo made it clear that "we have no account of the transformation of an ape into a man, nor any account of the commencement, progress or ending of such transformation . . . Let us therefore abide by the sacred history of the creation and descent of man."[123]

Crucial throughout was the distinction Nemo pressed between the appearance of earlier pre-adamite human species and the genesis of what he called "*the special* Adam," or "our" Adam, the father of the Jewish people,

ADAM'S ANCESTORS

whose creation "was consummated nearly six thousand years ago."[124] Each of these had its own biogeographical domain, with an associated range of settlement, fauna, and flora. The former world was inhabited by wild animals, the adamic creation by domesticated creatures. And there was no doubt about the superiority of adamic humanity over earlier human species, some of whom survived the local Noachian flood and continued in existence to the present day. The persistence of anthropometric characteristics bore witness to this history, and Nemo called upon the testimony of Poole to support his notions of racial fixity. All of this was reinforced by detailed exegesis of various Old Testament texts with the aim of defending scripture from the higher criticism of Bishop Colenso. The pre-adamite thus rescued Genesis from a two-pronged assault. First, it relieved readers of the need to reconcile the two accounts of Creation in the Mosaic narrative—a task that had attracted "great ingenuity," unfortunately misdirected, over generations.[125] Second, it undercut the suggestion that the two stories were the products of different ancient traditions, expressed in the use of different divine names—Jehovah and Elohim—which had been spliced together by an editorial redactor. As an apologetic device, the pre-adamite theory, rather ironically given its earlier role, was now being deployed to preserve scriptural reliability in the face of textual deconstruction:

> The difference between the two creations are pointed out by Colenso, who detects discrepancies between (as he thinks) two different accounts of the same transaction, instead of investigating them as the detail, first, of a general creation of mankind, and next, that of a fuller account of a particular race, made distinct from all the others by the Almighty, for a particular purpose. Where Colenso fancies he finds contradictions existing between the two chapters condemnatory of the historical veracity of the narrative, others deem them confirmatory of the mosaic revelations in a remarkable manner.[126]

℘ In their encounter with the new geology and the emerging human sciences, religious believers turned to the idea of a pre-adamite earth and pre-adamite races in order to retain solidarity between scientific knowledge and theological creed. To be sure, that formula was not universally welcomed, but for those who believed they could detect in historical remains and present-day bodies the echoes of pre-adamite ancestors—a creation before Creation—the theory enabled them to take science seriously and to preserve the integrity of the Genesis narrative. It facilitated the revelations of deep time that the geologists were exposing even while preserving a relatively recent date for the Mosaic story of Adam; it allowed room for the archaeological excavation of primitive artifacts and, later, human remains; it offered a means of untangling the complex genealogy of human languages; and it provided an explanation for racial difference without postulating an extended biblical chronology. Securing these gains took many forms. For some the pre-adamite earth was uninhabited, a world devoid of human life, a stage in preparation for the adamic drama yet to come. For others entire pre-adamic civilizations had populated the primeval earth; sometimes they took angelic or demonic form, sometimes they were human, and often they were considered inferior specimens of humanity. According to some, these residents were entirely wiped out before the advent of Adam; others were sure that pockets of pre-adamites survived to intermarry with Adam's progeny.

Whatever the emphasis of partisans, two hermeneutic tactics tended to predominate. One was to postulate a gap of indeterminate length between the initiating act of Creation and the story recorded in the early chapters of Genesis into which geological strata, archaeological artifacts, and anthropological diversity could be slotted. Another was to exploit differences between the two accounts of Creation in the early chapters of the book of Genesis, reading one as the origin of the human species, the other as the account of an individual, Adam, variously seen as the father of the Jews, the Caucasians, the Aryans, or some such.

Why was all this industry so important? In large part, of course, it was motivated by a passion to retain good faith with religious heritage in the face of scientific challenges. But it also sprang from the fundamental importance of the adamic picture to the Christian West's sense of its own identity, culture, and worth. It is therefore not surprising that the language of adamites and pre-adamites continued to feature, at least for a time, in other arenas as well. It permeated the science of ethnology at midcentury, for example, and also underwent transformations induced by Darwin's intervention and the advent of human evolutionary biology. Tracking the theory's fortunes through these writers will occupy our next two chapters.

5 ✌ ANTHROPOLOGY

Adam, Adamites, and the Science of Ethnology

THE PRE-ADAMITES HAD CERTAINLY come to dwell with theologians and writers of popular apologetics. This does not mean, however, that they did not occupy other cultural spaces too. They showed up, for example, in works with little concern for accommodating theology to scientific developments. Indeed, the language of pre-adamism could function as nothing more than a means of challenging the august authority of the Old Testament in matters of anthropological history and to that degree at least perpetuated something of the skeptical impetus associated with the theory's earlier incarnations. At the same time it also continued to feature in more dedicated scientific locations among those whose empirical inquiries remained, to one degree or another, fixed to biblical moorings. The diversity of opinions expressed on human origins in these arenas allowed for a variety of reactions to the pre-adamite proposal from different religious perspectives—with some espousing it, others dismissing it out of hand, and yet others allowing it as a possibility should the anthropological and archaeological evidence move decisively in favor of polygenism. This chapter charts this stretch of the story among those for whom pre-adamism remained necessarily polygenetic. We leave the transformation, effected to some degree by Darwin, of polygenetic pre-adamism into a monogenetic form until the next chapter.

Before turning to the ways in which it persisted among professionalizing ethnologists, one particular use of pre-adamite discourse merits comment; in this case it simply served as surrogate for primordial polygenism and the existence of ancient cultures predating recorded Hebrew history. Just the year before McCausland's rendition of pre-adamism first appeared, another book-length work on the subject—this time written from a markedly different spiritual address—crystallized. A rambling, unwieldy, and repetitious compendium, *Pre-Adamite Man* first appeared in 1863 under the pseudonym Griffith Lee. In fact, it was the work of Paschal Beverly Randolph (1825–75), a medical doctor, human rights champion, mystic spiritualist, and supreme grand master of the American branch of the Rosicrucian Order (Fraternitas Rosae Crucis), a confraternity that continued to attract those with esoteric interests in alchemy, Gnosticism, magic, and the Paracelsian tradition. Randolph was a controversial figure, and his teachings on occult sexuality—perceived at the time as free love and, later, as the precursor of ceremonial sex magic—resulted in a brief period of imprisonment in 1872, before he was reportedly acquitted of all charges.[1] His literary output was extensive; he was the author of some fifty books and pamphlets on health, philosophical arcana, *magia sexualis,* and the occult.[2]

In all likelihood *Pre-Adamite Man*'s initial pseudonymity was intended to protect its author from direct censure, for, as Randolph observed in the preface to the fourth edition, "the book has had mountains almost insurmountable in its path," although—in his own opinion at least—it had "outlived adversity [and] become a standard authority in the world of letters."[3] Others didn't think so. As one reviewer bitingly put it: "It is not worthy of perusal by Christian, Jew, or infidel. It is a work of great pretensions, but of no originality or merit . . . The thoughts . . . are exceedingly crude and disjointed; the statements even of pretended facts, are unreliable, and most of the quotations are inaccurate, and the style is awkward and often ungrammatical. As to the plan, there is none."[4]

These drawbacks notwithstanding, the book's general impulse is readily discernible. Its purpose was simply to lay out what Randolph took to be the standard evidence from linguistics, anthropology, archaeology, paleontology, and ancient history for the existence of humans long before the time frame that the biblical chronology for Adam could allow. Like other polygenists, Randolph rejected climatic accounts of racial differentiation. Commenting on the black race, for example, he noted: "Climate . . . therefore, had nothing to do in making him the hue he is; for in Africa, beneath the torrid sun, and in cold Icelandic regions, he and his children retain the same complexion . . . He must, therefore, be a distinct species of the *genus homo;* for, differing so totally from the sons of Adam, he could not have descended from the same source, and is consequently, and beyond all doubt, a Pre-Adamite Man!"[5]

This characterization, however, did not imply any inherent inability to progress; indeed, Randolph insisted that the black race was destined for "power and greatness" and dedicated the first edition of the work to Abraham Lincoln (a close personal friend and fellow Rosicrucian), though he did wax lyrical in his praise of European cultural superiority, "a superiority of structure and function, a superiority of temperament and material, a superiority of blood and race."[6] No doubt we discern here, too, a complex psychology in operation, for Randolph vehemently denied having any black ancestry when his enemies hurled at him the accusation that he was of Negro descent.

What is also notable about Randolph's intervention is that he was not at all motivated by any concern to use the language of pre-adamism as a strategy for reconciling science and the Bible. In fact, he took issue specifically with the harmonizing tactics of scriptural geologists. Attempts to read the days of Creation as geological epochs did nothing but ensure that the "grand and simple story is perverted, defaced and stultified by interpretations and exegeses, as violent as they can be, and such as would drive the writers of Genesis to distraction."[7] At every turn his passion was to ensure that theology gave way to science: "It is for Theology to suit herself to the requirements of advancing knowledge . . . She may struggle for a while, only to yield in the end."[8] Announcing his unorthodoxy at several points, he anticipated the survival of ethical Christianity, or what he called "*practical* Christianity," while welcoming the demise of its theological, or "*mystical*," components, and rejected the Incarnation, disavowing as incredulous the belief "that the Infinite, Awful, Eternal One, ever had a child by any woman in the physical sense, virgin or matron."[9] Besides, Randolph happily welcomed the most recent installment of Bishop Colenso's critical reading of the Pentateuch—one of a set of treatises uniformly excoriated by conservative believers—which he deemed "another eloquent protest against the historical inaccuracies of the book of Genesis."[10] He elaborated, too, on the fictions of chronology, both biblical and classical, through its typical resort to genealogical conventions, calendrical peculiarities, and numerological fixations, and insisted that the Genesis text—no more than one of a range of oriental cosmogonies—was the collation of various documentary fragments.

In sum Randolph was not troubled about whether or not Adam was a myth, though he was inclined to believe that there had been an ancient adamite people. His passion was simply to demonstrate the antiquity of the human race. As he himself put it: "The Adam of Genesis may be a myth; he may have been a miraculously created man; he may have been an immigrant from other parts of Asia, or he may be the Ideal Hero of the poets who sung [*sic*] of Auld Lang Syne when this world was six thousand years younger than it is today. In either case it matters little, so long as we can prove man's antiquity to be immeasurably greater than is usually believed."[11] For Randolph pre-adamite vocabulary was merely a surrogate label for an ancient humanity entirely separate from the line charted in the legends of the Hebrew people.[12]

Pre-Adamite Anthropology

If Randolph's pre-adamism advertises the esoteric fringes where the theory found a foothold, it would be mistaken to think that this was its natural environment or that it only surfaced in works of theological apologetic and thus had no place at the hub of the intellectual establishment. In fact throughout the middle decades of nineteenth-century Britain, the idea of a pre-adamite humanity thrived as part of the conventional discourse of the new sciences of anthropology and ethnology. This is plain, for example, from a reading of the proceedings of the Ethnological Society of London, which had originally come into being in the 1840s, at the instigation of Thomas Hodgkin, as an offshoot of the Aborigines Protection Society, though concerning itself rather more with the scientific study of the natural history of the human species than with humanitarianism.[13] But the society had gone into decline during the 1850s and was later reinvigorated by a group of anthropologists and archaeologists representing the newer trends in their subjects. Soon, however, cracks began to appear, and friction with the older Quaker-humanitarian element in the Ethnological Society resulted in the formation of the breakaway Anthropological Society in 1863.[14] The details of this institutional chapter in the history of British anthropology need not detain us here, save to say that the new society represented a tougher polygenetic and racialist line of thinking. As we shall later see, the writings of certain key members, such as James Hunt and Robert Knox, display a much more secularized polygenism that should not be conflated with religiously inspired pre-adamism. But others who participated in the society's scientific inquiries into human antiquity and diffusion at midcentury still found sustenance in the idea of Adam's predecessors.

Crawfurd's Linguistic Polygenism

During the life of the Ethnological Society the name of the aging John Crawfurd (1783–1868), a close associate of the eminent geologist Roderick Murchison, featured prominently.[15] Crawfurd, who played a major role as president and vice president, was an Indian army doctor turned orientalist, an East India Company resident at Singapore, and from 1818 a fellow of the Royal Society. Author of the very successful three-volume *History of the Indian Archipelago* (1820)—in which he elaborated in extenso on the spectrum of cultural forms, from anarchy to despotism, that he had discerned during his time in tropical Asia—he steadfastly promoted polygenism from a variety of perspectives ranging from physical anthropology to philology. "That the many separate and distinct races of man . . . are originally created species, and not mere varieties of a single family," he told his fellow members of the Ethnological Society in 1861, "there are many facts to show."[16] Predictably, he accorded a minimal role to the influence of physical and social environment on human physiology: neither form nor stature nor complexion were subject to climatic modification or to the effects of diet.[17] Race, simply, was the product of creation, not of climate, and still less of evolution.[18] And he was entirely content to leave his explanation there. It had simply "pleased the Creator—for reasons inscrutable to us—to plant certain fair races in the temperate regions of Europe, and there only, and certain black ones in the tropical and sub-tropical regions of Africa and Asia, to the exclusion of white ones." What was certain was "that climate has nothing to do with the matter." The story of human origins, he told the readers of the Ethnological Society's *Transactions*, went something like this: "When man was first called into existence, we may conjecture that he was planted over the earth in many small families or groups, often consisting of distinct species, or when not so of a single species, adapted to a variety of climates . . . In such an isolated state, each little group would construct its own separate language . . . and hence the multiplicity of tongues, and the variety of words and structures."[19]

This stance was reinforced by his profound antipathy to the Aryan theory of language origins of which Max Müller was the chief architect. While Müller in fact argued for the separation of ethnology and philology, there still remained many good reasons to join them.[20] The earlier tradition of linking racial monogenism with language remained attractive to some, such as R. G. Latham, author of *Man and His Migrations*, who, while acknowledging that some ethnologists had made too much of linguistic evidence for racial unity, was nonetheless sure that anatomists, archaeologists, and physical anthropologists had made too little of it.[21] Darwin's own suggestion that linguistic diversification was a fruitful way of thinking about "a genealogical arrangement of the races of men" further confirmed the association.[22] Crawfurd retained the connection, but his conclusions were radically different from the monogenist cast of Darwin's transmutationism. His own comparative analysis of numerous words in different vocabularies confirmed his judgment that "the theory which makes all the languages of Europe and Asia . . . to be of one and the same race of man, is utterly groundless and the mere dream of very learned men."[23] Language, because it was acquired rather than innate, was simply not a safe guide to race. So committed was Crawfurd to this scheme that he dismissed James Cowles Prichard's monogenism as a "monstrous supposition" hardly worthy of "serious refutation."[24] Throughout his *Transactions* essays and elsewhere Crawfurd's descriptions were punctuated with periodic commentary on the physiological and cultural inferiority of particular races, and yet he remained steadfastly opposed to slavery—a stance revealing that polygenism was not invariably cast as scientific underwriter to the ideology of slavocracy. Nevertheless, by placing different races in different regional locations, the Creator had suited racial culture to climatic circumstance. Climates were sufficiently powerful to keep races in their proper places. "Mere intemperance of climate," he informed his readers in 1863, "is sufficient to prevent man from making any advance towards civilization, and to hold him permanently in the savage state," and the "races of man which inhabit Africa correspond with the disadvantages of its physical geography."[25]

Pre-Adamites among the Anthropologists

Crawfurd was certainly not alone among members of the Ethnological Society in his belief in the creation of non-adamic humans. On the basis of his findings in Egyptian ethnology, for example, Reginald Stuart Poole insisted on the persistence of racial form and explicitly endorsed the double-adamism of his uncle in a paper read before the society in June 1862. Convinced that "permanence of type involves at least duality of origin," he nonetheless wanted to ensure his audience that he was concerned to rebut any charge that he "set no value upon the Biblical indications of the origin of man."[26] He also hastened to dissociate himself from the nasty racialist implications that American southerners had drawn from polygenism and the belief in racial hierarchy. "No ethnological theory or fact can shake the first principles of Christian morality," he insisted, wryly adding that belief in the common ancestry of the human species had not prevented good monogenists in America's southern states from abusing African Americans. For that matter "belief in the unity of man" had done nothing to deliver "the Negro in the Northern States any social recognition." In fact, he went on to insist that "the very knowledge that an inferior race had a separate origin, should rather arouse our generosity than awaken our dislike, and be an incentive to liberality rather than to tyranny."[27]

The following year, 1864, in February, the Indian-born Reverend Frederic W. Farrar (1831–1903), novelist, philologist, and later dean of Canterbury, supported George Gliddon in challenging the idea that the "universality of the Deluge, and the Unity of the human race," enjoyed the supporting witness of non-Semitic traditions, as was often suggested.[28] This paved the way for the November meeting of the Ethnological Society, at which Farrar presented the same audience with what he considered superabundant scientific testimony to the fixity of type and the impotence of climate, urging his readers to draw from his catalog "such inferences as appear . . . to be most truthful and logical."[29] His work on primitive language brought him to the admiring attention of Darwin, who proposed him for fellowship of the Royal Society in 1864. Farrar later repaid the compliment by arranging, controversially, for Darwin's burial in Westminster Abbey and preaching the funeral sermon.[30]

The inferences to be drawn from Farrar's challenge, of course, were hardly uniform, and different members of the society reacted in different ways. On the one hand, there were those who continued to believe that even if the idea of polygenism were to be verified, it would leave the question of the specific unity of humankind entirely untouched; on the other, there were those who pressed a more secular polygenism into the service of the harshest of racial ideologies. For the moment the salient point is that debate rotating around the issue of multiple Adams and divine creations of distinct human types were part of the conventional discourse of anthropology at midcentury. Farrar himself, for example, reviewed the evidence from philological history and the distinctions between monosyllabic, agglutinate, and inflectional languages for the Ethnological Society in May 1865 in order to support his claim that the "scriptural tradition which referred the diversity of language to a direct miracle shews a recognition of the fact."[31] Language families were primordial. Taking Reverend Farrar as emblematic, it is clear that the idea of non-adamic humans was firmly embedded in certain strands of both anthropological and theological discourse.

Even among the more materialist and secularist Anthropological Society membership, the theological pre-adamite did not disappear from the scientific scene. Much was still made of La Peyrère and his role in the intellectual prehistory of polygenism. Thomas Bendysche, for instance, one of the society's vice presidents, outlined his crucial contributions in his memoir on the "history of anthropology," dwelling on La Peyrère's role in the development of polygenism.[32] A subsequent contributor, writing as "Philalethes," set out to rescue La Peyrère no less from oblivion than from ignominy and compared his perfidious treatment at the hands of the self-designated orthodox to the virulence that the American physician and natural scientist Samuel George Morton (1799–1851) had suffered from critics such as John Bachman and Thomas Smyth, who vigorously held to the unity of the human race. For Philalethes, La Peyrère was to be understood not as a theological saboteur, as his latter-day critics alleged, but as a strategist in the cause of honest biblical interpretation unconstrained by despotic dogma. La Peyrère was to be revered both because he "broke through the meshes of a groundless traditional prejudice, and proved that even in Scripture there are no decisive evidences of man's descent from a single pair," and because he showed that "there are distinct indications of non-Adamite races." His "acumen, his candour, and his courage" were much to be admired.[33] A sympathetic portrait of this Galileo-like figure, and the denunciation he received at the hands of the guardians of the sacred flame, provided a historical lesson that contemporary enemies of science were too slow to learn. Philalethes had in mind those berating the radical polygenism of Morton, and his acolytes in the American South (whose racial ideology will detain us in chap. 7), and the archaeological revelations of Boucher de Perthes. Happily, Philalethes concluded, as each of these individuals had shown, "the true thought of the solitary thinker in his closet is stronger than priests and princes; is omnipotent even against the banded conspiracies of the whole world's prejudice and interest."[34] The strongly pro-science sentiments of Philalethes were themselves abundantly displayed again that same year, when, in a sympathetic account of Darwin's theory, he systematically demolished each of the features supposedly separating the human race from its animal counterparts. It was mistaken to seek to identify some discriminating trait, for there simply was "*no* such point of distinction; man does not form an order apart from the rest of the animal world."[35] The radical continuity between human and animal was thereby cemented.

The whole matter of Adam's language also surfaced within the new Anthropological Society. In a comprehensive review of Max Müller's *Science of Language* in 1863, Richard Stephen Charnock revisited several matters surrounding adamic speech. The historical record on the subject, he confessed, was bewildering. Some writers maintained that "the language spoken by Adam is lost," while others presented a baffling range of candidates;[36] Reading had argued the case for Abyssinian, Stiernhielm and Rudbeckius for Swedish, Skinner for Belgic, Lye for Icelandic, Hugo for Latin, and many more. His own inclination was to resist Müller's linguistic monogenism, and reflecting on African and American languages, which did not enter Müller's scheme, he observed that they displayed characters and idioms "of which it would be in vain to seek for traces in a primitive tongue."[37] His review of the adamic language debate and his elucidation of the linguistic taxonomy rooted in the account of Noah's sons—Shem, Ham, and Japheth—thus served to keep the question of Adam and non-adamic peoples in play.

A few years later the president of the Manchester Anthropological Society (and from 1871 vice president of the London Anthropological Institute), barrister and registrar of the bankruptcy court in Manchester George Harris (1809–90), elaborated on the "Distinctive Character of the Adamite Species" in an essay on "The Plurality of Races." The purpose here was to ascertain just how reliable the Bible was as an ally in the polygenist cause. Harris was not at all disposed to "dispute the authority or the inspiration of the Bible," and while acknowledging that its purpose was not to provide scientific instruction, he evidently did want to secure its benediction on the idea of humanity's plural origins.[38] The impression that Bible readers had that the sacred text taught universal descent from the adamic family was simply mistaken; plainly put, "the Bible has not been allowed to speak for itself."[39] In numerous passages the Genesis text simply presumed the existence of non-adamic peoples, and Harris rehearsed again the sort of standard evidence that had been in circulation since the time of La Peyrère, concluding that the subsequent narrative of the adamic line was only the story of the Hebrew people.

By the late 1860s the idea of pre-adamites as a reconciling tactic was obviously still circulating within the anthropological community. But it also served as a means of providing religious legitimation for the pluralist credo of a discipline that increasingly unsettled Christian assumptions about the unity of humankind and the moral imperatives that flowed from that doctrine. Commentators on theology and anthropology who did not hew to this polygenist line were less happily received, as George Moore—physician, polymathic author, Baptist layman, and grandfather of the philosopher G. E. Moore—bluntly found out.[40] Even though Moore was willing to speak of the inferiority of certain races in his 1866 book on *The First Man and His Place in Creation,* because he had attributed the emergence of the black races to the agency of climate rather than to original constitution, he found himself on the receiving end of James Hunt's scorn in the pages of the October 1866 issue of the *Anthropological Review,* in which he took up what he described as "the remarkable fact that some Darwinists are Monogenists, and, what is still more remarkable, that some Darwinites in this country are even now teaching as a scientific induction, that there is, at the present day, but one species of man inhabiting the globe."[41] To Hunt, Moore's "reckless speculations" about that "interesting creature 'The First Man'" enjoyed both "charming simplicity"—and no evidence.[42] Another reviewer for the *Popular Magazine of Anthropology* apparently considered Moore as a "malicious, incoherent, scribbler" whose efforts were fit only for "elderly *females of both sexes,* and Sunday-school children."[43] Moore replied testily. The exchange reveals that while Peyrèrean-style theology was tolerated, no doubt because it conformed to prevailing anthropological orthodoxy in that quarter, monogenist doctrine was lampooned at best, denounced at worst.

Early in the following decade, in 1872, when the Anthropological Institute of Great Britain and Ireland brought together the earlier rival ethnological and anthropological factions, Charles Staniland Wake (1835–1910)—Egyptologist and student of marriage, kinship, and ancient symbol worship—also took as his subject "The Adamites" for the readers of the first volume of the new institute's journal. A decade or so earlier he had argued, against Huxley and various materialists, that human superiority over the rest of the animal world resided in "a superior spiritual nature," and not just in humankind's intellectual or linguistic capacities.[44] Physiologically, in terms of the nervous system and emotional expression, there were certainly continuities between human and animal; by virtue, too, of their capacity to experience emotion, thought, and mental action, animals were possessed of a certain principle that "we call the soul";[45] it was only the spiritual powers of mental abstraction and generalization that definitively separated the human and animal creations. And having mapped that boundary to his satisfaction, Wake now wanted to engage in other forms of anthropological cartography. The burden of this later presentation was thus to find scientifically robust measures by which to discriminate effectively between adamites and pre-adamites. His own belief was that the dolichocephalic, or "long-headed[,] pre-Adamic stock" was essentially different from the brachycephalic "Turanian and Aryan peoples of Asia and Europe, with the Hamite and Semitic peoples of Western Asia and Northern Africa."[46]

Wake's interlocutors did not concur. Carter Blake, then lecturer in comparative anatomy at the Westminster Hospital School of Medicine, secretary of the Anthropological Society, and severe critic of T. H. Huxley's stance on human evolution,[47] certainly found Wake's paper arresting, but he felt it suffered from the fatal flaw that "the Hebrews, who claimed especially to be descendents of Ad-am, were undoubtedly dolichocephalous."[48] What was

ADAM'S ANCESTORS

not challenged was the whole project of seeking reliable means of dividing humanity into a bipolar adamite–pre-adamite taxonomy.

Adamic Anthropology

The multiple-Adam thesis that was beginning to find favor with some Christian apologists was supplemented by advocates among members of the growing community of anthropologists. But this certainly does not mean that adherents to traditional adamic anthropology had disappeared. Some rejected the pre-adamite proposal out of hand, though others, while refusing to embrace it, held it as a provisional option, allowing it as a potential harmonizing policy should the scientific evidence move conclusively against monogenism. Religious resistance, empirical assault, or anticipative apologetics were choices available to detractors.

Prichard's Defense

Opponents of polygenism could still call upon the towering, if fading, authority of James Cowles Prichard (1786–1848), whose celebrated critique of Lord Kames was still widely read. His *Researches into the Physical History of Man,* which first appeared in 1813 and was expanded into five volumes by its third edition, and *The Natural History of Man* of 1843, which had gone into its fourth edition by 1855, remained influential for many. That was the case at least until the advent of E. B. Tylor's evolutionism during the 1860s, despite those critics such as Gustave d'Eichthal, Arthur de Gobineau, and Paul Broca who thought of him as emblematic of old-fashioned biblicism. Prichard had progressively moved toward environmentalism in order to preserve his monogenist stance in the face of ever more assertive polygenetic challenges from British, European Continental, and American sources. Having grown up as a Quaker and later felt the influence of evangelicalism, Prichard, too, was profoundly committed to universal descent from Adam and thus found himself looking more and more to climate to explain somatic differences among racial groups. After all, he demonstrated his commitment to the truth of the Mosaic chronicle by seeking to accommodate the Genesis Days of Creation to geological periods and in his insistence that religion had not naturally developed but was, rather, a divine revelation.[49]

Early in his career Prichard had considered civilization as a race-forming factor, arguing that it stimulated variation in a way rather akin to the processes of domestication. But he increasingly moved toward identifying a correlation between climate and physical type while remaining ever sensitive to the excesses to which environmentalism had been put and repudiating the "absurd theory" held by the otherwise "ingenious" Stanhope Smith and, more reprehensibly, Johann Friedrich Blumenbach, who attributed black skin to bile secretions induced by the heat of the sun.[50] By resorting to the influence of natural environment, the door was opened to the drawing of analogies between human and animal development. Accordingly, Prichard cited the opinion of M. Poulin in support of the view that "acclimatisation . . . consists in certain permanent changes produced in the constitution of animals, which bring it into a state of adaptation to the climate." And while Poulin and Prichard stopped short of species transformation, the Lamarckian core of their views on hereditary transmission is plain in Prichard's reporting of the following "facts" established by Poulin: "Permanent changes or modifications in the functions of animal life, may be effected by long-continued changes in the habitudes which influence these functions . . . Hereditary instincts may be formed, some animals transmitting to their offspring acquired habits."[51] Certainly, Prichard never wholeheartedly embraced Lamarckism, but as Stocking points out, he increasingly retreated, like Darwin, toward it.[52]

If the climatic adaptations of animal species could be interpreted naturalistically, there was little reason to exempt the human species from the selfsame agents of change. Prichard's, like Smith's, racial account was thus inherently "evolutionary" rather than "creationist" and drew attention to his view of race as being far from fixed. As he put it, "All the diversities which exist are variable, and pass into each other by insensible gradations."[53] For Prichard, however, one component of his anthropological system disquieted many. Deeply convinced of progressive development from savagery toward civilization (although taken overall, as Stocking makes clear, his thinking is possibly better typified as diffusionist rather than developmentalist) and what seemed to him its self-evident corollary, progress from black to white, Prichard suggested that "the primitive stock of men were Negroes."[54] Whether or not he meant that Adam, in the state of perfection, was black has been queried, but however that question resolves, Prichard shared the idea of primitive blackness with the Cambridge physician John Elliston and the physiologist John Hunter.[55] Regardless, his authority lingered among certain factions of the ethnological community, despite the assault of several strategically placed individuals, not least because of his concerns about the racial abuse that colonists were inflicting on native peoples. At the British Association for the Advancement of Science in 1839, for example, he provided a grisly litany of what he termed "The Extinction of Races" as part of a recruiting campaign for the Aborigines Protection Society, whose motto AB UNO SANGUINE underscored humanity's essential unity.[56] That influence extended to the leading advocate of monogenism in archaeology, the Scottish president of the University of Toronto, Sir Daniel Wilson (1816–92), who for decades issued critical commentary from his Canadian vantage point on the militant, secular polygenism being peddled in the American South and argued the case for the physical and moral excellence of mixed races.[57]

Adam's Advocates

Back at the Ethnological Society, when the Havanna judge James Kennedy examined the question of the origin of the American Indians in March 1854, he expressed profound disquiet at those who countenanced the idea that "there were originally various distinct creations of beings of the human race"; such a stance was "contrary to our faith that 'God hath made of one blood all the nations of the earth.'" In his judgment the available data tended "to prove the correctness of the account given us in the Mosaic history, taken merely as history."[58] Kennedy's opinions, which drew sustenance from the "eminent" inquiries of "our late respected president, Dr. Prichard,"[59] were reinforced by those of Darwin's *Beagle* captain, the hydrographer and meteorologist Robert FitzRoy, now admiral, who told the society in 1861 that racial variation could occur very rapidly under certain conditions, mostly through intermarriage, and that this, together with patterns of geographical migration, could fully account for the nature and distribution of human occupancy of the earth. These circumstances, alongside observed affiliations between language groups, provided him with what he considered to be overwhelming evidence "in exact accordance with the historical account in the Bible" and, in particular, with the narrative of Noah's children, Shem, Ham, and Japheth. Part and parcel of this web of conviction was FitzRoy's belief, too, that when he was cursed "the mark set on Cain and his descendants was negro blackness."[60]

Or take the case of Robert Dunn (1799–1877), a clinical psychologist, member of William B. Carpenter's informal physiological psychology network, and fellow of the Royal College of Surgeons, who repeatedly argued the case for the unity of mankind from anthropometric measurements of human crania and cerebral physiology, from the tegumentary differences between races, as well as from psychology. As might be expected, Dunn granted considerable influence to the role of environment in modifying skin color and hair character and went on to adduce unitary arguments from hybrid fertility. He could not let the moment pass, he confessed, without "advert-ing to the confirmation which the revelations of the microscope have given to the dicta of Holy Writ"—that "*God had made of one blood all the nations of the earth.*"[61] Yet he did make it clear nonetheless that, while "the unity of the species by physiological and psychological evidence has been established, the '*quaestio vexata*' still remains: Have there been more creations than one of the same genus, more Adams and Eves than one single pair?" He himself sided with the monogenism of Edward Forbes, R. G. Latham, William B. Carpenter, and Prichard against Louis Agassiz; the supposition of the cre-ation of several distinct "*protoplasts*," one for each region of the globe, was not required to account for the geographical distribution of race.[62] "In these sentiments and in this conclusion," Dunn was pleased to report, "I think we must all cordially agree with Dr. Prichard."[63] Indeed, he repeated this same contention a few years later when he told the Ethnological Society that the "one common bond of universal brotherhood" stemmed from the fact that "God made of one blood all the nations of the earth and endowed them all with the same animal, intellectual, moral, and religious nature."[64] Divine ori-gin, not necessarily lineal descent from an original single pair, was the war-rant for a common humanity. But that conviction did not mean racial parity. The "intellectual inferiority of the Bushman, the Australian, and the Negro to that of the Indo-European" was "striking."[65]

Continental Reinforcements

Besides the long-standing authority of Prichard, supporters of human de-scent from an adamic pair could now call more and more on the scientific monogenism—even if not necessarily of the adamic variety—being champi-oned by the French naturalist Armand de Quatrefages (1810–92). His con-viction that racial difference was climatically induced exerted a profound influence on Isidore Geoffroy Saint-Hilaire and the French tradition of accli-matization.[66] Even while president of the Société d'Anthropologie, Quatref-ages stood out against many of the French anthropologists who favored Paul Broca's polygenetic views of race and who accounted for organic modifica-tion purely by interbreeding. Ever since observing the remarkable variability of domestic animals, which caused him to reject the belief that black and white humans belonged to different races, he relentlessly pressed the mono-genist case with the zeal of the converted. Significantly, Quatrefages traced the history of polygenism back to its roots in La Peyrère's theology, noting that "polygenism generally regarded as the result of Free Thought was biblical and dogmatic in origin." The history of monogenism was no less ideologically littered, and thus to avoid the taint of sectarian bias he hastily added that "monogenists are guilty of seeking in religious doctrines arguments in favour of their theory, and anathematising their adversaries in the name of dogma." As for himself, Quatrefages took seriously the findings of natural history and used the writings of figures such as Georges-Louis Leclerc de Buffon, Jean-Baptiste de Lamarck, and Georges Cuvier to support his critique of Agassiz's centers of creation. Moreover, he acknowledged that stances taken up on such scientific questions as variation, migration, and acclimatization were invariably conditioned by commitments to monogenism or polygenism.[67] Later the geographer and anthropologist Oscar Peschel was to argue that the polygenist belief that different human *species* "were at once sown broadcast by the Creator, in numbers as vast as swarms of bees," was a departure from sound scientific principle. "Any explanation of the present by the past is thus abandoned," he went on, "although it lies deeply rooted in human nature not to rest satisfied with observed facts until they have been reduced under some law of necessity."[68] His objection, to put it another way, was that polygenism departed from sound inductive inquiry.

Sound inductive inquiry, for the distinguished anatomist and physiologist Rudolph Wagner (1805–64), Blumenbach's successor at Göttingen, meant scientific agnosticism on polygenism. Wagner was well-known as a vigorous opponent of German materialism—in particular, Carl Vogt—and in a famous address to the Assembly of German Naturalists and Physicians, in 1854, he threw down the gauntlet and aroused the fury of materialists by postulating the existence of a soul substance and arguing that for ethical reasons science needed to retain belief in God. Vogt snapped back, insisting that true science had inescapable anti-idealist and atheistic consequences. His response served to polemicize the relationship between science and religion by focusing it around a sequence of bipolarities—science versus obscurantism, materialism versus idealism, spontaneous generation versus creation.

Parts of Wagner's original presentation were republished nearly a decade later in the newly founded *Anthropological Review* for 1863, though with an editorial note informing readers that several "passages touching on the supposed connection of the science of Man with historical Christianity and Revelation have not been translated, as these subjects have nothing to do with Anthropology."[69] What is significant in the present context, however, is that Wagner had made it clear that he was chary of the inclination of too many anthropologists to elaborate more and more Adams and paradises, some of whom had got up to as many as "fifteen or sixteen Adams." This "result ... no doubt appeared highly satisfactory to slaveholders," Wagner wryly noted, but in fact the "mode by which races have been formed is perfectly unknown" because it reached back to a primordial time "perfectly inaccessible to science." His own inclination, "quite irrespective of my religious convictions," was that all races could indeed "be reduced to one original existing, but only to an ideal type, to which the Indo-European type approached nearest."[70] What also reinforced this belief was Wagner's sense that the advent of Darwinism had reversed the image that the theory of descent from an original pair was scientifically antiquated.[71]

There was, of course, nothing inevitable about the link Wagner presumed between monogenism and abolition, as James Reddie made clear in his discussion of slavery for readers of the *Anthropological Review* the following year, 1864. The proposition that "all men are created equal," Reddie began, was widely assumed to be undeniable; the trouble was that it was just "utterly false."[72] Whether there were any implications in this realization for the practice of slavery was a subject on which Reddie affected to keep an open mind, claiming instead that his aim was to lay before his readers what was "really best for mankind as a whole."[73] Religiously, he noted, there was little to be said in favor of abolition, at least from a reading of the Bible. Indeed, it was notable that though "the writers, both of the Old and New Testaments were unacquainted with any other than a monogenous theory, we do not find a syllable in any of their writings condemnatory of slavery."[74] Religious arguments against the institution, therefore, were "utterly untenable and groundless."[75] Besides scriptural testimony, Reddie believed that he could point to manifold evil consequences that had flowed from manumission in the United States; it was far from obvious that the abolitionists had ethics on their side. On the scientific front he found repugnant a recent American treatise advocating miscegenation; a newly available rebuttal of this "filthy theory" hoping to restore "the normal relation of the Races" had, unfortunately, "spoiled a very tolerable thesis, by ... Yankee extravagance."[76] In any case Reddie's concern was to detach the slavery question from "the monogenist and polygenist theories of the origin of mankind."[77] Whether "all men are descended from a single Adam or not" was simply irrelevant to the question of racial equality because the Bible itself bore witness to the permanence of racial features and passed no judgment on the practice of slavery.[78] He himself felt drawn to "the older tradition of the descent of mankind from a single pair" but was nonetheless convinced that some races were substandard.[79] The only question to be posed was thus: "*While the inferior grades are actually inferior, how is it best for the higher races to deal with them?*"[80] This species of monogenism was attractive to some slaveholders in the American South who held tenaciously to the adamic story precisely because biblical literalism could secure a scriptural justification for slavocracy.

Refusal

Pronounced . . .

The continued survival of scientific monogenism among individuals like these not only provided reinforcements for those skeptical about polygenism on empirical grounds but also convinced many that its pre- and co-adamite versions occupied forbidden theological territory. The doctrine of the unity of the human species that underlay the monogenism of Stanhope Smith, Prichard, Blumenbach, and Quatrefages came to the surface no less in theological works than in scientific discourse. In Smith's own institution, Princeton, the Presbyterian tradition steadfastly hewed to the monogenist line. In 1829, for example, Archibald Alexander (1772–1851), who drew up the blueprint that would guide Princeton theology for a century and through whom Old School Presbyterianism received its baptism in Scottish Common Sense philosophy, turned his attention to the relationship between scripture and nature. Here he dismissed the idea "that there were men before Adam" as "destitute of all shadow of proof," suggested that the varieties of the human species were due to "the great difference of climate and other circumstances of the nations of the earth," and declared as a "prejudice without foundation [the belief] that the colour of the whites was that of the first man."[81] Alexander speculated—like others after him—that Adam was neither white nor black and that it was "much more probable that our first parents were red men or of an olive or copper colour." He based this judgment on his understanding that in Hebrew "the radical signification of *Adam* is *red*."[82] Later, also at the Princeton Theological Seminary, Charles Hodge (1797–1878) similarly defended the unity of the human race, as did Benjamin B. Warfield, who, as we shall see, was able to integrate it both with evolutionary theory and with a newer form of pre-adamism. Hodge, for example, despite his admiration for Louis Agassiz, would not compromise on the question of humanity's common origin and called upon the researches of his Princeton colleague William Henry Green on the nature of biblical genealogies to extend the traditional time available for racial differentiation since the adamic Creation.[83] To him the virulent polygenism of Gliddon and Josiah Clark Nott's *Types of Mankind* (1854) was subversive of genuine religion and social polity alike. Such matters were infinitely more threatening than Copernican astronomy or Lyellian geology, for to Hodge they struck at the very heart of Christian theology and human identity and thereby compromised the moral nature of humankind.[84]

On the other side of the Atlantic, in Britain, conventional adamic anthropology continued to be represented in both popular and more serious theological works. In a lecture delivered in Exeter Hall in December 1848, for example, Reverend William Brock directed his YMCA listeners to the subject of "The Common Origin of the Human Race." Patching together a pastiche of popular philology, physical anthropology, and psychology, the rhetorician urged the common origin of mankind upon his young hearers. With exuberant enthusiasm he concluded his homily with an appeal to human solidarity as the basis for evangelistic endeavor.[85] It was a similar story for Elisha Noyce, too, whose *Outlines of Creation* (1858) for juvenile readers likewise held up monogenism as the only respectable scientific option.[86]

In Scotland evangelical Calvinists were certainly no more inclined toward polygenism. In 1850, for example, Hugh Miller, though widely applauded by pre-adamite partisans for his scheme of harmonizing Genesis and geology, made it clear that he would have none of the fashionable polygenetic talk. To him the whole soteriological structure of Christianity rested on the assumption of the specific unity of the human race, and any notion that "*Adams and Eves* [were] many" had to be ruled out simply because, as he put it, the "second Adam died but for the descendants of the first." Miller's essay was in fact an extended commentary on Thomas Smyth's volume on *The Unity of the Human Races*—a work berating those polygenists, notably Agassiz, whose claims were circulating in fashionable circles in the American South—and while he took the same anthropological line, he felt that Smyth had perhaps been too precipitate in arguing that both blacks and blue-eyed Goths could *not* have descended *naturally* from a common origin. To be sure, supernatural intervention could not be ruled out, a priori, but with a typically Calvinist suspicion of multiplying miracles gratuitously, he proposed that the horns of the dilemma would be removed if it were assumed that Adam was neither black nor white but, rather, a "mingled Negroid and Caucasian type." If so, then "neither the Goth nor the negro would be so extreme a variety of the species as to be beyond the power of natural causes to produce."[87]

Miller's antipathy to the plural origins of humankind was shared by his Free Church of Scotland colleague John Duns (1820–1909), professor of natural science at New College, Edinburgh, from 1864 until 1903, as successor to the distinguished John Fleming. The relationship between science and faith was Duns's scholarly passion, and it was only to be expected that the challenge of the newest anthropology would sooner or later come to the surface in his writings. In *Science and Christian Thought* and *Biblical Natural Science,* both of which appeared in the mid-1860s, he thus tackled the question of the antiquity of the human species. In the tradition of Calvinistic epistemology Duns made it clear throughout these works that his science was grounded in Christian presuppositions and that the claims of the biblical worldview should be taken on the principle of initial credulity, at least until "its insecurity is clearly and triumphantly shown."[88] The scriptural record, taken prima facie, plainly indicated a monogenist position, and Duns only found this confirmed as he perused the arguments and sifted the data of natural historians, philologists, and anthropologists alike. Moreover, Duns found himself revolted no less by the evolutionary musings of Carl Vogt than by the ethical relativism of Continental materialists and by the racist undercurrents among the American Mortonites, and he thus set himself the task of refuting the underlying naturalistic philosophy.

Duns's survey of the pertinent scientific literature was prolix, but his familiarity with the sources of the newest science did nothing to weaken his resolute adherence to the adamic world picture. Again following the well-worn path of Johann Friedrich Blumenbach, James Cowles Prichard, and R. G. Latham, he looked to the "power of *habitat,* climate and the like" to produce as great variations among the peoples of the world as among the lower animals, thereby paradoxically connecting up animal and human in a single explanatory network.[89] "Is there greater unlikeness between the head of the negro, the aboriginal Australian, the European, and the Hindoo," he rhetorically asked, "than there is between the head of the grey-hound and that of the mastiff or the bulldog?"[90] As for human language, he called on the authority of Wilhelm von Humboldt, Max Müller, and C. C. J. Bunsen to confirm that all linguistic variations could well have sprung from a single source: "The very highest authorities in the science of language," he reported, "unite ... in bearing testimony to the original unity of language and the unity of man."[91] The sciences, it seemed, were on the side of scripture—as was morality, for Duns slyly remarked that it was only to be expected of the Morton-Nott-Gliddon brigade that in their portrait of universal race history they "would assign a foremost place to the families to which they themselves belong"; they might have done this, Duns further sniped, "without attempting to un-soul others farther removed from the early centres of civilization."[92] Interwoven with the fabric of Duns's diagnosis, of course, were theological threads. And so, while scrutinizing archaeological and geological evidence of human antiquity (from Danish *kjökken-nöddings* [kitchen middens], lake dwellings, peat deposits from Scotland, and swamp cypress in Mississippi as well as other palaeobotanical data) and consistently reinterpreting it to shorten the time scale, he spelled out in detail the biblical objections to the pre-adamism of the recently published *Genesis of the Earth and Man* and *Pre-Adamite Man*. Chief among these were Saint Paul's theology of original sin, which presumed that all humankind had descended from the Old Testament Adam, and what he considered to be the "gratuitous and unwarranted" postulation of a double creation narrative in Genesis.[93]

Victorian Catholics were just as allergic to pre-adamite speculations. Indeed, according to William Astore, polygenism was substantially more troubling to the Catholic mind-set at the time than either geological time or evolutionary theory.[94] Original sin was the pivotal dogmatic tenet that could not bear the weight of the multiple origin theory. Nicolas Patrick Wiseman—later cardinal and first archbishop of Westminster—for example, took up the whole subject in his *Twelve Lectures on the Connection between Science and Revealed Religion* delivered in 1835. Wiseman's intervention showed a remarkable familiarity with the state of play in a range of disciplines. Beginning with the study of language and ethnology, he moved on to the earth sciences, history, archaeology, and critical hermeneutics. His excursus addressed the larger question of the general relationship between science and Christianity, but it also got down to detailed questions such as numismatic challenges to some biblical texts, hermeneutic queries surrounding the prophecy of Isaiah, and how the recent discovery of certain ancient monuments confirmed a passage in Ezekiel. In his third and fourth lectures Wiseman turned to anthropology. And it became clear that, while he was happy to read the Genesis record through geological lenses, he was certain that any tampering with the doctrine of universal human descent from Adam endangered the very fabric of Catholicism, which was rooted in "the deep mystery of original sin, and the glorious mystery of redemption."[95] Having in his first two addresses demonstrated to his own satisfaction the possibility of deriving all languages from a common source and having secured the supporting testimony of "modern ethnographers, that the language of men was originally one,"[96] he now called on the authority of both Prichard and Blumenbach in his anthropological defense of adamic monogenism. These writers were aligned "on the side of truth";[97] on the opposite flank stood the skeptical French naturalists—J.-B. Bory de Saint-Vincent, Julien-Joseph Virey, Jean-Baptiste de Lamarck. Wiseman abominated them. Virey's works, for example, were "even more revolting" than Bory de Saint-Vincent's, not least on account of his suspecting "a certain fraternity between the Hottentot and the baboon";[98] Lamarck had gone further still by arguing that a graduated, evolutionary chain existed throughout nature. Their collective demotion of the adamic family to merely one of a number of early human communities was deeply troublesome. For his part Wiseman thought climate mainly sufficient to explain racial difference and outlined an explanation remarkably reminiscent of the natural selection mechanism: "The different families or races among men, may owe their origin . . . to the casual rise of a variety which, under the influence of favourable circumstances—the isolation, for instance, of the family in which it began, and its subsequent intermarriages—became fixed and indelible in succeeding generations." To this he added commentary on the moral unity of humankind, confirming the existence of a moral sensibility "identical through the entire race."[99] All this confirmed that adamic mongenism was "more consoling to humanity than the degrading theories of Virey or Lamarck."[100] In many ways Wiseman's intervention mapped out the route that Catholics would adopt on the polygenist question for several generations.

In Germany, too, this self-same position was promoted by the University of Bonn's professor of Catholic theology Franz Heinrich Reusch (1825–1900) in his *Bibel und Natur,* which first appeared in 1862. The question of the unity of the human race preoccupied him for three chapters in this work on the "Mosaic history of creation in its relation to natural science." Culling the writings of numerous scholars—notably, Prichard, Humboldt, Quatrefages, Blumenbach, and Camper—he reported in detail on various anthropometric measurements, insisting throughout that polygenism was unwarranted. A number of considerations confirmed him in his judgment: the many different ways in which the human species could be carved up depending on the anthropometric measurement being employed; the relative insignificance of physiological differences compared with the "characteristics which are common to all mankind"; the observed prevalence of interracial fertility; and the "intermediate forms" and "mutability of species" that Darwin (whose findings were "favourable to the theory of the unity of mankind") had established.[101] His conclusion was simply that the "doctrine of the unity of mankind does not contradict any certain physiological conclusions."[102]

In the United States Clarence Walworth, a Catholic convert, Redemptorist missionary, and friend of John Henry Newman, took up the subject in his 1863 book entitled *The Gentle Skeptic.* This work, one of exceedingly few on science subjects by Catholic authors in the period, employed a dialogical style to raise scientific questions and provide suitable Catholic responses.

It became clear that the mouthpiece for Walworth found Augustine's figurative understanding of the Genesis Days of Creation persuasive and was pleased to report that this interpretation was confirmed in the writings of Origen and other Fathers. On the geological challenge and indeed on the Darwinian question (which he simply rejected as scientifically speculative and thus inadequate) he was entirely sanguine, but the multiple Adams and Eves of the anthropologists were profoundly troubling. The unity of the human species lay at the bedrock of the Catholic Church, and any compromise on universal adamic descent would simply rip apart "the whole fabric of the Christian faith."[103] To account for racial divergence, Walworth proposed that differentiation was effected both by the accumulation of small, naturally occurring variations and by sudden saltations operating under the influence of providence.

It was the same for Orestes Brownson (1803–76), another fervent convert, who found Darwinism to be incompatible with his politics on account of its conception of self-development—although this was nothing compared to the challenge of pre- and co-adamism, which could erode the very foundations on which the social order rested. The consanguinity of the human race was a political necessity, for it was the foundation on which the entire edifice of human rights was erected. Explicitly rooted in the Genesis narrative, Brownson's commitment to the "solidarity of the race" was foundational both to "social rights, the rights alike of society in regard to individuals, and of individuals in regard to society," and to international law and was the only bedrock on which opposition to slavery could coherently be grounded.[104] For Catholics, of course, Brownson insisted that Adam's universal fatherhood was no less a theological dogma: "Original Sin, the Incarnation, Redemption, Regeneration, indeed all that has hitherto been regarded as distinctively Christian, would have no meaning if the unity of the human race were not a truth."[105]

Yet this common humanity did not mean equality. Reusch did not hesitate to engage in racial evaluations, much to the detriment of aboriginal Australians and certain native Irish. In the former case the "physical and moral degradation of most of the tribes . . . their cannibalism, and the artificial deformities which they like," made them reminiscent "of the ape, the caricature of man";[106] in the latter their "miserable circumstances" induced "very degraded features," which were the "outward sign of low and barbarous conditions of life."[107] For his part Brownson happily, and much more directly, pronounced that the "colored races, the yellow, the olive, the red, the copper-colored, and the black, are inferior to the Caucasian, have departed further from the norma of the species, and approached nearer to the animal, and therefore, like animals, have become more or less subject to the action of the elements."[108] All humans might share universal reason, but they evidently shared it in different degrees.

Until well into the post-Darwinian era, then, the confessional certainties of both Protestant and Catholic disciples of adamic monogenism on both sides of the Atlantic drew sustenance from those anthropological writers who, whether committed to Christianity or not, repudiated an increasingly fashionable polygenism and held resolutely to human descent from a single stock. By the same token the aggressive advocacy of racial polygenism in anthropological circles, and indeed the persistence of a substream of ethnological work retaining the language of adamism and pre-adamism, did raise doubts in the minds of some apologists. Should polygenist anthropology come to be vindicated, these commentators pondered, the pre-adamite strategy might yet salvage scriptural authority from profane science.

. . . and Provisional

Even while monogenists were continuing to question polygenism's theological and scientific legitimacy, there were those who could see the functional value of pre-adamism as a last line of religious defense against skeptical science. Certainly, as often as not, such observers remained uncommitted to non-adamic humanity, but they wanted to leave the scheme as a viable strategy should science's darker speculations come to be verified. In the 1870s, especially in the wake of Lane and McCausland's proposals, some theological conservatives thus began to toy with making their peace with the pre-adamite.

To be sure, such anticipatory tactics were not altogether new. John Pye Smith (1774–1851), Congregationalist clergyman, theologian at Homerton College, and harmonizing geologist, had mapped out as early as 1839 the theological path to be followed should polygenism come to be vindicated. His election to fellowship of the Geological Society of London in 1836, and, four years later, to the Royal Society indicates something of the regard in which his scientific work was held. But he also had a significant reputation as a biblical scholar who was thoroughly acquainted with Continental critics such as Wilhelm Martin Leberecht de Wette, Abraham Kuenen, and Karl Heinrich Graf and who had reworked the traditional understanding of inspiration to allow, as John Calvin had done, for the language of divine revelation to be accommodated to the cultural condition of the biblical writers.[109]

In 1838 he brought out a short volume aimed at a youthful audience on *The Mosaic Account of the Creation and the Deluge, Illustrated by the Discoveries of Modern Science,* and the following year he published his magnum opus, *The Relation between the Holy Scriptures and Some Parts of Geological Science.* This was the printed version of the Congregationalist lectures he had delivered the previous spring in which he defended a local deluge, a long earth history, and the likelihood that animal life had been created in several different regional centers.

Geological questions were certainly to the fore in this analysis, as he dealt with traditional points of conflict—the age of the earth, the fossil record, the Days of Creation, animal death before the Fall, Noah's flood, and so on—but Smith also kept a close eye on anthropological developments. In his discussion of human origins he acknowledged that it was certainly not "an *impossibility* that the Almighty Creator should have seen fit to bring originally into being duplicates, triplicates, or other multiples of pairs, formed so alike that there should be no specific difference between them." But he declared that he himself thought this "antecedently improbable."[110] He did confess, however, that naturally observed variation had yet to show how a black complexion could be derived from a white forebear. The absence of an explanatory mechanism that dealt adequately with "the physiology of the mucous membrane," together with historical evidence of racial fixity since ancient times, left him wondering, and he eagerly anticipated the third volume of Prichard's great work in hopes of finding an answer. Hitherto Prichard's research had provided good "physical, philological, and historical" evidence "in proof of the *unity* of the human race, by derivation from one ancestry."[111] Still, there were niggling doubts. As he went on: "But, *if* the progress of investigation should bring out such kinds and degrees of evidence, as shall rightfully turn the scale in favour of the hypothesis that there are several RACES of Mankind, each having originated in a different pair of first ancestors,—what would be the consequence to our highest interests, as rational, accountable, and immortal beings?"[112]

Some strategy was evidently needed in such an eventuality, and pre-adamism remained an acceptable option. "If the two first inhabitants of Eden were the progenitors, not of all human beings, but only of the race whence sprung the Hebrew family," Smith mused, "still they were formed by the immediate power of God." Pre-adamism did not deny Creation; it confirmed it. Besides, the theological significance of Adam as "a figure of Him that was to come" remained in place; "the spiritual lesson would be the same."[113] Indeed, Smith told his readers that Isaac La Peyrère and Edward King had already shown how "some difficulties in the scripture-history would be taken away; such as—the sons of Adam obtaining wives, not their own sisters;—Cain's acquiring instruments of husbandry, which have been furnished by miracle immediately from God upon the usual supposition."[114] And needless to say, pre-adamism would handsomely accommodate the sorts of racial distinction that Morton was so enthusiastically promulgating on the other side of the Atlantic. The troublesome issue of original sin, so critical to Catholic rebuttals of non-adamic humanity, was not a source of significant concern to Smith either. To him universal sinfulness was simply an empirical fact—there were frankly too many "painfully demonstrated proofs" to deny it. And as for the transmission of Adam's transgression to all humanity, Smith went on: "The way and manner in which moral corruption thus infected all men, under their several heads of primeval ancestry, would be an inscrutable mystery (—which *it is now;*—) but the need of divine mercy and the duty to seek it would be the same."[115] As an anticipative tactic, should Prichardian monogenesis succumb to polygenist incursion, pre-adamism offered a way out: "Thus, if this great question should ever be determined in the way opposite to what we now think the verdict of truth, the highest interests of man will not be affected."[116]

Some thirty years later on, another Congregationalist clergyman, Joseph Parrish Thompson (1819–79), adopted a similar anticipative strategy. An editor of the *Independent* and cofounder of the *New Englander* and successively minister in New Haven and New York City, Thompson developed interests both in oriental scholarship, particularly Egypt, and in the relations between science and faith. In 1870 the publication of his *Man in Genesis and Geology* added to his corpus of writing—many in pamphlet form—on theological subjects, politics, biography, and a photographic survey of Egypt.[117] Later, in 1874, he provided a demographic analysis of the ethnic composition of the United States for the Glasgow meeting of the Association for the Promotion of Social Science, in which the antislavery sentiments he had worked out nearly twenty years earlier reasserted themselves.[118]

In *Man in Genesis and Geology,* which began as a sequence of lectures to his own congregation, Thompson announced his preemptive policy on the anthropological challenge. While "the doctrine of Perriere [La Peyrère] of Bordeaux, in the seventeenth century, in his famous treatise on the Pre-Adamites, which has been revived of late by an anonymous English author [Poole's presentation of Lane's account]," Thompson noted, was "open to some serious objections, it serves to show one possible way in which the Bible and Science may yet be harmonized upon the question of the antiquity of man and the unity of the race. It may prove eventually that there is in this brief record in Genesis a margin for all the discoveries of Science."[119] For the moment, however, Thompson was not quite prepared to throw his weight behind the proposal until the scientific evidence forced him into a corner. This was a familiar stratagem. Throughout the book Thompson characteristically accumulated subjunctives. *If* the date of ancient monuments could be fixed, *if* the age of excavated human artifacts could be determined precisely, *if* the historical differentiation of languages could be elucidated, then—and only then—certain steps might be taken.

For the present he repeatedly called upon the witness of those exploiting scientific uncertainty: Carl Vogt's insistence that rates of peat deposit, from which ancient tools were being excavated, remained unknown; John Cleghorn's challenge to the vast antiquity claimed by Samuel Laing for prehistoric remains in Caithness; the *Anthropological Review*'s admission that "every scientific opinion is speculative" and that there was "no opinion current among scientific men . . . that is not essentially *provisional*." But readers could not rely on scientific doubt to preserve the Bible. For Thompson conceded that the data did indeed "call for an extension of time considerably beyond the computed chronology of the Bible."[120] Even a conservative dating of Egyptian monuments and the "unchanging appearance of leading types of mankind, as far back as we can trace these in history," forced him to anticipate what "principles of adjustment between Science and the Scriptures" should be put in place.[121] Well-established facts about philological diversification and distribution and fossil evidence of human existence during the last glacial age only confirmed his judgment. La Peyrère's pre-adamites waited patiently in the wings.

The year following the publication of Thompson's book, 1871, a volume entitled *The Beginning: Its When and Its How* made its appearance. It was authored by Mungo Ponton (1802–80), a fellow of the Royal Society of Edinburgh; author of a variety of works on optics, the electric telegraph, and devotional subjects; and a photographic inventor. Quite appropriately, *The Beginning* was spiced with a series of luscious paleontological illustrations. Like Thompson, Ponton came close to bringing the pre-adamites onto the theological stage. "The whole subject of the unity and antiquity of the human race," he maintained, "is still involved in too much obscurity to admit of the formation of any decided opinion." But in his exposition of what he called the sixth creative epoch, he affirmed that "should the evidence then ultimately prove sufficient to establish the existence of races of men older than the Caucasian, it would not be difficult to reconcile the Hebrew narrative with such a state of affairs." There was always the pre-adamite safety valve: "There may have been a Negro Adam, a Mongolian Adam, and perhaps two or three more besides the Caucasian Adam," he affirmed, and they may have been formed either at the same time or in a sequence of separate creations. Ponton did not explicitly endorse such a revisionist scheme. "All that has been attempted here," he commented, "is to show that there is, in the Hebrew narratives, nothing inconsistent with the hypothesis that there may have been several primitive races of men, each starting from an Adam— a man formed from the ground by some gradual process analogous to those by which, even at the present time, perfect organisms are slowly elaborated from the most minute and rudimentary organic elements."[122]

Even if these writers could not yet bring themselves to welcome the preadamites onto the platform of history, they were happy to keep them on backstage retainer should their presence eventually be called for. In one form or another pre-adamism thus continued to cast its shadow as much over anthropological science as theological apologetics. Whether in the writings of those we examined earlier who sought religious accommodation to geology, archaeology, and linguistics; in the discourse of mid-nineteenth-century ethnology and anthropology; or in the theological writings of those who kept a close eye on this new "science of man," the question of Adam's identity and lineage, and the possibility that he might not have been alone, persistently reasserted itself.

꒐ While Ponton steadfastly opposed Darwinism and elaborated a sequence of objections to the principle of natural selection,[123] hints contained in his cryptic comment about the formation of humanity through "some gradual process" gesture toward, perhaps even anticipate, the possibility of an evolutionary rendering of the pre-adamite scheme. The writers whose works have been canvassed in this chapter all routinely denied Darwinian evolution, whether they were professionalizing anthropologists inclined either to monogenism or polygenism or theological commentators repudiating nonadamic humanity or toying with it as a potential intermediary in the conflict between Moses and the ethnologists. The theological apologists examined in the previous chapter who openly espoused the pre-adamites were the same. The episteme—to use Michel Foucault's terminology—within which they all worked was still that of a fixist creation.[124] Pre- and co-adamism were just as straightforwardly creationist as the traditional interpretation of the Genesis account. The coming of Darwin had a significant impact on the structure of the theory, introducing the possibility of a departure from the polygenism with which pre-adamism was commonly associated, and bringing a monogenetic pre-adamism into view.

6 ❧ ANCESTORS

Evolution and the Birth of Adam

I N POPULAR CONSCIOUSNESS, if not in scholarly opinion, the theory of evolution finally killed off Adam. Widespread reports of "The Death of Adam," as several books have been entitled, have cemented this impression. John C. Greene's *Death of Adam,* for example, is subtitled "Evolution and Its Impact on Western Thought," and the first port of call in *The Death of Adam* by the Pulitzer Prize–winner Marilynne Robinson is "Darwinism." In fact, both of these authors are much too sophisticated to attribute to Darwin the assassination of the idea of Adam. Greene's account is really a subtle and striking narrative of the decline of static creationism since the time of Sir Isaac Newton;[1] Robinson's is a critical philosophical engagement with how the new Darwinians conceive of human nature and in particular their denial of human exceptionalism in the state of nature.[2] Nevertheless, the sense that Adam was finally exterminated by Darwin and his disciples is widely shared. In place of that image I want to suggest that for a significant body of opinion the coming of evolution meant the birth, not the death, of Adam. The thought here is that for those wanting to retain continuity with the heritage of adamic theology and yet are persuaded by evolutionary accounts of human origins, the suggestion that Adam was born of pre-adamite parents—that he had a direct genealogical lineage—found considerable support. Adam had a navel, for Adam had ancestors.

Evolutionary Pre-Adamism

This chapter in the story, then, modulates the narrative that has been unfolding in several more or less significant ways. First, at least in the hands of some, it marked a departure from the conventional association of pre-adamism with polygenism. By identifying the adamic family as the direct descendants of forebears and as a relatively late arrival on the stage of natural history, it preserved the unity of the human family. Not that common descent inevitably implied racial equality; evolutionists such as Thomas Henry Huxley and Darwin himself, we should recall, saw no reason to deny racial inferiority and superiority because humanity's common ancestor was pushed so far back into evolutionary history that it predated the emergence of the human species. It simply confirmed the consanguinity—a favorite expression at the time—of the human family. A monogenetic form of pre-adamism was thus now available in the marketplace of ideas. How its champions negotiated a range of doctrinal questions springing from this scenario will attract our attention in the pages that follow. Second, and again this applied more to some than others, these newer arrangements raised metaphysical questions about the human constitution and, in particular, about how the dualism of body and soul was to be handled. One strategy was to suppose that Adam's ancestors had undergone physical evolution to the point that a hominid body was suited to receive a human soul. This version allowed for both an evolutionary *and* creationist account of the appearance of Adam at the same time. Third, advocates of an evolutionized pre-adamism sometimes, but not invariably, allowed for the continued existence of pre-adamite stock and speculated on the relationship its members sustained with Adam and his family. In some cases the translation of other hominid groups into modern humans was countenanced, in some cases not.

Whatever the precise shape this revisionist pre-adamism assumed, the consequences for science, theology, politics, and history spiraled off in new and unpredicted directions. Such, of course, is the way with all harmonizing systems. Designed as they are to retain continuity between scientific discovery and theological beliefs, in fact they are mutually transformative; they alter both the scientific and the theological architecture of their patrons. Reconciling schemes are not to be thought of as neutral, disinterested joints holding together two limbs, or simply as bridges between two independent domains. They are, rather, more like a solvent that dissolves two substances to produce a new compound. Once adopted, monogenetic pre-adamism—like its polygenetic predecessor—channeled the intellectual and political energies of its champions along certain explanatory axes and obliged them to reorient both their theology and their science in one direction or another.

The suggestion that Adam's body had been subject to evolutionary transformation but that direct divine intervention was required to create his "soul," owed a good deal to the writings of the English biologist and Catholic convert St. George Jackson Mivart.[3] A critic of Darwin's theory of natural selection to explain the interim stages in the development of a useful structure—a criticism that Darwin took with great seriousness and sought to answer in later editions of the *Origin of Species*—he was nonetheless convinced that evolution had taken place. And in the same year, 1871, in which Darwin's *Descent of Man* appeared, he brought out his own account of *The Genesis of Species*. Here he spelled out what he considered to be a scientifically robust, and theologically acceptable, account of human origins. Mivart was certain that Adam was born, not directly created, at least in the physical sense.

It was not until the twelfth and final chapter of this volume that Mivart turned his attention to such matters. By this stage he had already laid out his own account of the evolution of species, highlighting his central critique of natural selection and advocating his own teleological conception of evolutionary development, which drew on the writings of the Irish amateur naturalist John Joseph Murphy and emphasized the correlated actions of internal and external forces.[4] So far as the human species was concerned, Mivart insisted that the human body "was evolved from preexisting material (symbolized by the term 'dust of the earth'), and was therefore only *derivatively created,* i.e., by the operation of secondary laws"; the "*soul* . . . was created in a quite different way, not by any preexisting means, external to God Himself, but by the direct action of the Almighty." This was as true for the first man, Adam, as for all other humans. As Mivart put it: "That the first man should have had this double origin agrees with what we now experience. For supposing each human soul to be directly and immediately created, yet each human body is evolved by the ordinary operation of natural physical laws."[5] This strategy allowed for empirical interrogation of the evolutionary history of the human physical form but exempted the soul from such scrutiny, for, he assured his readers, "physical science, as such, has nothing to do with the soul of man, which is hyperphysical."[6] Mivart's dual origin hypothesis opened the door to the possibility of genuine adamic ancestors, who were cast as subhuman in some critical dimension. Although Mivart does not seem to have resorted to the language of pre-adamism to depict them, others, as we will presently see, certainly would.

His proposal, unsurprisingly, was not uniformly welcomed. Huxley, for example, despite his friendship with Mivart, was soon dashing off to the St. Andrews University Library to ferret out a copy of Suarez to pronounce that Mivart simply could not be at once a faithful soldier of science and a loyal son of the Church.[7] Mivart thus found himself marooned, as Adrian Desmond puts it, between "the reactionary Pope and the militant Darwinians"; Huxley was only too happy to use the occasion to vent his spleen on every last trace of scholasticism and to stage himself as science's defender against papal dogma.[8] And later, spurred on by the critical attack launched on him by the Irish priest Jeremiah Murphy, Mivart found himself on the receiving end of ecclesiastical censure. His defense of scientific freedom just went too far, and several of his later writings on theological subjects were placed on the Holy Office's list of forbidden books.[9] Nevertheless, at the time when *On the Genesis of Species* appeared, expressions of support for his understanding of human evolution were certainly heard. His book received agreeable reviews in several Catholic periodicals, for example, and despite his personal antipathy to evolution, Pius IX was willing to confer on Mivart an honorary doctorate.[10]

Whatever supporters or detractors thought of his dualistic device, Mivart demonstrated at least to his own satisfaction how to retain the historical significance of Adam and yet allow for evolutionary transformism. Plainly, Darwin had not obliterated the adamic family. Nor for that matter did the advent of Darwinism sound the death knell to polygenetic speculations. In fact, as we will see in chapter 7, polygenism lingered long in the thinking of post-Darwinian anthropologists, who found a number of ways of denying the common origin of humanity. Among the set that congregated around the Anthropological Society in London, for example, a racialized polygenism continued to thrive, and it was similar in the American South. Indeed, Darwin's own writings could be read through polygenetic spectacles, for as George Stocking reminds us, "he granted that a naturalist confronted for the first time with specimens of Negro and European man would doubtless call them 'good and true species.'"[11] Simply put, race formation was so remote that today's racial groupings had been in existence for an immensely long period of time. For all intents and purposes modern races were polygenetic.

But what is also readily detectable is that among many of those who still held onto the idea of an original Adam and yet were persuaded by evolutionary theory, the development of a pre-adamite theory that was severed from its polygenetic moorings became rather attractive. These writers argued that Adam had human, or proto-human, predecessors, from whom he was descended, and that all were traceable back to a common point of origin. A monogenetic rendering of pre-adamism thus began to exert appeal.

Fig. 16. Alexander Winchell.

Alexander Winchell and Monogenetic Pre-Adamism

Perhaps chief among those who ruptured the ties between pre-adamism and polygenism and helped pave the way for the cultivation of a monogenetic variety of pre-adamism as a strategy for harmonizing biblical anthropology and human evolution was Alexander Winchell (1824–91), a leading American geologist and Methodist layman whose *Adamites and Preadamites* appeared in its first incarnation in 1878 (fig. 16). Winchell's efforts to mediate between science and religion earned him at once considerable acclaim and notoriety, including dismissal from Vanderbilt University. In two critical respects Winchell materially departed from much conventional thinking about human origins and theology. First, while other apologists often flirted with the pre-adamites as simply an anticipative tactic, Winchell actively embraced them as the only means of maintaining the credibility of biblical chronology in the face of scientific revelations. Second, his thoroughly monogenetic version stood in marked contrast to the polygenism of the American school of ethnologists, which was spreading like wildfire, though as we will see in the next chapter, it was no less virulent in its racial typecasting.

Winchell enjoyed a distinguished career in American science both as a geologist and as a reconciler of science and religion, and throughout his life he retained a typically Arminian enthusiasm for natural theology. His academic career was largely spent at the University of Michigan, where he variously served as professor of physics and engineering and later of geology, zoology, and botany. For a short time he was first chancellor of the University of Syracuse, having been appointed to that position in 1873, and then in 1876 he moved to a chair at Vanderbilt, which only required him to be present in Nashville for a couple of months each spring. It was summarily abolished in 1878, and he returned to Michigan in 1879 to serve as professor of geology and palaeontology.[12]

Besides publishing numerous technical works on petrology and mineralogy, Winchell contributed regularly to the *Methodist Quarterly Review* and thereby assumed the role of purveyor of science to the Wesleyan connection. In this latter role he saw it as his task to keep his coreligionists abreast of the most recent scientific developments and to modify traditional harmonizing schemes as required. His rehearsal of the teleological argument, for instance, departed from the Paleyan norm as he resorted to the idealistic—or, better, homological—version associated with such figures as Richard Owen and James McCosh, which located divine design in archetypal plans rather than in specific organic adaptations.[13] In his *Sketches of Creation,* for example, which came out in 1870, he pointed to various scientific resources that could widen the remit of natural theology beyond its familiar horizons. Geological investigations, he noted, were uncovering evidence of the appearance of new organic forms related to, but different from, earlier ones yet revealing "an identity of fundamental plan."[14] At this point his espousal of the idea of "four fundamental types" of organism as the blueprints on which all life was modeled predisposed him to doubt evolutionary theory. But he soon moved toward theistic evolution in its neo-Lamarckian formulation, which allowed for the appearance of adaptive features by means of natural law, albeit within certain limits. Thus, when he brought out *The Doctrine of Evolution* in 1874, he contended that it was "clearly the law of universal intelligence under which complex results are brought into existence."[15] In this work he laid out a detailed conspectus of the various schemes of what he called organic derivation, discriminating between the proposals of figures such as Darwin, Huxley, Alfred Russel Wallace, Alpheus Hyatt, Albert von Kölliker, and Karl

Gegenbaur, all of which demonstrated an intimate familiarity with current proposals.

Like many of his contemporaries on both sides of the Atlantic, Winchell certainly expressed reservations about aspects of the evolutionary program. He could not see, for example, how altruism could be produced from a mechanism—the struggle for existence—that was inherently "selfish" and further argued that the principle of selection required the "action of intelligent will."[16] He worried, too, that Darwin's own proposals seemed to require a widespread simultaneous appearance of new adaptive forms for which no generating mechanism seemed available.[17] Nevertheless, he found the general principle of evolution compelling, and within three years he had declared his support for the neo-Lamarckian mechanism that was being promulgated by Edward Drinker Cope (1840–97).[18] The "author's present conviction," he told the readers of his *Reconciliation of Science and Religion* in 1877, "is that the doctrine of the derivation of species should be accepted; and that the most tenable theory of the causes, instrumentalities, and conditions of this derivation is that propounded, in 1868, by Professor Edward D. Cope."[19] Winchell's project now was to demonstrate evolution's compatibility with theism and to do so in a way that avoided both "dogmatism and denunciation."[20] And his strategy was to move away from teleology to homology, thereby locating divine design less in particular contrivance than in overarching plan. Theologically sculpted though his version was, Winchell's outlook was now a conspicuously evolutionary one. Nor did he hesitate to expose his sentiments to Methodist audiences. The April 1877 issue of the *Methodist Quarterly Review,* for example, ran an extended review of Huxley's 1876 New York lectures on *The Direct Evidence of Evolution.* Here Winchell, while registering dissent from some of Huxley's claims, confessed: "We are pretty strongly persuaded that the doctrine of derivative descent of animal and vegetal forms represents the truth . . . We now think it is far safer to accept the hypothesis than to reject it. It is safer for the scientist, it is safer for the Christian."[21] The palaeontological record, empirical evidence of species variability, and developmental embryology all conspired to render plausible the evolution hypothesis.

Prior to his appointment to Vanderbilt in 1875, then, Winchell had given voice to his evolutionary sympathies, and he continued to strengthen this allegiance after his arrival in Nashville. But during his time there he had found it necessary to keep a careful watch on his tongue, and when he did fall foul of the university's board of trustees, he noted in an explanatory epistle to the readers of the Nashville *American* on June 15, 1878, "I have always taken pains, in my lectures at Nashville, to avoid the utterance of opinions which I supposed were disapproved of by the officers of the University."[22] What he had in mind was evolution. As he saw it, this is what lay at the roots of what he described as his "dismissal from office on account of heresy." He went on: "This heresy consists in holding with the great body of scientific men, that a method of Evolution has obtained in the history of the world, but not in holding that *man* is the product of evolution."[23] The language of heresy hunting, of course, was a rhetorical jibe recalling the fate of Galileo some three and a half centuries earlier, for when the strategically placed Bishop Holland N. McTyeire quipped, "We do not propose to treat you as the Inquisition treated Galileo," Winchell spat out the rejoinder: "But what you propose is the same thing . . . It is ecclesiastical proscription for an opinion which must be settled by scientific evidence."[24] The most distinguished scholar on the Vanderbilt faculty at the time, his dismissal caused quite a stir,[25] not least because, as Paul Conkin caustically quips, in the person of Winchell, Vanderbilt actually "came close to hiring a person who flirted with originality."[26] Indeed, Andrew Dickson White, who later would publish his immensely influential *History of the Warfare of Science with Theology in Christendom* (1895), wrote to Winchell from Paris, saying that his name would occupy a prominent place in "so important a matter of history."[27] Winchell was now on his way to iconic status in the annals of science-religion history.

In the minds of many, Winchell included, evolution was his downfall. As he noted of his chief adversary: "Evolution! this is the bugbear so big and black that nothing else could be heard or read by Dr. Summers, whenever I employed tongue or pen."[28] But there are good reasons to suppose that evolution was not the only factor in Winchell's removal from office. Far from it. After all, his views on the subject must surely have been known prior to his appointment. Indeed, it turns out that other matters were implicated, matters having to do with human racial difference and non-adamic humanity. On this issue, too, Winchell had tried hard to manage his speech. Before coming to Vanderbilt, he had written two long essays for his fellow Methodists on the religious nature of savages and barbarians in which he surveyed a wide range of natural historical, archaeological, and anthropological literature.[29] He had already hinted at the likelihood of pre-adamite human life back in 1870, when he suggested that because the Mosaic records represented an advanced population occupying the western reaches of Asia "at a period two or three thousand years antecedent to our era," there must have been "a long interval of human history still anterior to this date."[30] Now, in 1878, it was one of his supporters, David C. Kelley, a local Methodist minister, a scientific progressive, and a member of the Vanderbilt board, who announced that Winchell would deliver a public lecture on the topic "Races" that spring. Winchell was evidently a little taken aback under the circumstances and scribbled rather cryptically in his diary, "Dr. Kelley has advertised me to give a lecture on April 5 on 'Races'—horse races perhaps."[31] But he can hardly have been entirely surprised. The race issue hung in the air not least because his fifty-four-page pamphlet, *Adamites and Preadamites,* had appeared in the early part of that year, 1878, just shortly after the publication of his lengthy entry on the subject for the *Cyclopaedia of Biblical, Theological, and Ecclesiastical Literature.*[32] The thought that his opponents might have been setting him up certainly crossed some minds, not least when Bishop McTyeire suggested that he might take up the subject of evolution for his commencement address. Given the tension in the air, Winchell refused, speaking instead on "Man in the Light of Geology." On the very day of the graduation speech McTyeire told him that his scientific views were becoming an embarrassment to the institution, and shortly thereafter he received formal notice that his appointment was now terminated.

His views on human origins and anthropological history were certainly now receiving a fairly wide airing. The 1878 pamphlet, for example, was reprinted from the *Northern Christian Advocate,* and within a couple of years it would be expanded into a monumental five hundred–page treatise, *Preadamites.*[33] In all of these outlets Winchell defended the claim that there were human races on the earth prior to the biblical Adam. To Methodist ears it sounded awfully like the polygenism circulating elsewhere, though as we will presently see, this was emphatically not the case. Nevertheless, his citing of certain of its well-known advocates—such as Samuel George Morton and Josiah Nott—and his use of various anthropometric measures typical of that school of thought, no doubt compounded the misapprehension. What was crystal clear was Winchell's conviction that the black races, which he set out to establish as physically, psychically, and socially inferior to whites, were *not* descended from the biblical Adam but predated him. As he put it in his encyclopedia entry, "The conclusion is indicated, therefore, that the common progenitor of the Black and the other races was placed too far back in time to answer for the Biblical Adam."[34] While Winchell insisted that because the whole issue was "a matter of scientific fact, we should unhesitatingly appeal to anthropology for a final answer," his fellow Methodists in the Vanderbilt community did not see it that way.[35] For them the central concern was the implications of Winchell's presumed polygenism for their system of theology. What is also clear is that, courtesy of Winchell's pronouncements, the board of trustees at Vanderbilt University now occupied a rhetorical space in which evolution and polygenism were inextricably intertwined. Under such circumstances racial polygenism became the lens through which the local Methodist community's dealings with Darwin were refracted. And the editors of the *Nashville American* seemed to agree when they commented on Winchell's demise: "'His development of evolution and polygenism' caused the abolition of his lectureship."[36]

Precisely what was so troubling about Winchell's pre-adamism remains unclear, though there is no dearth of potential candidates. Anxieties could have sprung from many sources. Perhaps it was because he began both *Adamites and Preadamites* and his entry on the subject for the McClintock and Strong *Cyclopaedia* by advertising the insights of the notorious "heretic" Isaac La Peyrère. Readers were told in no uncertain terms that La Peyrère was "a victim of the intolerance of the times" and that his insights were "far in advance of his age."[37] (Later they would learn that he "was less impious and mad than the bond slaves of dogma who silenced his tongue.")[38] Perhaps it was because his claim that the black races were of pre-adamic stock seemed to Methodists to place them beyond the scheme of redemption—an embarrassing state of affairs to a denomination currently putting strenuous efforts into church outreach to the black community.[39] Or was it on account of his insistence that Adam was descended from black ancestors and may himself have been of dark complexion? Even Methodist largesse might not stretch that far. Maybe it was simply the postulation of beings to which Genesis bore no witness that caused concern or Winchell's insistence that there was "no evidence of the universality of the deluge of Noah."[40] Whichever was the case, evolution and pre-adamism proved to be too heady a brew for the Vanderbilt Wesleyans.

Just what, then, was the precise shape of Winchell's pre-adamism? He reviewed the basic narrative in a couple of paragraphs:

It was many thousand years ago that the first being appeared which could be called a man. Whether descended from a being unworthy to be called a man, is a collateral question which rests on other foundations . . . That first of all men did not make his advent in Asia, nor in Europe, nor in America. He appeared either in Africa or in a continental land which stretched from Madagascar to the East Indies . . . His skin was probably black, and well clothed with hair. He had implanted within him the divine spark of intelligence. He listened to the voice of conscience and felt the claims of duty. If not indigenous in Africa, his descendants took possession of that continent. They spread over Australia and Borneo and the lesser islands of the sea. In the course of thousands of years, they disseminated themselves over considerable portions of Asia.

The time arrived, at length, when, under the law of progressive development, a grade had been reached nearly on a level with that of modern civilized man, in respect to native capacities. Now appeared the founder of the Adamic family. His home was in central Asia. Seth and Cain were either his sons or nations descended from him. He had also "other sons and daughters" in the same sense. The Sethites and Cainites and other tribes of Adam, as they spread themselves eastward, displaced and at length exterminated the Preadamites of Asia.[41]

To a considerable degree Winchell's interest in the whole subject sprang from his obsession with racial geography, both historical and contemporary. And his various writings on the subject elaborated, in ever more painstaking detail, on the limited geographical dispersion of what he called the Noachites—Hamites, Semites, and Japhetites—across the Near East (fig. 17), on the major racial divisions of humanity with plentiful illustrations of representative types (fig. 18), and on the divergence of the black races from adamic anthropology. Ethnic group after ethnic group came within his scrutiny as he traced the global diffusion tracks of the primitive pre-adamites and plotted their position on his map of world racial genealogy. Questions of Hebrew and Egyptian chronology also surfaced, as did matters of race history and prehistoric archaeology. Evidence available from both Neanderthal and Cro-Magnon skulls was also built into the scheme. Throughout his various pre-adamite writings Winchell protested the theological propriety of his proposals, but in order to preserve his doctrinal credentials, at least to his own satisfaction, some major structural alterations to the architecture of traditional pre-adamism had to be undertaken.

From the outset it is clear that Winchell intended his account—which postulated that Adam was the natural offspring of pre-adamite parents—to support monogenism. In 1878 he presented his readers with this version of pre-adamism as the only way to retain a traditional chronology and yet do justice to empirical findings. "If human beings have existed but 6,000 years, then the different races had separate beginnings, as Agassiz long since maintained, each race in its own geographical area," he observed. "But if all human beings are descended from one stock, then the starting point was more than 6,000 years back."[42] Because biblical chronology located the adamic

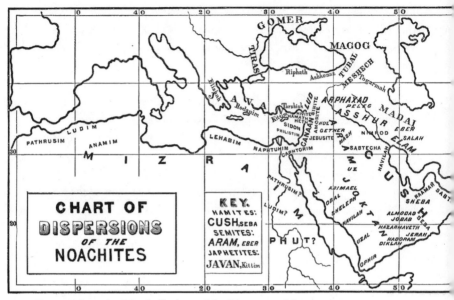

FIG. 17. Alexander Winchell's chart of the dispersion of the Noachites.

family somewhere around 4000 B.C., monogenists simply had no option but to entertain the thought that Adam had ancestors. After all, the evidence of persistence of racial type was very strong. The antiquity of race distinctions, for example, was confirmed from scenes depicted on ancient Egyptian monuments of the Twelfth Dynasty dating from about 1500 B.C. and from Pharaonic portraits of the same era (fig. 19). In the face of such records Winchell was sure that his pre-adamism provided chronologists with the temporal safety value they needed. But more, Winchell insisted that his scheme was theologically orthodox. In contrast to polygenism, it acknowledged "the unity of man . . . the possession of 'one blood' by all the races, one moral and intellectual nature, and one destiny."[43] Later in 1880, perhaps to rebut the misinformed charges of his adversaries, he reiterated the point: "I have not disputed the divine creation of Adam, even in maintaining that he had a human father and mother . . . I have not affirmed—even like M'Causland and other ecclesiastical polygenists—that mankind, one in moral nature, are not one in origin; since I believe the blood of the first human stock flows in the veins of every living human being."[44]

None of this implied, of course, that any racial group found itself banished beyond the arc of redemption, and Winchell worked hard to provide doctrinal warrant for rejecting the idea that descent from Adam was a perqui-

ADAM'S ANCESTORS

site for regeneration. Certain that heresy hunters were waiting in the wings, he sought to forestall their anxieties about the soteriological status of the pre- and co-adamites. In his view pre-adamism did "not interfere with current views of the catholic scope of the redemptive 'scheme.'"[45] Here Winchell worked by analogy with the theological responses to the earlier challenge of the plurality of worlds—the existence of sentient beings on other planets[46]—citing authorities such as Thomas Chalmers, David Brewster, and the Methodist Episcopal bishop E. M. Marvin, all of whom extended redemptive privileges to other worlds.[47] Winchell's inference was immediate: if redemption could encompass the farthest reaches of space, could it not reach back in time and thereby make "provision for the poor preadamites?"[48] To this view he added the testimony of the prominent Methodist theologian Daniel Whedon, who, though remaining skeptical about Adam's ancestry, had written on the retrospective benefits of the atonement and speculated that it could, in principle, incorporate pre-adamic humanity within its reach, should they have existed.

By severing the links between pre-adamism and polygenism, Winchell helped prepare the way for the later monogenetic version of pre-adamism

FIG. 18. Typical racial form from Alexander Winchell's *Preadamites*.

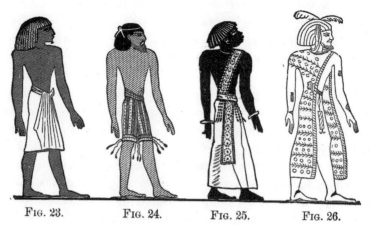

FIG. 23.   FIG. 24.   FIG. 25.   FIG. 26.

The four races of men known to the Egyptians. Fig. 23, *Rot* or Egyptian (*red*). Fig. 24, *Namahu* or Semitic (*yellow*). Fig. 25, *Nahsu* or Negro (*black*). Fig. 26, *Tamahu* or Mediterranean (*white*). Reduced from a portion of a painted relief of the Nineteenth Dynasty, about 1500 B.C.

FIG. 19. Ancient racial types from Alexander Winchell's *Preadamites*.

that would be deployed as a device for meeting the challenge of human evolution, even though he himself made the conscious decision not to stir up that particular hornets' nest. Nevertheless, he dropped hints of his scheme's evolutionary potential. For one thing he remained deeply critical of accounts of racial history that presumed black degeneracy from an originally perfect state; to him progression was the fundamental law of organic life, and this was reflected in his understanding of racial development: "Should it be claimed that the white man's Adam had descended from a common stock with the Negro, all nature cries out in assent. But should it be affirmed that the Negro is degenerated from the white man's Adam, every fact in nature shakes its head in denial."[49] What sustained this vision was his version of the pre-adamite theory. Pre-adamic monogenism allowed for progress; universal descent from Adam, understood as the first human being, required regression. Again, he noted that the "more we insist on the blood affinity of the races of man, the more we crowd upon attention the query whether a blood affinity may not exist between the lowest race and some type of being a little too low to be called human."[50]

At the same time, Winchell was careful not to show his hand too openly. "To assert that man has advanced from the lowest human condition," he reflected, "is not to assert that this condition was reached by advance from the brute. It is not necessary to assert this; and I wish the reader to note distinctly that none of the conclusions of this work rest on the assumption of man's derivation from a brute ancestor. Man may or may not have had such an origin; I do not trouble myself or the reader with that question."[51] And yet by calling attention to the inferences that were *not* immediately to be drawn from monogenetic pre-adamism, he was all the while raising in some minds the possibility that those very implications, while not self-evident, might nonetheless still be valid. Moreover, when he used evidence from forearm length, pelvic circumference and inclination, and cephalic type to suggest that certain branches of the human race showed signs of being intermediate between the adamite and the chimpanzee, he was approaching within a hairbreadth some theory of human evolution. His supportive references to the ways in which both Wallace and Mivart had exploited the body-soul dualism to allow for human physical evolution adds further weight to the suspicion that he was sympathetic to an evolutionary interpretation of the human species.

To summarize, then, Winchell emphasized the unity of the human species by tracing all human varieties back to an original pre-adamic stock. The descendants of this primitive group had dispersed from their point of origin across the face of the earth long before the adamic family had first appeared on the scene of history. Adam's significance, in this schema, was that he was "the immediate progenitor of the nations which figure in biblical history, and hence must not be expected to answer the requirements of the primitive ancestor of all mankind."[52] By the time of his appearance human society was already far advanced. Racially, Adam was the descendant of black forebears.[53] Indeed, rather like James Cowles Prichard and Dominick McCausland, Winchell interpreted the Hebrew *ADâMâH*—"redness or ruddiness of color"—to imply that Adam himself was, if not pure black, nonetheless "strongly colored."[54]

That Winchell found himself on the receiving end of Vanderbilt's disapproval does not mean that pre-adamism was entirely outlawed within the ranks of Methodist orthodoxy. The reception of his work among the upper reaches of the denomination was altogether more sedate. In fact, Daniel Whedon (1808–85), one of American Methodism's most distinguished intellectual leaders for nearly a quarter-century and editor of the prestigious *Methodist Quarterly Review,*[55] had already dispassionately reviewed McCausland's volume, and Winchell did not hesitate to quote generously from his liberal-minded review in his own book. Whedon quickly saw how McCausland's "ingenious, learned, and eloquent" proposals could prevent "the violation of the sacred text" by safeguarding Old Testament chronology and thus preferred it to the works of the Duke of Argyll. As for soteriological objections, Whedon insisted, as we have already noted, that "the justifying power of Christ's death . . . had a retrospective effect," anonymously citing La Peyrère's own argument that humanity's unity stemmed from moral identification with Adam, not lineal descent from him.[56] By allowing McCausland's, and indeed La Peyrère's, *polygenetic* pre-adamism onto the stage of theological acceptability, Whedon thereby gave Winchell the opportunity to justify his own monogenetic version as a good deal less open to charges to heresy.[57]

In the long term, however, Whedon turned out to be a wavering ally, and he ultimately drew back from endorsing Winchell's proposals, perhaps because in the interim he had been favorably impressed with Edward Fontaine's *Ethnological Lectures on How the World Was Peopled* (1872), which argued the case for the scientific legitimacy of pan-human derivation from an Edenic pair.[58] Fontaine, a southern Episcopal clergyman with distinct sympathies for scriptural geology, had accumulated evidence to challenge the antiquity of the earth, and Whedon welcomed his intervention as a serious challenge to Lyell's geology. It was a bitter pill for Winchell to swallow, not least because he had eulogized Whedon as "one of the noblest exponents of intelligent theology" and "too shrewd to be fooled by the shriveled old ogre of 'Orthodoxy,' who comes in the garb of Christianity, begging to be defended from the assaults of common sense."[59]

This was not the first occasion on which Winchell had cause to feel betrayed by fellow Methodists when he might have expected approbation. His earlier "Preadamite" entry for the McClintock and Strong *Cyclopaedia* is illustrative. The very fact that the compilers allowed Winchell this space is itself significant, of course, suggesting at least that Methodist debate on the subject remained open. And Winchell took the opportunity to pull together the data for human antiquity from cavern fossil bones, stone implements found in river drifts, human remains in loess and moraine deposits, volcanic tuff, peat bogs, kitchen middens, megaliths and tumuli, and lake dwellings, as well as contemporary anthropological research on "primitive" peoples. But the editors did not deliver the article into Winchell's hands carte blanche. At several key points they appended editorial footnotes contesting, questioning, and undermining his claims. Some of their quibbles were relatively routine, on precise points of information, for example, such as the inadequacy of employing certain astronomical data for dating Egyptian chronology.

But other complaints were rather more fundamental. They resisted, for instance, what they took to be Winchell's efforts to force scripture into harmony with science. Questioning the anthropological conclusions Winchell had drawn from the Hebrew meaning of the term *Adam,* they hastily affirmed that the "statements of Scripture must stand or fall by themselves, when fairly expounded by the usual laws of exegesis, and we are not at liberty to warp them into accommodation with discoveries in other fields." Again, when Winchell argued for Adam's descent from black ancestors on the grounds of progress from inferior stock, they dissented "toto coelo," especially from the "view that the Black races are in any essential point inferior to the others." To them black degradation, a state taken for granted, stemmed from "unfavourable surroundings . . . rather than . . . inherent lack of capacity." To the editors there was ample evidence, too, to confirm that the black races were numbered among the Cushite peoples and that this realization disposed "at once of the argument that Noah is the progenitor of the white races only."[60] All these editorial correctives were brought forward to preserve the traditional Mosaic narrative. But they do reveal the complexity of conceptual allegiances; the editors, biblical traditionalists as they were, clearly displayed their distaste for racist sentiments, whereas Winchell, the scientific luminary, based his rereading of the biblical text on a thoroughly racist anthropology. The links between religious and social conservatism are, on this issue, far from clear-cut.

Nevertheless, Winchell did find support from some quarters. An anonymous reviewer for *Appleton's Journal* commended the "step forward" that Winchell's pre-adamites had made in "the effort to effect a reconciliation between the conflicting claims of science and religion." With expansive citations this writer found Winchell's arguments to be "very strong" and predicted that they would "be influential with a certain large class of readers, chiefly because it accords careful and respectful treatment to beliefs and prejudices which are too often dismissed with contempt by those who approach these subjects from the scientific side."[61] And over a decade later, in a belated review of Winchell's treatise for the *Methodist Review,* Henry Colman conceded that "the theory of pre-Adamites conflicts with no biblical doctrine, and explains some otherwise difficult Scripture texts. It is not so connected with Scripture as to become a theological question, and must be decided by geology, ethnology, and history." While he himself considered Winchell's case as "not proven," on the grounds of his understanding of current scientific claims, and was pleased to report that "the facts of history, secular and scriptural, join in beautiful harmony with the faith . . . that Adam was the first man," he did mistakenly but illuminatingly refer to La Peyrère— Peyrerius—as "an orthodox Dutch [*sic*] ecclesiastic."[62]

Darwinism, Pre-Adamites, and Conservative Theology

Whatever his own personal fate at Vanderbilt, Winchell succeeded in keeping the idea of pre-adamite humanity before those concerned with the relations between science and religion in the new Darwinian era. Of course, such ideas retained little credibility for those using Darwinism to undermine religion or among those professionalizing scientists who privatized religious belief. Whereas secular versions of pre-adamism could prevail in the early nineteenth century because talk of multiple creations provided a shared language among ethnologists and anthropologists, the same simply did not apply to the post-Darwinian scientific scene. Once evolutionary theory, with its naturalistic overtones, had gained a firm foothold, the whole conception of pre-adamites only made sense to those who still attached some significance to the biblical Adam and who actively sought some rapprochement between evolutionary theory and theology. There were indeed those, such as Ernst Haeckel, who felt that the very idea of a first man simply had no scientific meaning. Haeckel himself was inclined toward monogenism, as he made clear in his two-volume *History of Creation* (1883), but he explained that the question of monophyletic or polyphyletic descent, if construed as the search for the first human being, was a fundamentally misconceived project. As he put it:

> While we for many reasons believe that the different species of speechless primaeval men were all derived from a common ape-like form, we do not of course mean to say that *all men are descended from one pair.* This latter supposition, which our modern Indo-Germanic culture has taken from the Semitic myth of the Mosaic history of creation, is by no means tenable. The whole of the celebrated dispute, as to whether the human race is descended from a single pair or not, rests upon a completely false way of putting the question. It is just as senseless as the dispute as to whether all sporting dogs or all racehorses are descended from a single pair . . . A "first human pair," or "a first man," has in fact never existed.[63]

Of course, attributing mythological status to aspects of Old Testament theology was not the prerogative of vituperative critics of religion such as Haeckel or Carl Vogt. For long enough there had been theologians, particularly but not exclusively in Germany, who took a higher critical view of scripture. For them the Creation story of Genesis, with its depiction of Adam's creation, probation, and fall, were truly "mythological" and could therefore be sifted of all scientific content to leave behind the moral message as a sort of spiritual residue.[64]

In light of these currents of scientific and theological thought, it is not surprising that the survival of the pre-adamite was left very largely in the hands of those with more or less conservative religious convictions. Recalling the skeptical origins of the idea and the odor of heresy that long clung to it, this is a notable change of fortune. At the same time, as I have suggested, the theory that remained was transformed. In general terms the pre-adamite came to be regarded either as a relatively sophisticated descendant of an earlier more savage humanity or as a near-human or subhuman predecessor of the fully human Adam, the first being to be graced with the *imago Dei*. As we shall see, such a conception could be interpreted in a variety of ways, but that it was refashioned into an evolutionary form is what most clearly demarcates it from earlier versions of the theory.

To illustrate something of this reconfiguration, an examination of several conservative theological engagements with ideas about human evolution during the final decades of the nineteenth century and up to the middle of the twentieth will serve to disclose something of the dimensions of this conceptual realignment. Because the apologetic legacy of these thinkers was so enduring, their use of "heretical" pre-adamite thought forms seems all the more significant.

Protestant Proposals

Because he occupied a "chair of the harmony of science and revealed religion" specially created for him at Princeton in 1865, the first of its kind in any American college, the pronouncements of Charles Woodruff Shields (1825–1904) on the subject of human origins inevitably invite our perusal.[65] Both his institutional affiliation and his self-appointed task of surveying the entire gamut of scientific and theological learning in the immediate aftermath of the Darwinian onslaught prepare us for a Presbyterian perspective on the question. Predictably, his monumental conspectus *Philosophia Ultima or Science of the Sciences* of 1888–1905, a work inviting comparison with the grand synthetic panoramas of Herbert Spencer or John Fiske, took up the disparate claims of rational (scientific) anthropology as compared with those of revealed (scriptural) anthropology.

Shields was clearly familiar with the pre-adamite theory both in its role as an anthropological thesis and as a strategy for harmonizing science and scripture, and his references throughout the work to the writings of figures such as

Giordano Bruno, Isaac La Peyrère, Lord Kames, Louis Agassiz, Josiah Nott, George Gliddon, Dominick McCausland, Reginald Stuart Poole, and a host of others displayed his considerable erudition. At this juncture Shields's aim was simply to lay before his public the range of scientific theories available, but it was already plain that he harbored some sympathy for the pre-adamite theory. Thus, he was discouraged to find, for example, that "the guardians of orthodoxy in our day are denouncing Agassiz and [Edward] Forbes for a theory of co-adamite races, which might really support their own doctrine of a high Adamic covenant, as distinguished from mere inherited sin." By this assertion Shields intended to intimate that Adam could be conceived of as *representative* of the entire human species irrespective of time or place and thereby relieve anthropology of the burden of commitment to universal descent from the adamic pair. As he put it, "Whether the human family be of one race or many races, the first Adam and the second Adam would still be their chief moral representatives."[66] So it was no surprise that in 1900 he would announce that the Genesis record dealt with the adamic-Caucasian race as a typical or representative segment of the entire human family and the bearer of salvation to the world—a view clearly assuming the continued survival of both pre- and co-adamites across the reaches of history's stage.[67]

Thus far, Shields's affirmations did not go beyond an endorsement of standard polygenetic pre-adamism. But there are hints that he was moving toward an evolutionary reworking of the concept. He had, for instance, no sooner introduced La Peyrère's system to his readers as "first a Protestant infidel, then a Catholic priest of the Oratory," than he immediately went on to link it both with the polygenesis of Agassiz and with Bishop Joseph Butler's treatment of "several questions of recent anthropology, such as the material origin of man, his development from an animal state, and his gradual predominance as the governing animal in our globe."[68] And having made this connection, he presently added that earlier students of biblical anthropology such as John Calvin and Theodore Beza were writing at a time when it "was too soon as yet to attempt any scientific verification of . . . dogmas, such as is beginning to be made, by associating co-Adamite and pre-Adamite theories of the savage and animal origin of man, with a special divine dispensation to Adam as the natural progenitor of the Caucasian race and federal representative of the whole human family." But Shields could foresee a time, not far distant, when "the secular evolution of Adam from the animal species shall be claimed to be as scriptural and orthodox as that of the animal from the vegetable races, or that of the organized planet from the inorganic nebula."[69]

At this stage he himself was rather hesitant to adopt too rapidly what he called the "eclectic spirit," despite the lead given by "Chancellor Winchell" and "St. George Mivart, the distinguished Roman Catholic naturalist," and he certainly was biting in his cryptic dismissal of the pre-adamism of *Primeval Man Unveiled* (1880), whose anonymous author (James Gall) "strangely blended" the "wildest delusions of angelology . . . with the latest speculations of the ethnologist."[70] For all that, drawing inspiration from "Dr. McCosh [who] from the first has led the creationists into alliance with the evolutionists," Shields had come to the opinion that from "present signs it would seem that the tide of controversy has turned in favor of Evolutionism in some form and degree."[71] As it turned out, he was remarkably prescient in his predictions, and his own resort to a covenantal conception of the Fall (which emphasized Adam's representativeness of the whole human species without reference to inheritance) as crucial to mediation between scripture and evolutionary anthropology was a move that later conservative theologians would make for precisely the same ends. Hence, it is not surprising to find Shields himself affirming, in 1900, in terms remarkably similar to Mivart that if the "creation of man" was to be regarded as "a process rather than as an act," it would be conceivable that "the animal organism of man" could have been endowed with "psychical qualities and divine resemblances . . . as new miraculous acts or subsequent processes during the historic period."[72]

If Shields sought an evangelical benediction on his evolutionary reading of the human story, he needed to look no farther than to two figures sharing his own campus: Reverend George Macloskie (1834–1920), who crossed the Atlantic from Ireland in 1875 to take up the new chair of natural history at the College of New Jersey, as Princeton was then called, and the recently appointed professor of didactic and polemic theology in the Princeton Seminary, Benjamin Breckinridge Warfield (1851–1921), who assumed that position in 1887. One or two comments on these voices, sympathetic as they were to Darwinism, will serve to illustrate some of the conceptual moves that were being advanced in the effort to reconstruct theological anthropology on evolutionary foundations.

Macloskie's appointment to his position was something of a last resort for James McCosh, the Princeton president, who had found it extraordinarily difficult to fill the post. By securing the services of his former Irish student, McCosh at last found a candidate with his own inclinations, both theological and evolutionary.[73] Over the years Macloskie assumed the role of purveyor of science to the Presbyterian fraternity both through various speaking engagements and his regular articles on scientific subjects for the denomination's magazines.[74] Thus, in the late 1880s and 1890s and into the early years of the twentieth century, he drew the attention of his fellow Presbyterians to a range of matters of pressing scientific and theological import: the role of speculation in science, accommodationist strategies, and theistic evolution.[75] As part of this enterprise, it is not surprising that the pre-adamites fell under his scrutiny, and in a listing of topics taken up for discussion on science and faith in 1892, they took their place alongside such other issues as "The Iconoclasm of Science," "Creation," "Evolution: The Right Attitude," "Creation of Man," and "The Deluge." Macloskie's notes disclose only the sketchiest of impressions of his approach to the subject, save that he identified La Peyrère as its source and portrayed John Pye Smith and Alexander Winchell as its latter-day representatives. But he did make it clear that "the theory of Preadamites does not affect the unity of the race" while at the same time noting that the "Bible seems to set forth Adam as our first father" and that Winchell "exaggerates the difference between negro & white."[76] From other writings, however, we can readily infer the kind of pre-adamism that is likely to have attracted Macloskie's interest.

It is not difficult to excavate the foundations on which the architecture of Macloskie's conception of human evolution stood. Two articles published around the turn of the twentieth century articulate his stance. In his account of "Theistic Evolution" in 1898 he paused to comment on the implications of the theory for human origins. "It has been recognized that man may have been both created and evolved," he wrote, "that his creation may have been effected under divinely directed evolution, either as a natural development, or possibly as a development with supernatural incidents and expedited." "We are satisfied," he went on, "that it is not right to hold the hypothesis of evolution as applied to the human race to signify that there could be no pair of first parents and no fall into sin. The most generally accepted views of human evolution will favor the monogenetic view, which signifies an ancestral pair."[77] Evidently, Macloskie entertained the likelihood that the biblical Adam was the descendant of some pre-adamic stock. And to some degree at any rate, Darwin could be called as supporting witness. Five years later Macloskie turned yet more specifically to the same subject, and here his espousal of a Mivart-style solution to the question clearly surfaced. Evolution, he gladly announced, "destroyed Louis Agassiz's theories that the different races could not be descended from common ancestors; and also the strange fancies that many of the animals and plants were created in multiple."[78] Darwinism had obliterated polygenism. In its place there was increasing recognition "and justly so, that man has somehow been evolved."[79] Where did that leave Adam? Macloskie was clear: "Evolution . . . will be held by the biblicist to be a part, the naturalistic part, of the total work of his making, the other part being his endowment miraculously with a spiritual nature, so that he was created in the image of God ... As a member of the animal kingdom, man was created by God, probably in the same naturalistic fashion as the beasts that perish; but unlike them, he has endowments which point to a higher, namely a supernaturalistic, order of creation. By present showing, man's evolution would probably consist in the production of a single pair of ancestors."[80]

Further support for this kind of maneuver was forthcoming from Macloskie's colleague at the Princeton Seminary, B. B. Warfield, an immense authority within late-nineteenth- and twentieth-century reformed theology and a supporter of evolution.[81] That Warfield had found the conceptual resources in the long-established distinction between primary and secondary causes to meet the challenge of Darwinian biology was crucial to the story he wanted to tell about the origins of humanity. And indeed, he had already expressed enthusiasm for Shields's efforts along these lines in *Philosophia Ultima*.[82] Before specifying the precise theological arrangements he envisaged, it is worth reflecting on his approach to the pre-adamite theory more generally.

In 1911 Warfield wrote an article for the *Princeton Theological Review* on the antiquity and unity of the human race. Because he felt that the mere age of the human race was irrelevant to theology, he moved on to discuss theories that bore on the unity or diversity of the species and to review some of the major scientific theories available. On turning to pre-adamism, he made a number of telling observations that throw much light on his own treatment of the subject. To him polygenism was to be equated with the "co-Adamitism" of writers such as Paracelsus, and not with pre-adamism, as was conventionally assumed. In his mind "co-Adamitism [was] the attribution of the descent of the several chief racial types to separate original ancestors," whereas "pre-Adamitism ... conceives man indeed as a single species, derived from one stock, but represents Adam not as the root of this stock, but as one of its products."[83] The writings of Zanini (Zaninus de Solcia), who is said to have been condemned in 1495 for believing in the existence of inhabited worlds before Adam,[84] Isaac la Peyrère, and Alexander Winchell, he reported, represented the monogenetic pre-adamite theory, while Arthur de Gobineau, Josiah C. Nott, George Gliddon, Louis Agassiz, and Cordonière were advocates of polygenetic co-adamism. Warfield himself warmed to neither of these versions. On the one hand, he loathed polygenism in every shape and form, as well as the racial pride that typically went with it, and on the other, he believed that Christianity's theological structure was bound up with Adam as the father of all humanity, not just of the Jews. Yet he did make reference to another "sort of pre-Adamitism [that] has continued to be taught by a series of philosophical speculators from Schelling down, which looks upon Adam as the first real man, rising in developed humanity above the low, beastlike condition of his ancestors." There Warfield left the matter, but his insistence that monogenism was "a necessary corollary of the evolutionary hypothesis" already raised the suspicion that this other "sort" of pre-adamism might have some validity.[85]

These inklings are not confirmed until we turn to Warfield's own views on the emergence of the human species. In an exposition of "Calvin's Doctrine of the Creation" in 1915 he made the case that Calvin had restricted the term *creation* to the initial act of *creation ex nihilo*. The only exception to this was the creation of the soul, for Calvin held to the Creationist, as opposed to the Traducianist view, that every human soul throughout the history of propagation was an immediate, not mediate, creation. Traducianists, by contrast, insisted that the spiritual component of humanity was transmitted through natural generation, not divine intervention. Accordingly, because all other modifications of what he termed the primeval "indigested mass" were by "means of the interaction of its intrinsic forces," Warfield could claim that Calvin's was a "very pure evolutionary scheme."[86] So far as the human species was concerned, it left open the possibility that the human body might have undergone a long history of evolutionary modification before receiving, by an act of immediate creation, a soul. Thus, when Warfield came to reviewing James Orr's *God's Image in Man* (1905), he allowed as a distinct possibility that the human body had been formed in emergent evolutionary fashion, "at a leap from brutish parents," and then fitted with a "truly human soul."[87] The compatibility of this conception with the alternative pre-adamite scheme that Warfield himself had identified is clearly apparent.

If Warfield left evolutionary pre-adamism as an inference to be drawn from his writings, other Protestant conservatives were explicit in their consideration of this possibility. Take, for example, A. Rendle Short (1880–1953), the Royal College of Surgeons Hunterian professor. Short was a distinguished surgeon and vigorous evangelical apologist, and in the latter capacity he took up the subject of the "Problem of Man's Origin" in a volume on science and the Bible that appeared during the 1930s. Here he hinted at the possibility that there "might conceivably have been pre-Adamite creatures with the body and mind of a man, but not the spirit and capacity for God and eternity." Certainly at this stage, he concluded that all this was "very difficult and speculative,"[88] but by 1942, when he produced *Modern Discovery and the*

Bible—a work destined to remain popular within British evangelicalism—he was rather more enthusiastic. Now he pointed out that it was by no means certain that *Homo neanderthalensis* was "to be regarded as man in the Bible sense of the word."[89] Rather, the Neanderthals could well be pre-adamite.[90] Indeed, it was just possible that the biblical Adam could have been formed by the infusion of spiritual qualities into some pre-adamic creature. As he explained: "They [the Neanderthals] may have been pre-Adamic, and Adam verily was a new creation, with spiritual qualities that they lacked. What sort of material the Creator used to make man, whether the dust of the earth directly, or the pre-existing body of a beast, we leave an open question."[91]

While Short thus allowed the possibility of human evolution, he was far from happy with the conventional Darwinian explanation and found himself much more inclined toward the mutation theory of Lev Simonovich Berg (1876–1950), the Russian geographer and ichthyologist. Accordingly, he spelled out in some detail the evidence that Berg had accumulated in support of what he called "nomogenesis." Its significance in the present context is simply that it kept open the possibility of an evolutionary jump from pre-adamite to adamite—precisely the sort of arrangement that Warfield had envisaged.[92] This option, as we will later see, was one that conservative Protestants concerned to maintain good relations between science and religion would continue to deploy throughout the second half of the twentieth century.

Catholic Considerations

On the whole Catholic thinkers have been rather less welcoming to the pre-adamites than their Protestant counterparts, though there is clear evidence here, too, of the infiltration of evolutionary pre-adamism into certain strands of the Catholic tradition. What further complicates the story are the cultural politics that long attended discussion about Darwinism in Catholic circles. The opposition to evolution among a group of Jesuit thinkers from the 1860s brought together by Pope Pius IX to combat the forces of modernity found expression in the pages of *La Civiltá Catolica*. G. B. Pianciani, for example, insisted on the need to retain the fixity of species against Lamarckian ideas of transformism.[93] Nevertheless, the official custodians of Catholic dogma did not take any action against Catholic evolutionists until the 1890s, when attitudes began to harden as the aging Pope Leo XIII—hitherto a force for moderation—lost ground to the increasing influence of Roman Jesuits and the machinations of the *Civiltá Catolica* coterie. The French Dominican priest Dalmace Leroy, author of *The Evolution of Organic Species* (first published in French in 1887), and the American John Zahm (1851–1921), who played a significant role in the development of the University of Notre Dame in Indiana and published *Evolution and Dogma* in 1896, particularly came under fire.[94]

Both these writers protested the theological acceptability of evolution, albeit within certain limits. Leroy's apologia, erected foursquare on Mivart's foundations, was directed against both atheistic renditions of evolution and theological enemies of the theory. But because Leroy hinted that the human body might be the product of evolutionary forces, the cardinal prefect of the Sacred Congregation of the Index of Prohibited Books considered *The Evolution of Organic Species* to be unacceptable. Although the finding that his book was censured was conveyed to him privately, as the decree was not made public, Leroy was devastated, and he speedily retracted it in March 1895. It was much the same with Zahm. While he had earlier condemned Darwin's *Descent of Man* as the cause of social inequity, his purpose in writing *Evolution and Dogma* was to demonstrate its theological acceptability. Likewise inspired by Mivart's perspective—the "brilliant author of 'The Genesis of Species,'" as he styled him—Zahm turned to what he called "the Simian Origin of Man" toward the end of his analysis.[95] Here he engaged in detailed patristic exegesis to justify a teleological rendering of evolution and to claim that the animal origins of Adam's body were compatible with the theology of Augustine and Aquinas.[96] Defending the Creationist—as opposed to Traducianist—explanation of the human soul as a direct divine creation for each and every human individual, he argued that Aquinas's understanding of the creation of Adam was in keeping with the generation of the human soul ex nihilo and the evolutionary production of "a *substratum* . . . into which the Creator breathed the breath of life."[97] In sum: "In the development of man, as in that of the lower animals, there is an ascending succession of substantial forms, by means of which that which is destined to become a human body, acquires a proper structure and receives the necessary disposition for becoming the receptacle of a rational soul."[98]

The book was an international success, and an Italian translation rapidly appeared, but readers of a fuming *Civiltá Catolica* were assured that it was all a tissue of lies, redolent with reckless assertions and hopelessly compromised by dubious assumptions. The Congregation of the Index denounced it in 1897, focusing very largely on the matter of human origins. Like Leroy, Zahm reportedly withdrew his efforts despite the favorable review he received the following year from the English bishop John Cuthbert Hedley, who, for his pains, was presently condemned by *Civiltá Catolica*.

Various factors played their parts in these intrigues. A fear of "Americanism"—the tendency on the part of American Catholics to display intellectual independence and to adapt Catholicism to an American context—was one such force. Another suite of concerns rotated around the routine conflation of Darwinism and modern scholarly trends—notably the new biblical criticism. Reactions to evolutionary biology were thus often all of a piece with responses to wider challenges to established ways of thinking. Critical, too, was the matter of intellectual authority and who had the right to interpret the tradition's received canon. Those who found evolution in the writings of Aquinas and Augustine were arrogantly failing to pay due deference to the history of neoscholastic commentary. Later on, fears of Darwinian-inspired eugenic policies played their role in shaping Catholic evaluations of evolution.[99]

Yet despite Huxley's efforts to proscribe Mivart's Catholic Darwinism, the difficulties that Leroy and Zahm experienced at the hands of the Congregation of the Index, and the more general anxiety over modernizing trends in culture, a succession of Catholic scholars have continued to defend the theological propriety of a circumscribed Darwinism and to rethink the status of the biblical Adam in light of evolution. And it was at this fulcrum point between evolutionary theory and exegesis of the Genesis text that the pre-adamite theory manifested itself. On at least one element in the nexus of pre-adamite ideas, however, the Catholic tradition has remained uncompromisingly hostile, namely, the belief in the continued existence of pre-adamites alongside the adamic family. This version of co-adamism was branded as heretical by Christian Pesch in 1910, while the papal encyclical of 1950, *Humani Generis* (to which we will later return), reinforced the stance that "Christ's faithful cannot embrace the opinion that after Adam there existed on this earth true men who did not take their origin through natural generation from him."[100]

In the interim Humphrey Johnson's hearty defense of monogenism during the interwar years prompted him to query the once fashionable suggestion that "neolithic man was 'adamite' and palæolithic man 'pre-adamite.'" To him that maneuver was gratuitous. "Provided that we do not demand any high degree of mental development in the first true man," he affirmed, "there is . . . no very grave difficulty in deriving all rational beings who have ever inhabited our planet, from one pair."[101] Thus, Johnson's rejection of polygenetic pre-adamism was orchestrated by pushing the date of Adam's first appearance farther and farther back in time. In order to sustain his reading of race history, moreover, he had to couple his monogenetic stance with a belief in retrogression. In fact, he later argued in 1950 that the discovery in Palestine and Moraira, Spain, of fossilized human remains exhibiting a combination of modern and Neanderthaloid features suggested that the two racial types were less divergent than at first believed; moreover, "the fact that those Neanderthal skulls in which the pithecoid traits are most accentuated are chronologically the most recent [was] an indication that retrogression has been at work."[102] Such a stance, of course, had ideological possibilities, particularly when linked to the outmoded theory of ontogeny recapitulating phylogeny. So, for all his commitment to the organic unity of the race, he openly declared that because "the history of the individual recapitulates the history of race we may infer that the backward races of mankind, such as the Negro and the Australian, are to some extent mental degenerates."[103]

The monogenist cast of Catholic thought, however, needs to be interpreted with caution, not least because monogenism is entirely compatible, as we have seen, with at least certain versions of pre-adamism, and, as if to confirm that very point, O. W. Garrigan noted in 1967 that for Catholics "any preadamite hypothesis must allow in some way for the doctrine of original sin and the unity of the present human race."[104] The appearance of pre-adamites from time to time in post-Darwinian Catholic literature is not unexpected. Thus, the geologist N. Joly, in his 1883 *Men before Metals,* which appeared in Appleton's *International Scientific Series,* cited a certain Abbé Fabre as asserting that "prehistoric archaeology and palaeontology may, without running counter to the Scriptures, discover in the tertiary beds and in those of the early part of the quaternary the traces of *pre-Adamites.* Since it disregards all creations anterior to the last deluge but one . . . Bible revelation leaves us free to admit the existence of man in the grey diluvium, in pleiocene and even in eocene strata."[105] Again in 1913, A. J. Mass allowed for the possibility of pre-adamism when he asserted that "the existence of a human race (or human races) extinct before the time of Adam . . . is as little connected with the truth of our revealed dogmas as the question whether one or more of the stars are inhabited by rational beings resembling man," and he called on the supporting testimony of Catholic dogmaticians such as Domenico Palmieri and Jules Fabre d'Envieu. Nevertheless, although Mass did not outlaw pre-adamism, he felt that neither the scientific evidence nor La Peyrère's theological system were compelling, nor was what he called the "political-social Preadamism" of McCausland, Poole, and the American polygenetic ethnologists.[106]

These brief admissions of the pre-adamites into Catholic conversation prepare the way for some further treatments of human evolution by Catholic writers during the first half of the twentieth century in the generation after the Leroy-Zahm controversies. The contributions of Henry de Dorlodot, priest, theologian, and director of the Geological Institute at Louvain University in Belgium, are a convenient point of access to these debates. Dorlodot made no secret of his evolutionary convictions and took up the whole subject for a series of lectures during World War I. A limited edition of them was printed in 1918, though they were apparently antedated to avoid German censure and appeared with a 1913 imprint under the title *Le Darwinisme au point de vue de l'Orthodoxie Catholique.*[107] In 1922 an English translation appeared as *Darwinism and Catholic Thought.* It was substantially a theological work devoted to an elucidation of the writings in the classical Catholic tradition in order to substantiate their compatibility with evolution. Accordingly, Dorlodot perused and commented on the relevant encyclicals, the biblical commission of Leo XIII, and the exegesis of the Hexameron, as well as the writings of the Fathers and the Aristotelian scholastics, and showed to his own satisfaction that they cohered with a natural evolutionary account of species transformation. Throughout the lectures he relied on the time-honored distinction between primary and secondary causes. As for the question of human evolution, Dorlodot resorted to Catholic principles of embryological development and cited Saint Thomas Aquinas to substantiate the view that "the human embryo before it reaches a state fitted to be animated by the intellectual soul must have been animated successively, not simply by a vegetative soul and then by an animal soul . . . but rather must have passed—*per multas generationes et corruptiones*—through a great number of substantial transformations."[108] For Dorlodot the only break in the chain of natural evolution was the creation of the human soul—a view much in keeping with that of his friend and colleague Mivart. Whatever feathers Dorlodot ruffled in Rome, the book was enthusiastically received by English-speaking Catholic reviewers.[109]

The English translator of Dorlodot's work, the Jesuit Ernest C. Messenger, also contributed significantly to the Catholic debate on evolution with his own *Evolution and Theology* (1931), which understandably drew on Dorlodot's historical groundwork, so much so that one reviewer quipped that the best parts of his book were those based on Dorlodot's analysis.[110] Here again much space was devoted to the elucidation of traditional Catholic sources—Saint Ephrem, Saint Basil, Saint Gregory of Nyssa, Saint Ambrose, and so on—all to support the legitimacy of the thesis that the creation of species was by natural causes. Then he moved on to a much more detailed examination of human origins, arguing that the emergence of the human race by way of phylogenetic descent was entirely compatible with tradition and that opposition from modern theologians stemmed from their overly literalistic reading of the Old Testament. At this point Messenger paused to introduce the pre-adamites. As much as anything else, they performed the useful function of "accounting for the many apparently imperfect types of humanity which recent archaeology has revealed." If they were fully human and had not died out by the time of Adam's appearance, he argued, the need for direct divine intervention in the emergence of Adam's body would have been unnecessary. If that was not the case, and some special divine activity were involved at this point to prepare the body for the infusion of the soul, Messenger was nonetheless certain that because the whole purpose of creation was to "lead up to man," it was "in every way fitting that God should thus have made use of secondary causes in the formation of Adam's body" and that other pre-adamic "creatures may have taken an active part in preparing for his coming."[111]

Messenger's work received "fairly wide distribution" in the years immediately after 1931, but stock copies of the book were destroyed in air raids on England during World War II.[112] Nevertheless, it certainly had provoked a spate of comment in the Catholic press, and so, in 1949, Messenger brought out a collection of these essays under the title *Theology and Evolution.* Many of his reviewers painstakingly interrogated Catholic tradition to determine the legitimacy of Messenger's reading, and many of the arguments and counterarguments addressed subjects no less central to conservative Protestant theology.[113] The nuances of the debate need not detain us further, save to note that when P. G. M. Rhodes, a sympathetic commentator, observed that it "may well be imagined that God so directed the course of evolution that animals which might be described as sub-human came into existence, possessing nearly, but not quite, the human configuration, without the rational soul," this was very largely the monogenetic pre-adamism that had emerged in the wake of Darwinian evolution.[114]

In his deliberations Messenger added a further dimension to the Catholic debate on human evolution that revolved around embryology and its analogical significance. After presenting a thumbnail sketch of the history of embryology and reprinting an address he had given on Aquinas's embryological thought, he used these reflections to vindicate the mediate animation theory that Dorlodot had championed. The argument centered on the question of when the soul entered the human fetus. Messenger and Dorlodot both urged that the idea of its infusion only *after* a certain period of development (hence, mediate rather than immediate animation) was thoroughly in keeping with tradition. The discussion soon moved off into questions of technical Catholic dogmatics, but its significance to the pre-adamite question stemmed from its bearing on the issue of recapitulation. To be sure, Messenger felt that Haeckel's "Law of Recapitulation," as originally formulated, had little to commend it. "But viewed in the light of the Thomist theory of the succession of forms in the human embryo . . . it is highly suggestive," he posited. "For, if a human being at the present time goes through first a vegetative and then a sensitive or animal stage, it is difficult not to think it likely that the human race as a whole may have had a similar history."[115] What this idea opened up was the possibility that in a natural succession of evolutionary forms, humanization took place at some point along the sequence—both for the individual and the human race. Or to put it another way, just as the soul entered the body after a period of natural development, so too the first complete human, Adam, might well have had pre-adamic hominid-like predecessors.[116]

⁂ During the final decades of the nineteenth century and the first half of the twentieth, pre-adamism underwent some significant modifications. The coming of Darwinian evolution raised, in a radically serious way, the possibility that the human species had experienced an evolutionary transformation. Religious believers, who still felt bound to the Genesis narrative of Adam as the father of all humanity but found themselves persuaded by the evidence for human evolution, developed strategies for accommodating their science and religion to one another. Typically, the idea promulgated by Mivart, that the human body had been subject to evolutionary change but that the soul was implanted in the first true human by divine agency, opened up the thought that there had existed pre-adamic hominids who approached, but had not yet attained, human status. Adam, to put it another way, was physically born of pre-adamite parents but was the first recipient of a truly human soul. By this arrangement traditional doctrines about the fall from grace and the inheritance of original sin could be preserved intact. Of course, apologists for evolutionary pre-adamism failed to enjoy the universal benediction of their coreligionists, and perhaps for this reason their advocacy of the scheme was, at least on some occasions, rather tentative. My focus here has been on Protestant and Catholic conservatives who were scientifically progressive. Whether similar tactics were developed within other religious traditions— notably Judaism and Islam—remains to be seen. But among those whose opinions we have canvassed here there was an assurance, to revisit the title of this chapter, that evolution, far from heralding the death of Adam, in fact confirmed the very opposite: the literal birth of Adam. Running alongside these conjectures was a different, though sometimes related, pre-adamite trajectory, one that exploited both polygenetic and monogenetic pre-adamism for the purposes of racial ideology and its political correlates. It is to this line of development that we now turn to begin to excavate this parallel deployment of pre-adamism, both secular and religious, in the interests of racial supremacism.

7 ✥ BLOODLINES

Pre-Adamism and the Politics
of Racial Supremacy

THROUGHOUT THE ERA when pre-adamism was undergoing
its own evolution from polygenism to monogenism in response
to Darwinian transformations, the theory's potential to serve the
politics of racial ideology was being fully exploited among advocates of white supremacy. Polygenism was eagerly harnessed in the cause
of racial apologetic by those of a more or less secular outlook, while pre-
adamism was embraced by some religious partisans intent on cultivating a
racialized theology. On both sides of the Atlantic, pre- and post-Darwinian debates on human origins were permeated with the cultural politics of
race relations. The question of human origins was not simply about science
and species; it was about society and sex, cultural identity and racial purity.
Bloodlines mattered—economically, politically, spiritually. Labor economics, voting rights, marriage laws, church membership—all were folded into
the passion to determine just who was, and who was not, of Adam's lineage.

Connections between scientific questions of race and the institution of
slavery were deep and lasting, as we saw in chapter 3. During the late eighteenth and early nineteenth centuries debates about slavery drew on scientific declarations about human origins. In the case of abolitionists arguments
were often mounted to refute the idea of racial inferiority and to underscore
the unity of humankind, while advocates of the slave trade, particularly in
the United States, called on racial science to justify their policies and the
hierarchical ranking of racial groups. Of course, there are no neat histori-

cal lines of causation to be drawn here. Many scientific proponents of racial inferiority rejected slavery, while champions of human consanguinity often defended it. As Peter Kitson shows, many slaveholders felt no need to call on polygenist anthropology to justify themselves. James Tobin, for example, writing in 1787, found the views of David Hume, Voltaire, and Lord Kames on the subject to be unconvincing and happily argued that the condition of his slaves was much superior to that of Britain's laboring classes.[1] Abolitionist writing, by contrast, often called on scientific claims about the unity and common origin of the human species in their crusades and wedded them to Christian universalism. Scientific questions about the beginnings of the human species, whether common or plural, along with the religious investment in origin narratives, were thus freighted with political cargo. The intensification of these preoccupations throughout the nineteenth century, and on into the early years of the twentieth, and the often contradictory inferences drawn from them are chief among the themes that will guide our scrutiny of Adam and his ancestors in the pages that follow.

Science, Race, and Human Origins

In both Britain and the United States a rather more secular polygenism was cultivated as a virulent strain of racialist politics during the Victorian period. And the citationary architecture of the anthropological writings of its devotees on both sides of the Atlantic reveals conspicuous bibliographical interweaving. To apprehend something of the ways in which theological pre-adamism of both monogenetic and polygenetic stripes was used to sponsor racial ideology at the time, a thumbnail sketch of its secular scientific counterpart in both locations will be useful.

Polygeny in Britain

In Britain the impetus behind the reinvigoration of anthropological polygenism sprang, at least in part, from the defense of a slave economy and its associated racial stereotyping, which manifested itself in critiques of Adam Smith's classical market economics. If David Levy's analysis is well founded, Thomas Carlyle—who dubbed economics the "dismal science" on account of its finding the secret of the universe in the all-too-banal relations of supply and demand—was critically implicated in these machinations through the way in which he "morphed racial slavery into an idealized feudalism."[2] Carlyle's critique of classical economics revolved in part around "the early utilitarian presumption that all should count equally in the calculus of social good."[3] To Levy this assumption was manifestly false, as two groups—Irish and black Africans—bore witness. While British evangelicals were campaigning on the principle that black slaves were both "men and brothers," Carlyle was lauding the virtues of social hierarchy and the idea of natural masters and declaring the fundamentally subhuman status of those races who refused to work. They were economic children, requiring careful tutelage—and serfdom was the answer. For Carlyle, we might say, *Homo sapiens* was *Homo economicus*. Thus, in his 1849 "Occasional Discourse on the Negro Question" he proposed re-enslaving freed Jamaicans and warned, in characteristically florid rhetoric, of the dangers of an alliance between evangelical abolitionism and classical economics: "These two, Exeter Hall Philanthropy and the Dismal Science, led by any sacred cause of Black Emancipation, or the like, to fall in love and make a wedding of it,—will give birth to progenies and prodigies; dark extensive moon-calves, unnameable abortions, wide-coiled monstrosities, such as the world has not seen hitherto!"[4]

The chain connecting Carlyle-style economic racism directly to British anthropological polygenism is not a long one. It is symbolized in one prominent link: James Hunt (1833–69). To Hunt scientific anthropology was the perfect substitute for democratic-utilitarian economics. It provided precisely the kind of graduated racial hierarchy that suited the needs of imperial economics. Hunt's role in the foundation of the polygenist-oriented Anthropological Society of London is well-known, and the ways in which this coalition differed from its predecessor, the Ethnological Society of London, have already been noted.[5] His significance at this point, as with that of Robert Knox (1791–1862), lies in the secularized polygenism that both promulgated. Like their American counterparts, Josiah C. Nott and George Gliddon, their vision of humanity's multiple origins was not pre-adamite in any theologically significant sense. Hunt, for example, was no more happy with those enamored of evolution than with those committed to Genesis. To him there was little difference between "a disciple of Darwin and a disciple of Moses—one calls in natural selection with unlimited power, the other calls in a Deity provided in the same manner."[6] Thus, in an evaluation of natural selection presented to the 1866 meeting of the British Association for the Advancement of Science, he expressed his amazement at those Darwinians, notably Thomas Henry Huxley, who insisted that there was only one human species and that all had descended from one primitive stock.[7] To Hunt races were persistent and primordial, not subject to evolutionary change. And so he continued to voice his opposition to any suggestion that adaptation to climate was involved in racial development. Earlier, in 1862, he had drawn on the demographic statistics of mortality compiled by medical practitioners such as Ranald Martin and A. S. Thomson to resist the notion of human cosmopolitanism.[8]

Hunt's diagnosis was informed throughout by the writings of Knox, who in 1850 had published *The Races of Men*.[9] Knox, an Edinburgh anatomist and former army surgeon, had turned to the study of ethnology in the 1840s in the aftermath of the infamous Burke and Hare scandal involving the procurement of cadavers for dissecting in class.[10] Here Knox sketched out his guide to a variety of races—Jewish, Coptic, Germanic, Celtic, Slav, African, and so on—wedding his diagnosis to a mélange of rigid anti-acclimatizationism, hereditarianism and polygenism and drawing all the while on the Continental tradition of transcendental biology. Archetypal plans, not external environment, best explained organic diversity. So firmly committed was Knox to the idea that racial groups were structurally suited to specific regional environments that he went so far as to claim that no race could permanently change its terrestrial zone without degeneracy. For his efforts Knox was rewarded with Ralph Waldo Emerson's grudging acceptance of his "unpalatable conclusions" as "charged with pungent and unforgettable truths."[11]

The pungent and unforgettable implications of polygenism in the Hunt-Knox mode are certainly not difficult to discern. Different races were separate biological entities, and there were profound physical and moral chasms between them. As Hunt put it in an infamous address to the newly formed Anthropological Society in 1863: "The skin and hair are not the only things which distinguish the Negro from the European, even physically; and the difference is greater, mentally and morally, than the demonstrated physical difference."[12] In a lengthy follow-up discussion with a large number of interlocutors, he paused to refer to Dr. Berthold Seemann's "observations on the dying out of mixed races."[13] So, despite speculating that some blacks had benefited from the infusion of European blood, Hunt's message was plain: race mixing was bad. Knox's credo is yet more crisply stated: "With me, race or hereditary descent is everything."[14] Plainly, racial stocks should remain separate. Bloodlines should be preserved.

In Britain support for secular polygenism characterized many of those circulating around the nascent Anthropological Society. In an anonymous piece on plurality of races, which can be taken as emblematic of this line of development, the reviewer of the English translation of the French physiologist Georges Pouchet's work made it crystal clear that theologians simply had no business dabbling in anthropological questions. It irritated him that "we are free to speculate on the age of rocks, and even to inquire into the succession of plants and animals; but man is a sacred, and, therefore, a forbidden subject." How pitiful it was that "his origin, antiquity, and special relationships have all been settled by a tribunal that laughs at induction, and treats opposing facts with derision." It was all well and good to examine differences between "brown and white bears" and "African and Asiatic elephants," the reviewer went on, "but an Esquimaux and a European, a Negro and a Persian, were to be invariably treated as of one species. Freedom of inquiry ceased with man."[15] It was for this reason that Pouchet's defense of polygenism, as an instance of hardy scientific freedom, was welcome. It displayed none of the British tendency to tone down hard truth to placate a timid public. But members of the Anthropological Society were made of sterner stuff: "They are afraid of no statements, nor do they stand in dread of any conclusion." Pouchet's *Plurality of Races* was thus pleasingly "offensive" in its very title and agreeably "antagonistic" in its spirit.[16] For the future any "attempted meddling" by theologians in questions of racial origins must be taken as "an impertinence that the dignity of science can afford to treat with the silent contempt it so richly deserves."[17]

Polygenist anthropology thus contributed significantly to the outlook of those Victorians congregating under the banner "race is everything." It took its place within a powerful arsenal of scholarly weapons that racial thinkers fastened upon to justify their case, including writings on national character, biological heredity, geographical location, human acclimatization, and phrenology. This development marked a signal break with what Nancy Stepan characterizes as an earlier "monogenist, ethnographic, diffusionist, environmentalist tradition," which, while not "necessarily non-racist," had nonetheless drawn heavily on James Cowles Prichard's "moral disgust for slavery, his belief in the essential humanity of the African, and his Christian faith in the psychic unity of all the peoples of the world."[18]

Champions of the American School

In the United States the development of polygenetic racial science was nowhere more conspicuously prosecuted than in the endeavors of the Philadelphia physician and anatomist Samuel George Morton (1799–1851), the Alabama surgeon Josiah Clark Nott (1804–73), and the English-born Egyptologist George R. Gliddon (1809–57), with support from Louis Agassiz. In *Types of Mankind* (1854), for example, a work of mammoth proportions dedicated to the memory of Morton, Nott and Gliddon declared their intention of supplementing the theological critique to which Blumenbach and Prichard had been subjected with their own scientific refutation. However noble Prichard's sentiments might appear, they were certain that his commitment to Scripture had subverted his scientific intentions, and they did not hesitate to identify instances of "special pleading" and the "suppression of adverse facts," which together left "little confidence in his judgment or his cause." "How few," they lamented, "possess the moral power to break through a deep-rooted prejudice."[19] Their own project was different. With little concern for harmonizing science with the Bible, they embarked on the construction of a thoroughly scientific polygenism in which the language of pre-adamites in the theological sense had no place. As H. F. Augstein puts it: "What distinguished scientific racialism from earlier assertions of polygenism is its final departure from theology and, in particular, the renunciation of the consanguinity of European and Semitic peoples. Nott and Gliddon dismissed both."[20] After all, Gliddon had long been lecturing on errors in the Bible and took a lot of pleasure in what he called "Parson-skinning," while Nott relished the blood sport of clergy baiting and privately dismissed ministers as "skunks."[21] Nevertheless, even as Gliddon sided with those who read the Creation narrative as a "paradisiacal myth," declared the "*historical individuality* of Adam" untenable, and happily spoke of a "mythic 'Adam and Eve,'" he was only too happy to sweep into the polygenetic fold the apologetic pre-adamism of the anonymous Edward William Lane, who, Gliddon noted, had essentially resurrected La Peyrère's formula.[22] In such cases he was keen to insist on polygenism's theological heterodoxy. Indeed, Gliddon was so infected with the *odium theologicum* that even Nott lamented his zeal when on yet another occasion he "pitched into the Bible & Parsons."[23]

The general shape of American polygenism—the American School of Ethnology, as it was called—is now well-known, thanks to the labors of a number of historians.[24] And the rather more secularized and distinctively racialized pre-adamism that flourished among this group throughout the middle decades of the nineteenth century took various forms. Morton, for instance, was fixated on human crania and put together the world's largest collection of skulls at Philadelphia's Academy of Natural Sciences, dubbed the "Morton Golgotha" by his friends. His 1839 *Crania Americana* was designed to present statistically reliable measurements of human cranial capacity and thereby to establish the cogency of the polygenetic thesis. Since 1843 Nott had used his anatomical expertise to argue for the infertility of racial hybrids and thus to combat interracial marriage. And the aggressively pompous Gliddon reported in his 1844 *Ancient Egypt* evidence for the fixity of racial difference stretching back to the ancient world so as to undercut appeals to the environment as the agent of racial differentiation.[25]

The collective endeavors of the American School, however short-lived their scientific standing, brought a variety of additional rhetorical devices into the discussion. Statistical measurement, visual imagery, and the cultivation of what might be called moral cartography were all conspicuous features of the project. Consider first Mortonite numerology. His statistizing practices were scrutinized by Stephen Jay Gould, who uncovered something of how what he referred to as an unconscious "finagling" of the data delivered findings perfectly fitted to Morton's racial tastes.[26] By ignoring dependent variables such as age and sex, generalizing from atypical groups, and so on, Morton could supply a convenient hierarchy, with Caucasians comfortably located on top, Native Americans in the middle, and Africans at the bottom. Moreover, within specific groupings there was a statistical hierarchy too. Among the Modern Caucasians the Teutons did best, averaging 92 cubic inches, whereas the Celts scored 87 and the Indostanic family 80. At the other end of the spectrum American-born blacks chalked up 82 cubic inches, while the Hottentots and Australians only achieved 75 cubic inches each.

At the Philadelphia Academy of Natural Sciences, Morton's successor, J. Aiken Meigs, librarian and professor at the Philadelphia College of Medicine, further perused the Mortonite data and issued his own lengthy inventory of "The Cranial Characteristics of the Races of Men." It was a sort of progress report on what he called "human cranioscopy" and constituted very largely a further statistical refinement and elaboration of his late mentor's endeavors.[27] If anything, Meigs was able to exploit the statistical massaging of the Mortonite data even more than had Morton himself.[28] He was able, for example, to reduce the sample of German skulls from Morton's eighteen to fifteen and thereby to end up with a higher mean score—95 rather than 90 cubic inches, a very considerable increase. Similarly, he was able to reduce Morton's average size for American blacks: whereas Morton's analysis of the twelve skulls of "American-born Negroes" yielded an average score of 83 cubic inches, Meigs's analysis of the same data delivered a figure of 80.8 cubic inches. These differentials, of course, might seem insignificant were it not for the moral weight they were made to bear. Moreover, Meigs was able to make moral virtue out of statistical vice, as his use of a biased sample of five Anglo-Saxon skulls reveals. He felt constrained, for example, to exclude item number 62 because it was the "skull of a lunatic Englishman." Of the remaining set the largest was number 991, which attained 105 cubic inches; it had belonged to an English soldier. The second largest—weighing in at 99 cubic inches—was number 59, a Mr. Pierce, who was described as a convict and cannibal. The others, numbers 80 and 539 were, respectively, Gwillym, an English convict, and James Moran, who had been executed for piracy and murder. The cluster hardly displayed a group characterized by moral excellence, a fact recognized by Morton and Meigs, who were nonetheless able to turn it to their advantage. Even though the sample was drawn from "the lowest class of society," it was nevertheless "remarkable" that they collectively delivered "an average of 96 cubic inches."[29]

Visual imagery also formed part of the Mortonite arsenal, manifesting itself in at least two forms.[30] First, there were the finely drafted illustrations of head forms and crania that graced the pages of the American School's publications (fig. 20). These drawings performed the rhetorical task of essentializing race images in the minds of readers and of conveying a sense of the empirical seriousness of their enlightened inquiry by deploying the techniques of the scientific illustrator.[31] Careful observation, the images implied, would confirm racial views—and thus the racial viewpoint of the authors' polygenism. Gliddon's side-by-side pen drawings of the heads of "an inferior type of mankind and a superior type of monkey" were thus designed to emphasize the "palpable analogies and dissimilarities" between them (fig 21).[32] And the same was true of his "Ethnographic Tableau," appended to *The Indigenous Races of the Earth*. Its first column was composed of Louis Agassiz's geographical regimes—the Arctic, the Australian, the Polynesian, and so on; the second contained Aitken Meigs's craniometric samples; the third, Gliddon's own graphic depiction of the different human racial groups; and the fourth, the linguistic evidence deduced from the work of Louis Ferdinand Alfred Maury. The whole visual undertaking was to establish "what Baron de Humboldt has so eloquently deprecated—and Count de Gobineau so strongly insists upon—viz.: the existence of *superior* and *inferior* races."[33]

Gliddon's construction of a racial chart depicting the world pattern of civilization constitutes a second venture into the realm of visual rhetoric, this time through the medium of anthropometric cartography. His self-named "monkey chart" of the "Geographical Distribution of the *Simiæ* in Relation to That of Some Inferior Types of Men" sprang from his interest in the zoo-geography of Louis Agassiz, who had remarked in several places on the parallels between the distributions of black orangutans and black humans. Gliddon's aim was to translate the polygenist theory of centers of zoological creation into the visual language of cartography. Encompassing fifty-four monkeys and six humans, the map's purpose was crystal clear: to provide a visual display of the claim that "within the black circumvallating line which

Koriak

Aleoutian

Tartar

Chinese

CUVIER

Bulgarian

FIG. 20. Typical head forms from the work of the American School of Polygenist Anthropology.

surrounds the zone occupied by the *simiæ,* no 'civilization' . . . has ever been spontaneously developed since historical times" and, second, that "the most superior types of Monkeys are found to be indigenous exactly where we encounter races of some of the most inferior types of Men" (fig. 22).[34]

For all its statistical and physiological finesse, the racial geography that the Mortonites constructed was fundamentally moral cartography. Time and time again, moral censure found its way into the "scientific" depiction of racial groups. On the basis of his numerical data Morton declared that the structure of the native American "mind appears to be different from that of the white man, nor can the two harmonize in their social relations except on the most limited scale"; and the "Greenland esquimaux" were pronounced

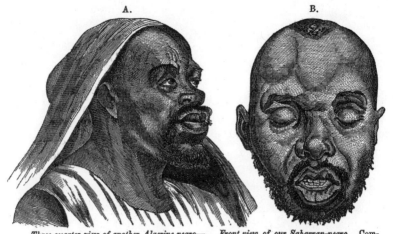

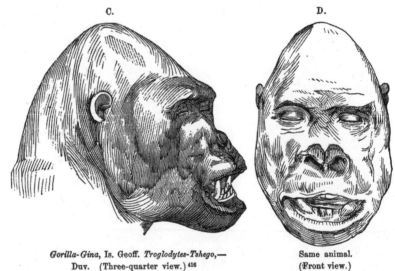

A.

B.

Three-quarter view of another Algerine negro— *Front view of our Saharran-negro.* Com-
"*Biskree.*" [415] pare his tinted *profile* in No. 26 of our
"Ethnographic Tableau,"—from B. de
St. V.'s plate.

C.

D.

Gorilla-Gina, Is. Geoff. *Troglodytes-Tshego,—* Same animal.
Duv. (Three-quarter view.) [416] (Front view.)

FIG. 21. Illustration from J. C. Nott and G. R. Gliddon's *Indigenous Races of the Earth.*

"crafty, sensual, ungrateful, obstinate and unfeeling, and much of their affection for their children may be traced to purely selfish motives . . . Their
mental faculties from infancy to old age, present a continued childhood . . .
In gluttony, selfishness and ingratitude, they are perhaps unequalled by any
other nation of people."[35] Racial mapping simply *was* moral geography.

What lay behind such judgments was a passion to keep racial bloodlines unadulterated. Thus, Morton took up the subject of "Hybridity in Animals, Considered in Reference to the Question of the Unity of the Human Species," in the mid-1840s and concluded that hybrids were contrary to nature.[36] Agassiz was also troubled by the thought that racial amalgamation would inevitably follow in the wake of emancipation. Nott's obsession with the same subject was part and parcel of his horror at the thought of racial intermarriage. After all, he was certain that everyone "conversant with breeding of Horses, Cattle, Dogs and Sheep, is aware of the effect of the slightest taint of impure blood." If, as he was sure, there were "several *species* of the human race" and if, as he was no less convinced, there was "a law of nature opposed to the mingling of the white and black races," then the mixing of Caucasian blood with "that of interior races" was of the deepest "Political, Moral and Religious import." The consequences were plain: "That the Negro and Indian races are susceptible of the same degree of civilization that the Caucasian is, all history would show not to be true—that the Caucasian race is deteriorated by intermixing with the inferior races is equally true."[37] As for Gliddon, he insisted that "mulattoes, produced by intercourse between exotic Europeans of the white race, with equally-exotic African females of the black, die out . . . in three or four generations."[38] Sexual regulation was needed to prevent racial transgression. The implications of their science for the politics of slavery, moreover, was never far from the minds of these poly-

FIG. 22. Gliddon's chart of the "Geographical Distribution of the *Simiæ* in Relation to That of Some Inferior Types of Men."

Bloodlines: Pre-Adamism and the Politics of Racial Supremacy

genists. As George Frederickson shows, Gliddon believed his research on Egyptian crania would provide strong support for the slave system, while Nott told James Henry Hammond, a South Carolina planter and politician, that "his work was designed to confound the abolitionists."[39]

The American School's Critics

Of course, American polygenist anthropology did not go uncontested. Throughout the middle decades of the century conventional adamic monogenism continued to be defended by the successors of Stanhope Smith and James Cowles Prichard. In America's southern states, for example, some ministers stood up against what they saw as the malign implications of Mortonite polygenism. John Bachman thus defended the unity of the human race on scientific and scriptural grounds by arguing that polygenism was born in infidelity and nurtured in skepticism.[40] Bachman, the leading scientific critic of the American School, focused on the question of hybridity and with impressive empirical detail refuted case after case that Morton had instanced in support of polygenism as well as advancing various anatomical arguments in favor of human unity. For Bachman polygenism threatened not only the authenticity of scripture but also the ideological fabric of Christian civilization. In a review of *Types of Mankind* he made it clear that he considered that Gliddon was embarked on a project to discredit Christianity by "heaping on the Holy Scriptures all manner of epithets of derision and contempt."[41]

Thomas Smyth (1808–73), a Belfast-born Presbyterian clergyman who spent nearly forty years in Charleston, South Carolina, also produced a robust defense of *The Unity of the Human Race* in 1850. Operating within the confines of an Ussherite chronology for the human race and relying on a vast range of authorities—among them Stanhope Smith, Prichard, and Bachman himself—he marshaled both scientific and scriptural arguments, bolstered with historical, philosophical, and linguistic findings, to undermine Agasssiz-style polygenism. Along with Nott and Gliddon, who "ridicule[d] with profane and vulgar buffoonery the doctrine of the unity of the races," Smyth had in his sights those claiming that "the record found in the Book of Genesis has reference only to the ancestors of the Sacred, or Jewish, or Caucasian race."[42] He was thus engaged in a mission to rescue Adam from partisan abduction and from the subversive tactics of the likes of Voltaire and La Peyrère. To him Adam's attributes were "those, not of any one race of men, but of the whole human family." Adam's bloodline was the patrimony of every human, and on this the entire structure of Christian redemption rested. "All the races of men . . . who are interested in Christ and in His gospel," Smyth proclaimed, "are, and must be of Adamic origin, seed, or blood . . . And hence, as it is expressly commanded to preach this gospel to every creature in all the world, all must be of the same original Adamic family and origin."[43] Smyth thereby garnered the forces of his considerable scholarship to mount a campaign against pre-adamism and American School polygenism alike. To him these were scientifically unphilosophical, morally reprehensible, and religiously unchristian ideas.

The Testimony of Modern Science to the Unity of Mankind, which appeared in 1859, added further support to the Bachman case. It was the work of the University of Virginia's professor of anatomy, physiology, and surgery, James L. Cabell (1813–89). Cabell's defense of the specific unity of the human race was grounded both in biblical conviction and scientific evidence. His target was twofold: infidel polygenists and scriptural revisionists. His own investigations had convinced him that organic modifications could be induced quite rapidly—by the influence of climate, for example—and that, Lamarckian-like, "these newly acquired characters may then be perpetuated by hereditary transmission."[44] Some of the varieties within species remained permanent, but their genesis in environmental causes meant that Cabell could happily reject the arguments of Agassiz and Morton on species and Gliddon on Egyptian monumental inscriptions. Besides, the "undoubted power possessed by the various races of men and by the domesticated animals, to undergo acclimation in every quarter of the globe . . . indicates the possibility that the former may have sprung from a common origin."[45]

From a different ideological location others such as the Princeton theologian Charles Hodge suspected that the currently fashionable polygenism, exhibited by anthropologists of Gliddon's stripe, was more than a little motivated by a passion to furnish "a satisfactory foundation for the perpetuity of African slaveholding."[46] Attempts to define the nature of the human species on a scientific basis was simply wrongheaded; they ignored the "higher bond of union in the identity of *pneuma.* The rational and immortal soul belongs to all, and it is the same in all . . . The rational soul of the Caucasian, of the Mongolian, and of the African, do not differ the one from the other, more than the soul of one Englishman differs from that of another. There may of course be a great difference in the mental endowments of different races of men, as there are among the different members of the same family. But this does not affect the question of identity."[47]

Yet none of this meant that advocates of the unity of the human race were committed to egalitarianism, still less abolitionism. The idea of black inferiority was just too ingrained for that. Bachman, for example, staunchly defended southern slavery and argued, on the basis of the biblical curse on Ham, that the black races were designed, and destined, for servitude. He considered the "Negro [to be] a striking and now permanent variety" who might improve through intermarriage with whites—a morally repugnant price to pay for racial enhancement.[48] Smyth, too, for all his vigorous protestations that there were no innate differences between the intellectual capacities of blacks and whites, did not use his assault on polygenism to attack slavery. Far from it. In fact his quarrel with the American School was in part because he felt the entire undertaking would actually undermine the traditional biblical basis of southern slavocracy. As he put it:

> The introduction in the South . . . of this novel theory of the diversity of races, would be a declaration to the world that its institutions could no longer rest upon the basis which has always been hitherto assumed, and that this theory has been adopted for mere proud, selfish, and self-aggrandizing purposes . . . It would remove from both master and servant the strongest bonds by which they are united to each other in mutually beneficial relations . . . Neither would this theory be less dangerous in its bearing upon all the interests involved in the proper and Scriptural course to be pursued towards the slave population which, by the providence of God and the coercive and mercenary policy of Great Britain, has been intrusted to her management. God is in this whole matter . . . The relation now providentially held by the white population of the South to the colored race is an ordinance of God, a form and condition of government permitted by Him.[49]

Put simply, scientific anthropology bestialized slavery; adamic theology sanctified it. For Smyth slavery had the benediction of scripture, and any undermining of biblical authority was thus socially subversive. For generations polygenism's spiritual infidelity was well-known; after all, it "was for this purpose the theory was introduced by Voltaire, Rousseau, and Peyrere." But now its seditious politics were laid bare. Christian slavery, to Smyth, did not mean that slaves were chattel; it gave rights and privileges; it was providentially ordered. But it was slavery all the same. No wonder polygenism was "impolitic to the South."[50] It would wreck the building blocks on which southern Christian civilization had been raised. Nevertheless, Smyth's personal efforts at slave reform and his involvement with fellow Southern Presbyterian clergyman John Adger and John Girardeau in establishing the Zion Presbyterian Church for slaves resulted in his being vilified by some Charleston planters as an abolitionist.

Among southern clergy at the time this viewpoint was widely shared.[51] A plain, unadorned reading of the Bible seemed to sanction the slave system, and there was no need to turn to secular science or unorthodox readings of Genesis to support it. To them such perfidious projects would only defile a laudable biblical institution and weaken the foundations of southern patriarchal communalism. Scriptural arguments in support of slavery abounded.[52] Reverend George Armstrong pronounced in *The Christian Doctrine of Slavery* (1857) that abolitionism had sprouted from the infidel breed of philosophy that had inspired the French Revolution.[53] George Howe, professor of biblical literature at the Presbyterian Seminary in Columbia, in an unrestrained attack on the polygenist lectures of Josiah Nott, repudiated efforts to justify slavery in the language of biology, preferring instead the vocabulary of the Bible, under which "the slaveholding patriarch, the slaveholding disciple of Moses, and the slaveholding Christian lived, protected and unrebuked."[54] Robert Dabney, doyen of the southern Presbyterian theologians, high-ranking Confederate officer, and well-known opponent of public education on account of its trend toward social leveling, excoriated abolitionism as the product of atheistic theories of human rights.[55] On polygenism and abolitionism, as with a multitude of other issues, Dabney considered the struggle to be between the forces of rationalism and revelation. Not surprisingly, with a primal belief in white supremacy, he juxtaposed what he called "scriptural politics" with "new-fangled republicanism" and "ultra-democratic" Jacobinism.[56] According to Mark Noll, after the South's defeat he "offered only intuitive tribal sentiment, especially on the spectre of black-white intermarriage," throughout the first half of a speech in which he opposed the ordination of African Americans.[57]

All these figures were allergic to any modernizing trend and joined together to repudiate evolutionary theory.[58] Other similar cases could readily be enumerated. Reverend George Junkin, then at Washington University, delivered biblical arguments against abolitionism to the synod of Cincinnati in 1843; thirteen years later the Virginia Baptist Thornton Stringfellow presented *Scriptural and Statistical Views in Favor of Slavery;* James Henley Thornwell detected in slave labor the merciful hand of Divine Providence.[59] To all of them the Bible sanctioned slavery, and abolitionists and polygenists alike were undermining its supreme authority. Humane Christian slavery, they believed, was under attack from two radically different sources: an opportunistic abolitionism fueled by northern greed and economic self-interest; and a degenerate anthropology that would dehumanize whole races.

Like Bachman, proslavery propagandists who did not countenance polygenism typically found justification for their stance in the Old Testament story of the curse on Noah's son Ham. Armstrong's *Christian Doctrine of Slavery* (1857) located the origins of slavery in the narrative. So did Dabney's *Defence of Virginia* (1867). The Mississippi Presbyterian James A. Sloan's *Great Question Answered* (1857), the Baptist J. L. Dagg's *Elements of Moral Science* (1860), and the Tennessee clergyman Samuel Davies Baldwin's *Dominion; or, the Unity and Trinity of the Human Race* (1858) all likewise turned to the story of Ham to underwrite the institution. Stephen Haynes has elucidated the twists and turns of this widespread maneuver.[60] What he has shown is that in the antebellum South standard interpretations of the curse, which often found sexual transgression at the heart of the matter, were routinely ignored in favor of a reading that underscored family dishonor. Ham's failure to act honorably by laughing at his father's drunken nakedness, Haynes argues, resonated with the premium that the South put on an honor code, the violation of which justified condemnation to slavery.[61] Whatever the motivation, the Ham narrative provided a widespread, though not universal, warrant for slavery without any need to posit multiple human origins.

Of course, there were dissenting voices who interrogated the Bible for its abolitionist possibilities. Some of them ferreted out passages mandating the release of slaves after a set number of years; others pointed out that pro-slavery theology conveniently ignored practices such as polygamy that enjoyed Old Testament sanction every bit as much as slavery; still others dwelled less on specifics, preferring instead to make a Christian moral case against slavery's inhumanity. What became clear, however, was that biblical inductivism was not of itself sufficient to resolve the issue and that hermeneutics were simply shaped by the cultural conditions and political stance of commentators. As Brooks Holifield has shown, the slave issue precipitated a move away from what he calls a Baconian hermeneutic by introducing a historical consciousness that insisted on the need to locate biblical texts in the time and place of their writing.[62]

Needless to say, the African-American voices that entered the fray found Mortonite polygenism entirely objectionable.[63] Frederick Douglass, former slave and one of the foremost leaders of the abolitionist movement, chose as his subject "The Claims of the Negro, Ethnologically Considered" for his commencement address at Western Reserve College in 1854. Impressing on his hearers the moral weight of the question of race relations, he took up the cudgels against Morton, Nott, Gliddon, Agassiz, and other polygenists, arguing that "the credit of the Bible" was at stake in their anthropological corruption. Everything hung on the outcome of the question of human origins, for it was "connected with eternal as well as with terrestrial interests."[64] He thus proceeded to demonstrate with forensic precision the "ever conspicuous" Mortonite "contempt for negroes" and to impress on his hearers that the Pauline declaration that all nations were "of one blood" provided sufficient grounds to reject the multiple origin theory and to confound those who denied human status to the black race.[65] Adopting an environmentalist account of racial derivation, he insisted that "only about one fifth of all the inhabitants of the globe are white; and they are as far from the Adamic complexion as is the negro."[66]

Several other black writers added their support. In two issues, for example, of the 1861 edition of the *Christian Recorder*—the organ of the African Methodist Episcopal Church— an anonymous writer addressed his readers on the subject of ethnology. Descent from multiple sources was roundly rejected in favor of "the Jewish and Christian Scriptures," which "plainly teach, that mankind, of whatever race, family, or tribe, have descended from a single pair of progenitors, and the history of Adam, as given in the first book of the Pentateuch . . . furnishes the ground-work of belief upon this subject."[67] This self-same stance persisted for decades after the end of the Civil War. Martin Delany, abolitionist, explorer, and first African-American field officer in the U.S. Army, presented his defense of the unity of the human race in his 1879 tract *Principia of Ethnology: The Origin of Races and Color.* In his view differences in skin tones were traceable to the different complexions of Shem, Ham, and Japheth—the sons of Noah.[68] In 1886 the Baptist clergyman, historian, and first African-American graduate of Newton Theological Seminary, George Washington Williams devoted the first chapter of his *History of the Negro Race in America* to the subject of the "Unity of Mankind," with the aim of refuting "the absurd charge that the Negro does not belong to the human family."[69] Universal descent from Adam provided the sheet anchor of Williams's interrogation. Just a few years later, in an 1898 rebuttal of the pre-adamite speculations emanating from white Methodists, the American Methodist Episcopal bishop Benjamin Tucker Tanner bluntly declared: "The Negro is a man. He is of Adam. He is of Noah. The Negro is a brother."[70] Here Tanner, as with other African-American writers at the time, considered the human race as Hamitic.[71] Indeed, as late as 1906, W. O. Thompson happily used Darwinian science to support the Bible's monogenetic account of human descent because "both accounts agree that all races descended from a single primitive stock."[72] Surveying the range of black responses to evolution, Eric Anderson makes the telling observation that even "black religious leaders seldom directly attacked Darwinism or competing theories of evolution, preferring, instead, to make traditional, biblically based appeals for 'the brotherhood of man.'"[73]

Pre-Adamism and Supremacist Ethno-Theology

During the later decades of the nineteenth century advocates of pre-adamite theology faced opposition on at least two flanks. On the one hand, polygenetic anthropologists on both sides of the Atlantic labored long and hard to cleanse their terrain from theological interference. On the other, conservative theologians turned their guns on those who sought to reread the adamic story through the lens of scientific claims. So far as questions of racial supremacy were concerned, the polygenists believed they had secured scientific warrant for their belief in race hierarchy and the white domination of supposed inferior peoples. Theological apologists for slavocracy, particularly in the American South, had no call for secular science to justify an institution that they believed enjoyed biblical legitimation. Nevertheless, the potential of theological pre-adamism, both monogenetic and polygenetic, to serve the interests of white supremacy was sufficiently powerful to attract the attention of several writers during these years, and it is to their efforts that we now turn.

Adamic Repugnance

We return first to Alexander Winchell, whose efforts to find some theological rapprochement with evolution have already attracted our attention. His monogenetic rendition of pre-adamism, we recall, was in large part designed to preserve post-adamic biblical chronology intact even while allowing for the possibility of evolutionary change. At the same time he did not hesitate to marshal his pre-adamites in the cause of white supremacy.[74] In the scenario he advanced, Winchell was certain that "no such racial contrast existed between the family of Adam and the nonadamites as to originate a racial repugnance."[75] But this racial compatibility applied only to Adam and the pre-adamic Dravida in his immediate genealogical line—the very group among whom Cain dwelt on his banishment from Eden and from among whom he procured his wife. Lineages disaggregated much earlier in the course of racial history were a different matter (fig. 23).

A right understanding of Adam and his ancestors thus had direct consequences for contemporary race politics in the United States, and Winchell's monogenetic pre-adamite scheme was just as racially serviceable as polygen-

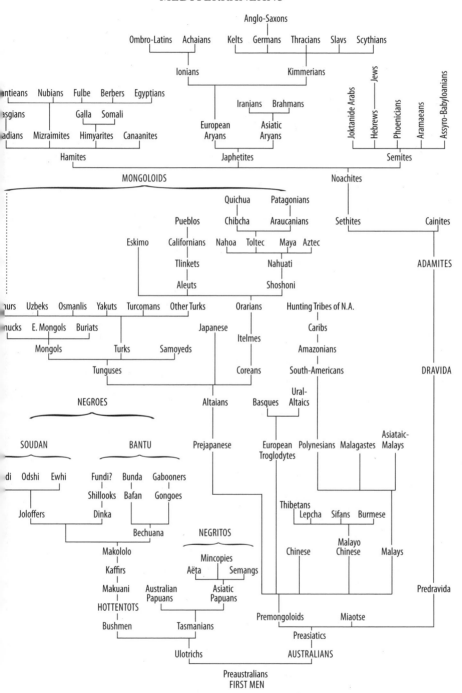

Fig. 23. Alexander Winchell's genealogical tree of the types of mankind.

etic pre-adamism. The portraits of the "pre-adamites" that graced the frontispiece of his book plainly advertised the direction in which his thoughts
inevitably moved (fig. 24). "The inferiority of the Negro is a fact everywhere
patent," he uncompromisingly declared in 1878, and this condition was emphatically *not* to be attributed to any substandard environment. Africa's interior, he judged, was such that "the salubrity of the climate, the fertility of
the soil, the vast hydrographic system of lakes and rivers, have all conspired
to give the interior of the continent natural conditions surpassed by those
of the site of any civilization which ever existed."[76] Environmental modification of the human species had proceeded apace, to such a degree that there
were now several physically, psychically, and linguistically distinct types of
humanity. This splintering of races was to be accounted for by natural means,
not by divine intervention, for Winchell's pre-adamism, in contrast to earlier
renditions, was thus thoroughly antagonistic to the notion of fixity of type.
But differences could become ingrained, embedded, entrenched, whatever
their origin, and Winchell was convinced that the racial types observable in
the present had remained unchanged since ancient times. "While we cannot deny that organism has been coadapted to environment in the progress
of ages," he noted, "it is true that characters finally acquired persist with a
wonderful degree of changelessness from age to age, and under the broadest
diversity of physical conditions. From the date of the earliest records the Jew
has been a recognizable Jew, the Negro has been distinctly a Negro, and the
Egyptian, and the Aryan and the Abyssinian have stood forth as completely
different as they appear to be at present."[77]

Thus, Winchell's description of black inferiority, and the intensity of his
disgust at miscegenation were as exuberant as any racial polygenist might ex-
hibit. But we should recognize, too, that they were not dissimilar to Huxley's
belief that the supposedly lower human races were closer to the ape than to
the highest exemplars of the white race.[78] Indeed, Huxley, repeating Morton's statistical findings, did make it clear that it was entirely possible to "admit that Negroes and Australians, Negritos and Mongols are distinct species,
or distinct genera and you may yet, with perfect consistency, be the strictest
of Monogenists." For himself he was convinced that it was Darwin who succeeded in "reconciling and combining all that is good in the Monogenistic
and Polygenistic schools."[79] A lengthy chapter devoted to "Negro Inferiority"
in Winchell's treatise thus dwelt in unrelenting detail on the race's presumed
anatomical, physiological, and psychical inadequacies. Cephalic index, forearm length, brain weight, mathematic competence, aesthetic judgment, racial inertia—all these were marshaled, along with side-by-side etchings of

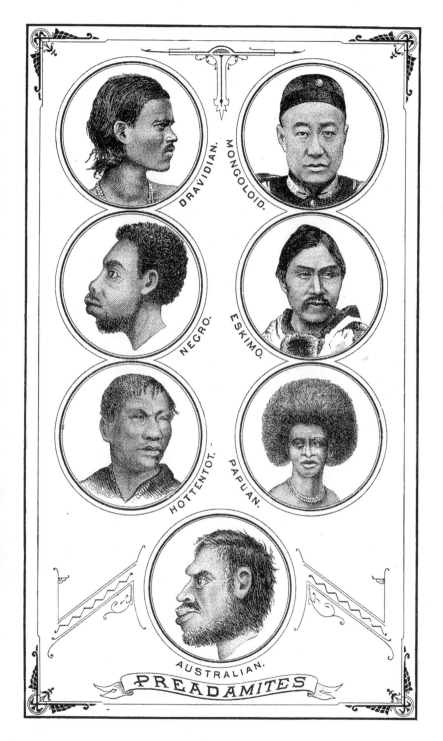

FIG. 24. Frontispiece in Alexander Winchell's *Preadamites*.

the facial profiles of a "female hottentot" and a "female gorilla," in support of his declaration that "just as far as the African diverges from the style of a white man, he approximates the lower animals."[80] In the light of this litany of denunciation, the need to preserve the adamic bloodline was an absolute moral necessity:

> I allow myself to pause here briefly, for the purpose of protesting against the policy of North American miscigenesis [*sic*], which has been recommended by high authorities as an eligible expedient for obviating race collisions. It is proposed to consolidate the conflicting elements by a systematic promotion of the interfusion of the white and black races. It is proposed, in short, to cover the continent with a race of Griquas. The policy is not more shocking to our higher sentiments, nor more opposed to the native instincts of the human being, than it is destructive to the welfare of the nation and of humanity.[81]

Advocates of racial intermixture were thus the special objects of his scorn. Wendell Phillips was castigated as "in danger of acquiring the title of 'most eloquent platform virago,'" that is, if his "sex did not protect him." Bishop Gilbert Haven's sentiments were dismissed as "painful." Canon George Rawlinson was berated for having "added his name to this cluster of self-appointed conspicuities" by advancing proposals that would lead to the careless destruction of "ethnic pearls."[82] All came under the whiplash of his tongue for their open advocacy of ethnic amalgamation. Their policy, which would only result in the widespread production of mulattoes, was the blindest folly. Drawing on Stanford Hunt's somatic measurements of United States volunteers during the Civil War, he reported that the brain weight of the average pure Negro (1,331 grams)—so much lighter than that of whites (1,424 grams)—was nonetheless greater than one-quarter-white hybrids (1,319 grams). When supplemented with his own "table of states of hybridization" and his natural law of "the mutual repugnance of races," the implications for national policy were obvious:[83] the continued production of "freckled, blotched and mottled complexions, uncouth extravagances of features, short life, infecundity and general sanitary feebleness, [which were] common characteristics of mulattoes" was both "shocking" to the "higher sentiments" and "destructive to the welfare of the nation and of humanity."[84]

Winchell's pre-adamism was thoroughly monogenetic; he insisted that his version left "the blood connection between the White and Black races undisturbed," and he repudiated the Hamitic hypothesis as groundless myth. Nevertheless, he could speak about the way in which racial groups had been "adulterated with Negro blood" or elevated (at too great a cost) with the "infusion of Caucasian blood."[85]

Eve's Seduction

Offensive though Winchell's scientific declarations no doubt are to modern ears, they were nothing compared to the startling proposals advanced during the second half of the nineteenth century by a suite of theological writers whose project was literally to dehumanize Africans as nonhuman pre-adamites. The central trope in their armory of degradation was the projection of what Mason Stokes calls a "racist Edenic mythos," which recast the fall from grace as a *sexual* offense on account of Eve's miscegenating liaison with a black pre-adamite symbolically cast as a serpent in the Genesis record. As Stokes observes, "This reading of the temptation of Eve worked to buttress hysteria over the issue of . . . amalgamation . . . Thus miscegenation becomes the reason for the Fall."[86] Eve's sin lay in polluting the pure adamic bloodline. And what was illicit in Eden was criminal in the United States. The segregationist bio-theology these writers assembled has had a lengthy afterlife, serving as a potent countercurrent to the religious impulses that later shaped the Civil Rights movement. Over the past century and more, as Jane Dailey observes, "Narratives such as these had two key pedagogical aims: to make the case for segregation as divine law, and to warn that transgression of this law would inevitably be followed by divine punishment."[87] Something of the shadows these nineteenth-century racial zealots have continued to exert on white nationalist communities up to the present day will be sketched in the next chapter. For the moment we turn to the twists and turns of their supremacist sexualizing of pre-adamism during the second half of the nineteenth century.

Chronologically, what Colin Kidd calls "the bizarre contortions" of Samuel Cartwright (1793–1862) provide a suitable point of departure.[88] A Louisiana physician and pro-slavery advocate, he identified in 1851 what he called "drapetomania" as a black psychotic condition to explain the inclination of slaves to escape captivity.[89] Later he further developed his anatomical prosecution of black inferiority in an 1860 article on the Mosaic record in which he summoned into existence two primordial "races of intellectual creatures with immortal souls, created at different times." Then, noting that the Bible commentator Adam Clarke had been "forced to the conclusion that the creature which beguiled Eve was an animal formed like man, walked erect, and had the gift of speech," Cartwright took the next step of identifying Eve's tempter—Nachash, or the serpent—as a "negro gardener" belonging to the pre-adamite race.[90] The word *Nachash* allowed him by free association to elaborate a litany of serpent-like qualities as racial allegations: "The history of the creature is enclosed in the name, under the cover of a bundle of ideas . . . the serpent—. . . watching closely—prying into designs—muttering and babbling without meaning—hissing—whistling—deceitful—artful—fetters—chains—and a verb formed from the name, which signifies to be or to become *black*. Any good overseer would recognize the negro's peculiarities in the definition of *Nachash,* and the verbs connected with it, if read to him from a Hebrew lexicon."[91] Moreover, the perversion that Eve triggered did not stop at the gates of Eden. Her initial act of gross interracial sexual union was evidently a morbid psycho-moral condition that afflicted the adamite bloodline after Cain's murderous rebellion and resulted in "a very general amalgamation . . . between the Adamic and the inferior races."[92] According to Cartwright, their depraved hybridity came under divine judgment at the time of the Flood. As Frederickson tellingly observes, such "fantastic theories cannot . . . be dismissed as the curious musings of an isolated crackpot," for "Cartwright's views were respected in the [American] South, probably more than Nott's."[93] The idea's subsequent history bears out that impression.

While Cartwright did not deny the fully human status of Eve's black paramour, others did. The propagation of the Evite sexual saga was perpetuated in the 1867 work of a Nashville publisher and clergyman, Buckner H. Payne (1799–1883), writing under the pseudonym "Ariel."[94] But this time the illicit encounter was embellished with an added exoticism: the black tempter did not have an immortal soul; he was a subhuman beast. The scenario was depicted in the briefest compass toward the end of Payne's tortured booklet on *The Negro: What Is His Ethnological Status?* In fact, this piece of virulently anti-black propaganda carried a sequence of interrogative subtitles: "Is he the progeny of Ham? Is he a descendant of Adam and Eve? Has he a soul?" and so on. Payne's strategy, initially, was to undermine the conventional view that the black races were the progeny of Ham's cursed son Canaan, and he worked hard to retrieve the dignity of their racial offspring by charting their diffusion into Mizraim—Egypt—where archaeological ruins indicated a culture of "surpassing magnificence and grandeur."[95] Having established to his own satisfaction that this was a white civilization possessing "all the arts and sciences, manufactures and commerce, geometry, astronomy, geography, architectures, letters, painting, music etc," he used these accomplishments as the foil against which to disparage black history "as blank."[96] If the theory of the Hamitic origin of the black race was false, it followed that representatives of it must have been in the ark because the Flood was universal, and Payne inferred that "the negro entered the ark *only as a beast*." His stark conclusion was plain: "the negro is not a *human* being—not being of Adam's race"; "he was created *before* Adam."[97]

The final component in Payne's convoluted racial theology preoccupied him for the remainder of his attack: interracial mixing. So absorbed was he with bloodlines that he read the Old Testament with what can only be described as a miscegenation hermeneutic. Event after event in the biblical chronicle of divine judgment was interpreted as punishment for one crime and one crime only—"the crime of *amalgamation or miscegenation of the white race* with that of *the black—mere beasts of the earth*."[98] The banishment from Eden, the Noachian flood, the condemnation of Sodom—all were swept into this obsession. To Payne this was "a crime, *in the sight of God, that can not be forgiven by God*—never has been forgiven on earth, and never will be. Death—death inexorable, is declared by God's judgments on the *world* and *on nations*" who committed the horrible crime of "amalgamating, miscegenating, with the *negro-man-beast, without soul—without the endowment of immortality*."[99] The present-day reverberations could not be missed. As he worked his way toward the conclusion of his jeremiad, Payne asserted: "And *you,* the people of the United States, are upholding *this profanity*."[100] The reenslavement of African Americans or their deportation back to Africa was the only way that the nation could cleanse itself. For the meantime his inflammatory warning was stark: "A man can not commit so great an offense against his race, against his country, against his God . . . as to give his daughter in marriage to a negro—a *beast*—or to take one of their females for his wife. As well might he in the sight of God, wed his child to any other beast of forest or of field."[101]

No sooner was Payne's diatribe made publicly available than its claims were contested point by point by Robert Anderson Young (1823–1902), a minister of the Methodist Episcopal Church, South, and sometime president of the Wesleyan University in Alabama. Young's critique had been provoked by a request he had received to reply to Ariel's charges from a group of women supplying artificial limbs to Nashville's Confederate wounded. Young was evidently appalled at what he read and variously castigated it as "*balderdash*," a "weak, wicked, and infidel publication," written to "impress the vulgar mind."[102] He was chagrined to report that the work "is before the public. Thousands have read it. Some believe it. A few have been damaged by it. If the negroes read it and believe it, they are ruined." Certainly, Young's rebuttal did not extend to egalitarianism. He confessed that he did not believe "in the social equality of the Negro."[103] But on page after page he contested Payne's grammatical infelicities, ill-informed ethnology, scriptural manipulation, and anatomical misapprehensions and called on the support of Bachman, Richard Owen, Humboldt, and Prichard. In frustration he cited the words of a colleague who, when told about Young's proposal to answer Ariel, could only sigh, "Do you propose to *reason* with a madman?"[104] His own account of racial evolution concurred with the "master-minds in Ethnology ... [who] have arrived at the conclusion that the Negro Variety of the Human Species has only been developed in the course of ages, within the African tropics, and was derived from Egypt and Assyria."[105]

To support his refutation Young had called on the opinions of a Tennessee Confederate physician, Dr. George S. Blackie, later cofounder of the medical journal the *Southern Practitioner* but currently running a school for young ladies—the Shelby Female Institute—in Nashville's South Cherry Street. A graduate of the universities of Edinburgh and Bonn, Blackie willingly lent his support, and provided a postscript, to Young's booklet, in which he presented the scientific case for Cuvier's insistence that "there is but one genus homo; and one species, homo sapiens."[106] The fact that the offspring of white and black unions was fertile was one indicator that the races did not belong to different species, Blackie reported. Moreover, he called on the authority of Blumenbach to point out that "the most dissimilar human skulls bear more resemblance to each other than the long head of the Neapolitan horse does to the short skull of the Hungarian." Human craniometry was no clue to species difference. The frequent occurrence of Melanism and Albinism in various regions of the globe further supported his case that in "the human species, varieties occur in one race approaching the characteristics of another."[107] Blackie was also sure that what was considered beauty was distributed across racial groups, and he commented on the "handsome and regular features" of the black abolitionist Frederick Douglass and of a black actor he had earlier seen in Paris.[108]

The Young-Blackie riposte did not, however, put an end to the Ariel controversy. Far from it. A further refutation was forthcoming, for example, from Harrison Berry, a literate enslaved artisan from central Georgia who wrote about slavery even while still in captivity; he is reputed to have published in 1868 *A Reply to Ariel,* in which he pinpointed the "interest of the great diabolical slave power, making the enslavement of the Negro justifiable on the hypothesis of his being a beast" as the motivating force behind Payne's tirade.[109]

That same year another reply appeared under the initials M.S. in which the author dismissed the original work of Ariel and the critique of Young and Blackie alike as "embodying some of the most absurd notions ever entertained by educated divines." A firm believer in polygenism and that the "negro is of a different *species*" of pre-adamic origin, he repudiated both Ariel's bestializing of the black race and Young and Blackie's advocacy of the unity of humanity and their naturalistic account of racial evolution.[110] To him there was no doubt about the human status of Africans; neither was there any doubt about their separate origin and inferiority. To bolster his views M.S. expanded on his understanding of hybridity, giving voice again to the absorption with bloodline purity that characterized so much writing on the whole subject at the time. Aware that the offspring of interracial unions were not sterile, he worked hard to argue that their fecundity was temporary and disappeared within four generations, and he took this as proof that the partners were of different species. Much of the argument rehearsed standard polygenetic science, particularly after the fashion of Agassiz: "We believe, with Prof. Agassiz, that all the different species of animals and plants were introduced upon the earth in groups, at different times, suited to their habits and natures, but that whole families were not; that the negroes and tropical Asiatic, Malays, West Indians, and Central American Indians were all introduced into those different centres or zones at one time, and long anterior to the Adamic race."[111]

Nevertheless, M.S.'s intervention did include one or two further novelties. First, M.S. insisted that different kinds of sexually induced diseases marked out a species border between black and white. While "unbridled, licentious, and indiscriminate sexual intercourse between the Caucasian or Adamic species produces . . . gonorrhoea—a disease which is not fatal," he wrote, the same behavior "between the white and the negro, or Indian, is known to generate *syphilis,* one of the most loathsome and destructive diseases known to man." It comes as no surprise to find that M.S. considered that "*amalgamation,* or, in modern parlance, *miscegenation,*" was "an accursed sin."[112] Second, while not dwelling in much detail on the whole matter of Eve's tempter, M.S. did pause to side with Adam Clarke's suggestion that the "serpent" was in human form and had the power of speech. But rather than identifying the tempter as black, the devilish instrument of seduction was branded as "most likely a Mongol of intelligence and pleasing appearance."[113] Third, M.S. resorted to the new scientific language of electricity, atomic energy, and molecular motion to explain the relations between matter, mind, and soul and to urge that it "is the larger and more perfect structure of the intellectual lobe of the brain, in the Adamic race, that characterizes them as a superior species to all the others of the *genus homo* . . . They receive and transmit those vibrations, or the principle called *motion,* in a higher degree than any other created beings."[114] This mélange of popular electrical images provided M.S. with the grounds for a white supremacist final pronouncement: "The soul of man, of the Adamic race, being tabernacled in a more perfect system of nerves and enlarged brain (intellectually), for the concentration of those progressive and infinite divine emanations which characterize him as of the ruling race, and the master-spirit of creation on this earth, is without limit."[115]

Another anonymous contributor, writing under the name Prospero, provided a sequel to Ariel's sketch. Using the different Hebrew names for "God" in the Genesis account, the author argued for two separate creations, pre-adamic and adamic. The pre-adamites were identified as Mongols and Negroes, together with their mixed progeny, and, like Ariel, Prospero located Eve's tempter among them. This circumstance, as Stokes indicates, allowed the writer to eroticize, perhaps more than Ariel, the moment of Eve's sexual downfall.[116] The scene materializes as the serpent, taking the form of an "African medicine-man, or conjurer," whispers "his diabolical temptation into the ear of unsuspecting Eve," and the fatal seduction is complete. For Prospero, of course, the implications of that Edenic obscenity ran on and on, and heaven's primordial outrage at the "contamination of Caucasian by inferior blood" found its "most recent illustration" in the United States. For the future, Prospero concluded, "this great Commonwealth will never achieve its destiny, so long as the negro is allowed to vote, or to exercise any political right or privilege whatever. To subject the Caucasian to the Negro is a higher crime against nature than to place the negro under the ape or the baboon."[117]

In these engagements with Ariel's provocation, the whole matter of Eve's transgression was usually only tangentially, if colorfully, advertised. Yet the identification of the beguiling serpent, or Nachesh, as a black pre-adamite enjoyed numerous supporters among Christian hyper-racialists. D. G. Phillips, a Louisville writer using the pen name "A Minister," for example, told the readers of *Nachesh: What Is It?* that Africans were not of Adam's race and that the tempting serpent was "a negro."[118] In A. Hoyle Lester's *The Pre-Adamites, or Who Tempted Eve?* however, the episode became the leitmotif of what Stokes describes as a "mini-romance" that sexualized the account in novel ways that granted "Eve an agency often denied her in other accounts."[119] Once this former slave owner had laid out his polygenetic reading of the Creation narrative—in which the African, the Malayan, and the Turanian or Chinese families were placed *ad seriatim* in their distinctive geographical provinces before the appearance of the adamite or Caucasian race—Lester moved directly to the scene of the Fall. The "fairest queen that ever graced the courts of the earth" is introduced as making "her *début* on the arena of life in the romantic shades of Eden." But soon she is "wearied with monotony," suffering from boredom and lustful for novelty. Lester portrays her wandering desire to slake "her thirst at gushing fountains where she dreamed no mortal had yet partaken."[120] It is here that she meets a "handsome stranger"—not this time a black pre-adamite but an enticing Mongolian—with whom she made repeated trysts and "lingered long in some solitary shade." The forbidden fruit is tasted: "Eve, poor woman, yielded to the evil machinations of this seductive deceiver. She rose from the mossy couch a wiser but fallen creature, and returned to the presence of her lawful companion disrobed of virtue, that precious jewel, the brightest ornament of her sex."[121] Cain, the "mongrel offspring," was the monstrous issue.[122] With this foundation laid, Lester rehearsed the standard obsessions of white supremacist bio-theology—the abominations of miscegenation and blood mingling, anthropometric measures of Caucasian superiority, a physico-moral inventory of racial groups, denial of the traditional Hamite account, and linguistic polygenism.

The casting of Eve's tempter as a pre-adamite creature of one stripe or another was a critical tool in the weaponry of one virulent strain of white supremacist ethno-theology during the nineteenth century, not least because it directly connected their social horror of bloodline mixing with original sin understood as interracial sexual transgression. The loss of paradise was the original price paid for Eve's illicit liaison; the consequences of contemporary amalgamation policies would be no less devastating for the United States.

Bestializing Variations

The power of pre-adamism, in one form or another, to serve the interests of race hatred and the language of gross racial abuse evidently acquired a considerable following during the second half of the nineteenth century. In many cases the same scientific and theological strains of argument reverberated through the books and pamphlets that partisans produced. Minor variations proliferated, and new themes surfaced from time to time. But the basic thrust of the scheme was to bestialize the African and to provide warrant for a fixation with blood purity. In a sequence of articles printed in the *Eastern Lutheran* and the *Chambersburg Democratic News* and drawn together into a book in 1898, Gottlieb Christopher Henry Hasskarl (1855–1929), a Pennsylvania Lutheran minister, culled a large number of scientific and religious works to confirm that the black race was not adamic and was therefore "inevitably a beast."[123] Hasskarl's articles were so laced with lengthy extracts from writers such as Winchell, McCausland, Ariel, and a host of others that it is difficult to extract his own thoughts from the quagmire of quotation marks. But this, in and of itself, is indicative of just how plentiful were those works dwelling on racial difference that could readily be swept into a white supremacist credo. Most of the usual self-confirmatory "evidence" surfaced in Hasskarl's invective, though he did pause to add his own observation that Africa—"nature's death-trap to heroic Christian missionaries"—disclosed the extent to which the continent was "the dungeon of a specie [sic] of a race of men without souls." This was hardly surprising, of course, for Hasskarl had managed to convince himself that the church's task was only to "'go' Christianizing those of the genus homo that are humanly constituted," that is, only those who were "true descendants of 'the First Adam.'"[124]

Needless to say, B. J. Bolding, pastor of an African Methodist Episcopal Church in Franklin County, Pennsylvania, did not see things the same way and put pen to paper to contest Hasskarl's claims. His own bishop, Alexander Walters, provided an introduction to Bolding's riposte, expressing his pain at the "bitter and malicious things" that had been said by the enemies of his race but noting that Hasskarl had struck "the hardest blow" and had displayed "the most malevolent spirit in such malicious business."[125] Bolding's response was remarkably measured in his reply to Hasskarl's "pet theory," and he patiently defended the Hamitic origins of the African, the scientific case for monogenism, the common structure of all human languages, and the impossibility of determining the skin color of Adam. He contested Hasskarl's diagnosis of African mission activity, too, arguing that, despite "its terrible climate, bad government and petty wars," the continent had actually been remarkably responsive to Christian activity.[126] But underlying all these assertions was Bolding's foundational conviction that all human races were "descendants of the one original human pair, Adam and Eve," and that the "theory of a plurality of species and of origin in the present races of men" was "unphilosophical, and contrary to right reason."[127] The structure of Bolding's entire defense depended on the speciousness of the pre-adamite scheme.

Charles Carroll was, if possible, yet more vicious than Hasskarl in two mammoth works published as the new century dawned. The title of the first, *The Negro a Beast* (1900), carried its message on its sleeve. No less chockful of extended cited extracts, Carroll quoted from figures such as Jean Louis Armand de Quatrefages, Paul Broca, Paul Topinard, Ernst Haeckel, Johann Friedrich Blumenbach, Alexander Winchell, and a library of others, including "Ariel," whose reputation he sought to retrieve from the "unmerited abuse of the negroized clergy." With these resources to hand he turned his guns on the theory of evolution by natural selection. It was only through this atheistic theory of descent that "the Negro obtained his present unnatural position in the family of man."[128] Darwinism's monogenism thus troubled him deeply, and he garnered every scrap of evidence he could find to confound it. Indeed, he attributed its delusions to fashionable notions about racial amalgamation that had in turn fathered atheism, the development theory, and a philosophical naturalism. His own book was a ferocious attack on the black race as "not an offspring of the Adamic family" and thus "not of the human family."[129] On page after page it displayed its author's deep-seated phobia about "mixed blooded tribes," who, like the children of Cain, were "soulless" because "they were of amalgamated flesh."[130] Such—by now standard—pre-adamite fodder was served up to Carroll's fairly wide readership, though his intervention was not without its own curiosities.[131] He suggested, for example, that Cain's murder of Abel was on account of the sexual jealousy aroused in him by the fact that "the beautiful Adamic woman, who . . . would have been the wife of Cain, would now become the wife of his brother."[132] As for the identity of the serpent in the Garden, he explained in the 1902 *The Tempter of Eve,* it was a pre-adamite female—a black woman who had become Eve's maidservant and confidante.[133] And he dangerously flirted with the idea of racial annihilation by reminding his readers that "God very seriously advocated extermination as a possible remedy for the very class of creatures which, in America, Bishop Nelson is pleased to term Negroes."[134]

It was much the same with William H. Campbell's *Anthropology for the People,* which had come out in 1891 and was presented as the work of "Caucasian." Blending, like many before him, scriptural commentary and scientific aspiration, Campbell produced what Paul Harvey has choicely dubbed a "truly miscegenated offspring"—namely, "a mytho-scientific racism that blended folklore, Darwinian science, and biblical exegesis."[135] In fact, Darwinism was not mingled into Campbell's rabid concoction, for he devoted a lengthy chapter to refuting it, largely on account of its assumption of common human descent. Instead, he held to the increasingly obsolete authority of scientific men such as Agassiz, Morton, Nott, and Cartwright to defend polygenism. The work was a restatement of typical pre-adamite supremacist convictions: the inferiority of pre-adamic races, the error of the Hamitic theory of black origin, the restriction of Christian promises to Adam's line, a bio-determinist reading of the black constitution that made it suitable only "for a low and subordinate sphere of life—that of a servant or laborer under the control of a superior," and an insistence that "God intended the Adamic race to be kept from admixture with any inferior blood."[136] As he began to wind up his account, moreover, the motivating forces of his pre-adamite anthropology came sharply into view. Tellingly entitled "Importance of Correct Views of Anthropology," chapter 15 turned to current conditions in the United States, and here the politics of pre-adamism displayed themselves in their true colors. Campbell sought to destabilize his audience by painting a menacing picture of political life in Haiti and Mexico, where the public sphere was corrupted by mixed-blooded "mongrels." To avoid the proliferation of such "mulatto criminals" in the United States, the lessons of pre-adamite anthropology must be learned: "If the polygenic or plurality theory is true, then the relations of races must be very different. In this case the races of mankind are not of the same origin, nor of 'one blood,' and it is plainly God's will that they should be kept separate and distinct. To preserve blood purity is a paramount duty, and the corollary of universal brotherhood is brutalizing and detestable."[137] This was the only way in which "White supremacy" could be secured, and Campbell drove home the point in the final pages of his conclusion: "Correct anthropological views would put an end to beastly miscegenation, not only between Caucasians and negroes, but between Caucasians and the yellow races . . . Blood purity is an absolute and imperative duty, and miscegenation a damning sin against nature and nature's God."[138]

∅ For supremacists haunted by fears of racial amalgamation in the aftermath of black emancipation, pre-adamism delivered a suitably flexible but sufficiently stable set of themes to satisfy their longings. They reread the doctrine of original sin through the spectacles of miscegenation, rejected the traditional resort to the Hamitic origin of Africans, and celebrated the judgment of scientific anthropologists who supported primitive polygenism. Pre-adamism provided just the kind of warrant they needed to convince themselves that their anxieties were scientifically and theologically justified. It was remarkably well suited to the social phobias of the era. Why clergy turned to pre-adamism from the standard Hamitic narrative that, for southerners, had long been sufficient to provide an account of African origins and a justification for slavery lies, I think, in pre-adamism's capacity to serve as a tool to combat interracial mixing in ways that the Hamitic account never could. By identifying distinct adamic and pre-adamic bloodlines, white supremacists could construct a bio-biblical dogma that allowed traditional loyalty to the Bible to draw on a mélange of scientific specialties.

8 ❧ SHADOWS

The Continuing Legacy of
Pre-Adamite Discourse

WHILE PRE-ADAMISM AS A SPECIES OF theological anthropology reached its heyday during the second half of the nineteenth century, it would be mistaken to think that it has since then become entirely extinct. To the contrary. It has continued to thrive in certain distinct ecological niches. To be sure, the idea only had purchase for those among whom the notion of a historic Adam retained some significance, religiously or scientifically. But this very fact points to a profound irony. Conceived in heterodoxy and fostered in profanity, the pre-adamites have come to reside among religious conservatives and fundamentalists. And yet the theory's versatility to perform different functions has remained a characteristic feature of the scheme. For some it has continued to be employed within an antievolutionary framework; for others, both Protestant and Catholic, it has enabled rapprochement with Darwinian biology and allowed for an evolutionary rereading of the Mosaic narrative; for still others its virulent racial potential remains attractive, and it has been recast as a pillar in the edifice of certain branches of contemporary white nationalism. In what follows we chart something of the way pre-adamism's shadow thus continues to fall on questions of human origins up to the present day.

Pre-Adamism and Antievolutionism

During the early decades of the twentieth century supporters of pre-adamism were to be found among those suspicious of evolution's metaphysics and concerned about the direction Darwinism seemed to be heading. Indeed, there is evidence of its survival among some of those considered to have played a critical role in the genesis of modern fundamentalism. One of the most remarkable, albeit terse, expressions of support for the theory in its traditionally nonevolutionary guise among the fathers of this party is to be found in the writings of Reuben A. Torrey (1856–1928). Torrey was, in many respects, a fundamentalist par excellence.[1] He had served his time as one of D. L. Moody's foremost chiefs of staff and took on editorial management of the last two volumes of *The Fundamentals*—a set of theological statements resisting modernizing impulses in theology. Whatever the revivalist overtones of the movement he found himself spearheading, Torrey's intellectual roots in the cultural ethos of the New England tradition committed him to a firm belief in the mutual reinforcement of science and scripture. He accordingly applauded James Dana's concordist reading of Genesis, an understandable enthusiasm, perhaps, given the fact that he had studied under Dana at Yale. And while he did not find it possible to negotiate a similar accommodation of scripture to Darwinian biology, though he acknowledged it was possible to be an evolutionist and still accept biblical infallibility,[2] he welcomed the pre-adamite as a peacemaker between biblical religion and archaeological science. Predictably, he claimed to have found the pre-adamite *within* the pages of Genesis, but he was delighted nonetheless that his discovery should match scientific findings so well.

Certainly, Torrey's observations were, as I have indicated, exceedingly brief. Nevertheless, he plainly stated that "all verses after the first verse of Genesis I seem rather to refer to a refitting of the world that had been created, and had afterwards been plunged into chaos by the sin of some pre-Adamic race, to be the abode of the present race that inhabits it, the Adamic race." And again: "It should be said further that it may be that these ancient civilizations which are being discovered in the vicinity of Nineveh and elsewhere may be the remains of the pre-Adamic race already mentioned . . . No one need have the least fear of any discoveries that the archeologists may make; for if it should be found that there were early civilizations thousands of years before Christ, it would not come into conflict whatever with what the Bible really teaches about the antiquity of man, the Adamic race."[3]

A much more sustained defense of anti-Darwinian pre-adamism in the early twentieth century is to be found in the writings of Sir Ambrose Fleming (1849–1945), fellow of the Royal Society, president of the Victoria Institute, first president of the Evolution Protest Movement, and for forty-one years professor of electrical technology at University College London[4] (fig.

FIG. 25. Sir Ambrose Fleming.

25). During a distinguished career he made pioneering contributions to the development of the telephone, radio, and television and was nominated by several leading scientists, including Guglielmo Marconi, for a Nobel Prize.[5] Now well advanced in years, he took on the presidency of the Victoria Institute in 1927 largely on account of his sense that higher criticism, a secularized Darwinism, and scientific materialism were eroding public confidence in the Bible. As the society's president, Fleming was head of a movement that had committed itself "to investigate fully and impartially the most important questions of philosophy and science but more especially those that bear upon the great truths revealed in holy scripture, with a view to defending these truths against the oppositions of science, falsely so-called."[6] His connection with the Evolution Protest Movement was a public demonstration of his stance on Darwinism, as was his book-length tract for the times, *Evolution or Creation?* which first appeared in 1933. All these efforts certainly associated him with antievolutionary sentiments, though in fact his real objection was to a naturalistic Darwinism that removed all metaphysical possibilities of divine design from nature.[7] His critique very largely revolved around his unease about both materialism and monism as adequate accounts of the nature of matter and energy, preferring instead what he labeled "Theistic Idealism, namely, that the Universe is not a Thing; it is a Thought, and Thought implies a Thinker. In other words, it is essentially an Idea or Thought, in a Universal Divine Omnipresent Mind."[8] Indeed, finding consolation in Darwin's reference to the "Creator" in the final pages of *The Origin of Species,* Fleming sought to distance Darwin's stance from that of his modern neo-Darwinian followers, whose outlook he found much less tolerant of the idea of Creation, and sided with such critics as the German comparative anatomist Albert Fleischmann and the Russian biogeographer Lev Simonovich Berg.

With his twin commitments to science and Christianity, it was not surprising that Fleming would sooner or later turn to the challenges to Christian theism arising from prehistoric archaeology and anthropology. The claim by Bishop Ernest W. Barnes in 1927 that there was scientific consensus on human evolution from an apelike stock, Fleming found scientifically objectionable and in "flagrant contradiction with the teaching in the Inspired Scriptures."[9] The boundaries of his own thinking were clear: the Genesis record required special acts of creation at three points: "first, in connection with the creation of *matter;* secondly, at the creation of animal *life,* and thirdly, at the creation of Man."[10] On Monday, January 14, 1935, he turned to the theme of human origins for his presidential address to the Victoria Institute.[11] The *Daily Telegraph* gave prominent coverage to the lecture in its columns the following day, and the equivocations over evolution that Sir Ambrose had given voice to provoked comments from various quarters. According to its report, Fleming had "delivered a fierce and heavy assault on the centre of modern scientific doctrine, the principle of the evolution of the universe, of the animals and man."[12] In its Saturday edition he replied to his critics, and here he spelled out his own strategy for reconciliation, a strategy subsequently elaborated in detail in *The Origin of Mankind,* which appeared later that year. It was, fundamentally, the pre-adamite theory of Dominick McCausland and was put forward as an alternative to "the Evolutionary theory of the origin of the Human Race over a vast span of past time, from animal ancestors by Darwinian natural selection."[13] Of course, since McCausland's time new archaeological and anthropological findings had become available. In 1891 Eugene Dubois claimed to have discovered fragments of *Pithecanthropus erectus* near Trinil in Java, and these remains were widely interpreted as the first hard evidence of a primitive hominid. In 1907 Mauer Sands had unearthed part of a jawbone with teeth of human type near Heidelberg, and thus was born *Homo heidelbergensis.* And most famous of all, Charles Dawson's 1912 discovery of human remains in a gravel bed at Piltdown in Sussex, which were subsequently pieced together by Smith Woodward of the British

Museum and Teilhard de Chardin, though later shown in 1953 to have been fraudulent, were enthusiastically seized upon by evolutionists at the time. Fleming was suspicious of all three and either argued, in typically creationist fashion, that illustrators had let their imaginations run wild by constructing both facial and skeletal profiles from the scantiest of fragments or generally questioned the extravagant claims being made on the basis of negligible data.

His approach to the evidence for "Neanderthal man" and the Cro-Magnons, however, was very different. He fully acknowledged that the Neanderthal skullcap and skeletal fragments excavated at Düsseldorf in 1856, in 1887 near Spy in Belgium, and in the years up to 1914 in Krapina, Croatia, and southern France were compelling evidence. But he went on to suggest that this Neanderthal race was replaced in Europe by the superior Cro-Magnons, complete skeletons of whom had been found in the Pyrenees and the Dordogne. The latter evidently possessed considerable ingenuity, artistic ability, and handicraft skills, as revealed in their bone and flint instruments and cave drawings. But while some archaeologists, such as Gustav Schwalbe, argued that the Neanderthal evidence supported a human evolutionary conception, Fleming resisted such an interpretation. As he put it, "The upshot of it all is that we cannot arrange all the fossil remains of supposed 'man' in a lineal series gradually advancing in type or form from that of any anthropoid ape, or other mammal, up to the modern and now existing types of true man."[14]

It was at this juncture that the pre-adamites came to his rescue. Quite simply, Fleming suspected that the Neanderthals were a pre-adamic stock, whereas the specially created Cro-Magnons were the adamic antediluvians of the biblical narrative. That he should turn explicitly to McCausland's polygenetic pre-adamism was thus doubly understandable, for he was just as concerned as McCausland to preserve a relatively literalistic hermeneutics of scripture. By this means Fleming could at once take seriously the patterns of early migration, crack those hoary exegetical chestnuts about Cain's wife and city, and yet again press the theory into the service of racial ideology.

Fleming's racial proclivities surfaced as a consequence of the way in which he built a more specifically "spiritual" component into his conception of Adam, alongside accompanying physiological and psychological traits. Adam's predecessors were certainly human in the sense that they possessed "powers . . . which separated them from animals before the arrival of the Adamic man at the approximate date 5500 B.C." But the "Creation of the Adamic man," he explained, "was the appearance on earth of a being more eminently endowed with psychical faculties of initiative, authority, and powers of intercommunication than before, and with special powers of intercourse with the Creator."[15] In Fleming's mind what confirmed his polygenetic pre-adamism was his belief that humankind could be divided into several distinct *species:* Caucasian, Mongoloid, and Negroid. The distinctive spiritual component he presumed to be the possession of the adamite line allowed him to consider that the Caucasians possessed some spiritual superiority over other racial stocks. It was thus bad enough that interracial miscegenation produced physically degenerate offspring; that it produced spiritual mulattoes was too great to bear. From the beginning, it seemed, ethnic intermixture was spiritual bastardy too. "Intermarriages between members of these great racial divisions are possible," Fleming averred, "but the progeny are usually feeble, not long-lived, and of poor psychical quality."[16] All this, moreover, had implications for his understanding of the Mosaic narrative itself. In his own words: "If we then look at it from the point of view of the Biblical account of the Origin of Man, we see there recorded that the failure of the Adamic race to keep its privileged position originally given to it, and the inter-marriages of the descendants of the Adamite with prior created human beings brought about a state of moral degradation and violence tending to destroy the Divine purposes for that Adamic race. Hence came the Divine decision to destroy by a flood the bastard races, and begin again with a selected and God-obeying representative of the pure Adamic race." And again: "What it [the Noachian flood] did imply was a wiping out of all the bastard cross-breed race between Adamic and pre-Adamic man, and the beginning of the Caucasian race from the sons of Noah, Japheth, and Shem. It is allowable to presume that sufficient of the Mongolian, Negro, and other human species survived to continue the population of Eastern Asia and middle Africa."[17]

Fleming's resort to the pre-adamite theory was not, needless to say, universally welcomed. It certainly rubbed the anatomist and anthropologist Arthur Keith (1866–1955) the wrong way. In point of fact it was largely because of Fleming's proposals that Keith put pen to paper to produce, in March 1935, a tract for the times entitled *Darwinism and Its Critics,* the first part of which had already appeared in the *Literary Guide.* This was not the first time that Fleming had fallen into Keith's line of fire. In 1928, while enlarging on the implications of Darwinism, he challenged Fleming's assumption that "the modern Darwinist is impelled by an anti-religious motive."[18] Deferential though he was to Fleming's scientific accomplishments, he found his resuscitation of Paleyite teleology entirely unconvincing, preferring to locate design within the inherent properties of living matter, not in an external divine agent. Now, as he turned his big guns on those anti-Darwinians who still held to the "impossible theory" of special creation, he picked out Fleming's pre-adamism as a target for special bombardment. It was scientifically dubious. How did Sir Ambrose know that the Neanderthals had no "spiritual nature" while the Cro-Magnons had? "Such acumen," Keith remarked, "is beyond the power of all anthropologists of my acquaintance." But more than that, it was biblically suspect:

> Sir Ambrose ventures the opinion "that there have been pre-Adamitic races of beings, whom I call hominoids in my address, but were not 'man' in the psychical and spiritual powers or possibilities in the Biblical sense of the word." He even ventures the opinion "that between true man and anthropoid apes there may have been some species of hominoids created." Is not Sir Ambrose taking an unwarranted liberty with the inspired word by introducing acts of creation and types of humanity of which there is no mention in the Mosaic record? Or would it not be more in keeping with scientific method to give up the theory of special creation, seeing that the truth has to be prosecuted to gain a verdict in its favour?[19]

Fleming's failure to convert Keith to his way of thinking hardly came as a surprise. The response from William Dickson Lang FRS, keeper of Geology at the British Museum, who took up Fleming's address in a review for *Nature* in 1935, was also critical, though without as much venom. Here Lang expressed his concern at the "less tolerant view" of evolution that Fleming had expressed, compared with those of Bishop Charles Gore, Christian Socialist and editor of *Lux Mundi,* who saw evolution as the method of divine action in the world. For him Fleming had too easily based his anti-Darwinian case on metaphysical grounds of the sort that actually lay beyond the remit of the practicing scientist. In sum Lang judged that "Sir Ambrose and his opponents speak different languages, and on terms on which agreement is impossible." For himself Lang could not see why a Christian could not "believe, with the early church, that the story in *Genesis* of man's creation is focused upon the spiritual truth that God created man, and is not to be regarded as a scientific account of the process."[20] Fleming, perhaps, had more reason to be disappointed at the response he received from fellow evangelical Christians. One reviewer, writing in 1936, for example, found it "refreshing to find a scientist of Fleming's standing arguing from the facts of Christianity for not only the possibility but the probability that Man's origin is due to a special creative act of God." But there the praise ended. Fleming's suggestion "not only that Adam was not the first man but that he was the ancestor only of the Caucasians" ran foul of the fundamental doctrine of the unity of the human race.[21] Again, in 1939, Louis Berkhof, an American Calvinist theologian of Dutch birth, felt the need to reaffirm the standard adamic narrative and to present the views of La Peyrère, Agassiz, Winchell, and particularly of Fleming as being utterly devoid of all credibility.[22]

The potential of pre-adamism to support antievolutionary accounts of human origins has surfaced yet more recently. In 1996 Dick Fisher, a Vietnam air force veteran with interests in science and religion, sought to retrieve the pre-adamite theories of La Peyrère and McCausland in his proposed solution to the origins question. Fisher judges that "theistic evolution as a bridge between science and the Bible drops us into a quagmire of interpretational difficulties" and is therefore "not a viable answer"; he opts for pre-adamism as a genuine alternative.[23] Holding to the special creation of Adam at around 5000 B.C. yet acknowledging the archaeological evidence of the existence of prehistoric hominids—precursors to *Homo sapiens*—he explicitly resurrects the views of La Peyrère, Lane, and McCausland to explain these artifacts. In this way he considers it possible to hold to "an enlightened, *literal,* interpretation of the first eleven chapters of Genesis," rather than succumbing to a "liberal, relaxed view toward the Old Testament."[24] Adam was a special creation, but prior to his existence the earth was inhabited by Neanderthal man and Cro-Magnon man. As for his contemporaries, Fisher considers that they remained outside the covenant enjoyed by the "Adamic bloodline" and "the ritual of animal sacrifice."[25] For him the critically central point is that Adam and Eve are to be understood not as "the first mammalian, bipedal couple, but the first covenant couple," standing at the inauguration of the Semitic "covenant bloodline," a family line into which all other peoples can be adopted.[26]

Evolutionary Theory and the Revival of Pre-Adamism

Whereas Fleming's pre-adamites occupied anti-Darwinian territory, during the second half of the twentieth century they began to appear as a feature of conservative theological writings seeking some accommodation between evolution theory and biblical theology. The general idea that unifies these endeavors here is that the *imago Dei* is given a theological, rather than psychical or physiological, meaning, thereby leaving room for the evolution of *Homo sapiens* prior to its transformation into the biblical Adam.

Because it was the first formal document to take up the question of human origins, the appearance of the encyclical *Humani Generis* of 1950 constitutes an important moment in the modern history of pronouncements from Catholic officialdom on the topic. Here Pope Pius XII opened the door to a free discussion of the question of the evolution of the human body from animal forebears. The direct divine creation of the soul was reasserted, but permission was granted to consider the possibility that the human physical form had been subject to evolutionary transformation. The canon had been prepared in view of the denial of monogenism by some ethnologists, particularly in France, and by some theologians who were flirting with the idea that Adam was not a single individual but, instead, a number of progenitors from whom the human race was descended.[27] It occasioned widespread comment in the Catholic media.[28] The tenor of the whole document was to affirm the unity of humankind and the monogenist account on the grounds of Catholic doctrine on original sin. But the very fact that—at least in the eyes of some commentators—*Humani Generis* skirted the issue of pre-adamism, which could be given a monogenist rendering, left room for considerable debate on the subject, particularly among French and Italian Catholics.[29]

Whether or not monogenism was specifically asserted, rather than just assumed, was a matter of considerable debate, but most commentators agreed that polygenism was condemned. Just what was to be made of this rejection of polygenism, however, was not so clear. Augustin Bea, for example, believed that the encyclical did not address the scientific side of the issue at all, whatever it affirmed for theology. He felt that whether there were forms of polygenism that could be made consonant with Church teaching was a question that had actually been shelved.[30] And other observers pointed out that the way in which polygenism was outlawed still left room for pre-adamism; still others felt that even if this were so, it mattered little, for pre-adamism was merely an antiquated seventeenth-century theory.[31] Yet time and again, the ritual rejection of pre-adamism in one mode was only prefatory to its reassertion in another. Thus, among those individuals partial to evolutionary science, its shadow was invariably cast over efforts to work out a theologically acceptable means of accepting that truly human types had developed from subhuman pre-adamic stock.

In the aftermath of the encyclical John O'Brien, at the University of Notre Dame, explored the whole question of religion and evolution, returning to the views of St. George Jackson Mivart, Henry de Dorlodot, and John Zahm, emboldened by the possibilities opened up by *Humani Generis*'s leaving "the development of the human body from other living creatures an open question, to be decided on the basis of the evidence."[32] The second edition of the volume carried an appendix by the Jesuit J. Franklin Ewing, bringing readers up to date on the question of human evolution. In one brief paragraph he turned to the pre-adamite question: "The possibility that there were true men before Adam and Eve, men whose line became extinct (in other words, Pre-adamites) is allowable. But the Pope does not mention them. In view of the fact that we have no evidence at all concerning their existence, we shall not mention them."[33] Evidently, Ewing understood the theory as referring exclusively to the possible existence of fully human, but now extinct, pre-adamic races. But an afterword dwelling on evolution's compatibility with Catholic faith by Achille Lienart, recently appointed cardinal, formerly archbishop of Lille, and a leading liberal at the Second Vatican Council, advertised the strategy that the human race had developed from pre-adamic hominids: "The animal origin of the human body is not an argument against the dogma of original justice and the preternatural gifts of the first man. If primitive humanity appears to have been physically very close to the animal world and but slowly to have acquired the perfection of its human features, the reason is that it began to be human only at the moment that it received a spiritual soul."[34]

A sophisticated treatment of just this possibility from the perspective of Catholic philosophical theology is to be found in the writings of the German theologian Karl Rahner (1904–84), who came under the censure of the Vatican authorities for his efforts.[35] In Rahner's mind pre-adamism, in the sense of fully human individuals existing before Adam, was so entwined with polygenism that he felt compelled to reject it. But he made it clear from a detailed scrutiny of the Church's official pronouncements that, in his judgment, polygenism could not explicitly be ruled out. What was crucial here, both to Rahner and to the spirit of Catholic officialdom, was a doctrine of original sin umbilically tied to "natural, biological generation" from the first ancestor.[36] By this account any acceptable polygenism would of necessity implicate its advocates in a theology of multiple original falls from grace. Thus, for Rahner to determine a viable stance on the *scientific* question of human origin was inextricably bound up with the theological conception of original sin. At the same time, it also depended on whether the human constitution was to be conceived of dualistically (as body and soul) or as fundamentally indivisible. Following the current drift of the theological tide, both Protestant and Catholic, Rahner expressed his reservations about crudely dualistic models that smacked of Greek infiltration.

With these twin commitments to original sin and to human psychosomatic unity, Rahner insisted that "in regard to the animal kingdom man is a

ADAM'S ANCESTORS

metaphysically new, essentially diverse species, not merely in the biological sense of the phenotype, not only in name, but in that ultimate root of his psychosomatic nature." And yet such "creationist" talk was to him entirely compatible with a belief that "biological development in the animal kingdom reached so advanced a stage of development in a number of exemplars that the transcendent miracle of 'becoming man' could take place in them."[37] What precisely Rahner meant by this assertion is not entirely clear, for he certainly did not envisage any naive plopping of a soul into an animal body. Nonetheless, he had in mind the "hominization" of pre-adamic creatures. What *is* significant is that when he came to teasing out just how the "creation of the spiritual soul" was accomplished, he resorted again to those embryological analogies already so serviceable to Catholic evolutionists:

> If the mediaeval doctrine is presupposed, and it is coming to the fore again, that the spiritual soul only comes into existence at a later stage in the growth of the embryo, several pre-human stages will lie between the fertilized ovum and the organism animated by a spiritual soul. These do not yet, therefore, stand in immediate and proximate potency to actuation by the spiritual soul . . . On that basis it is quite possible to say that an ontogeny viewed in that way corresponds to human phylogeny as present-day evolutionary theory sees it. In both cases a not yet human biological organism develops towards a condition in which the coming into existence of a spiritual soul has its sufficient biological substratum.[38]

With this is mind it is not at all surprising that Rahner should approvingly have cited Philipp Dessauer's comments that "shortly before the appearance of man, simian forms very close to man" had emerged. Having manual freedom, upright posture, and human teeth, they represented a "group of prehuman, animal forms" whose purpose was "to prepare the way for man."[39]

Even more recently the question of the relationship between polygenism and Catholicism was the subject of analysis by Augustine Kasujja (1946–), later archbishop of Uganda. Reviewing in some detail La Peyrère's original pre-adamite theory and linking it with the polygenism of Voltaire, Calhoun, Nott, and Gliddon, Kasujja moved on to discuss the significance of the discovery of fossil humans for the Catholic understanding of the origin of the human species. To advance his case, he drew on East African archaeological discoveries, in particular *Homo habilis,* to focus the question of whether monocentrism or polycentrism provided the best scientific explanation of human origins. What if, he pondered, advocates of polycentrism were to be vindicated? Would the magisterium and scripture allow for polygenism? For him the key question revolved around the transmission of original sin, and he argued that if it were to be conceived of in terms of solidarity rather than genetic descent, then the door would be open to a reconsideration of polygenism as a viable Catholic option.[40] Clearly, the Catholic tradition, despite appearances to the contrary, has not yet given its last word on polygenetic accounts of human evolution.

Similar maneuvers are also detectable within certain strands of mainstream Protestant theology, despite surface appearances to the contrary. Take, for example, the Swiss theologian Emil Brunner (1889–1966), for whom the whole enterprise of harmonizing the modern worldview of science with that of biblical cosmogony is a profoundly misguided project—and a fortiori the attempt to integrate biblical anthropology with the science of prehistoric archeology a myopic undertaking. "To equate the Neanderthal Man," he writes, "with 'Adam in paradise'—an attempt which is being made to-day, even by European theological university professors—merely produces an impossible bastard conception, composed of the most heterogeneous and incongruous views." And yet for all his denials Brunner remained convinced of a fundamental distinction between what he called *animalitas* and *humanitas:* "*Animalitas* here denotes a form of existence which achieves no personal acts at all, but merely acts of self-preservation and the preservation of the species; humanitas means, however, that in which—even if at first in a very rudimentary way—something personal, something which transcends *animalitas,* is achieved." So, when Brunner asserts that "even if man is descended from the animal world, as *humanus* he is something wholly new, not only in contrast to the ape, but over against the whole of Nature," we see the perpetuation of the essential threads of evolutionary pre-adamism woven into the fabric of neo-orthodox theology.[41]

Amongst more conservative Protestant thinkers concerned with taking seriously developments in evolutionary biology, this self-same strategy is also noticeable, though here the explicit language of pre-adamism and the critical importance of Adam as a historic individual are more conspicuously in view. In a standard conservative commentary on the book of Genesis that appeared in 1967, for example, theologian Derek Kidner explicitly returned to A. Rendle Short's pre-adamite theory to make peace with evolution. He suggested that if, as in all likelihood, "God initially shaped man by a process of evolution, it would follow that a considerable stock of near-humans preceded the first true man, and it would be arbitrary to picture these as mindless brutes." On the contrary, by disengaging the idea of the image of God from the notion of "rationality," Kidner was able to conceive of culturally sophisticated pre-ad-

ADAM'S ANCESTORS

amites who were "of comparable intelligence" to Adam but still lacking the spiritual *imago Dei*. Indeed, to avoid assigning "a stupendous antiquity for Adam," he countenanced the possibility "of the continued existence of 'pre-Adamites' alongside 'Adamites.'" More than that, he went so far as to suggest that after the "creation" of Adam, "God may now have conferred His image on Adam's collaterals, to bring them into the same realm of being." And by conceiving of Adam as the "federal" representative rather than genetic head of humanity, the knotty problem of original sin could be disentangled.[42]

Others have adopted and popularized this basic scheme. The evangelical statesman John Stott (1921), for example, writing for a youthful audience in 1972, observed that "my acceptance of Adam and Eve as historical is not incompatible with my belief that several forms of pre-Adamic 'hominid' may have existed for thousands of years previously. These hominids began to advance culturally. They made their cave drawings and buried their dead. It is conceivable that God created Adam out of one of them."[43] In a similar vein the British geneticist R. J. Berry (1934–) speaks of pre-adamic hominids with very substantial cultural acquisitions and intellectual prowess. To him the image of God is again given a spiritual meaning, thereby leaving room for the evolution of *Homo sapiens* prior to the race's transformation into the biblical Adam.[44] The Baptist theologian Bernard Ramm (1916–92) likewise kept the pre-adamite theory before the American evangelical mind as a means of retaining anthropology within the confines of theological acceptability. Ramm's *Christian View of Science and Scripture* came to achieve almost legendary status among those orthodox Protestants who wanted to keep an open mind about the scientific enterprise. And not without good cause. His survey of the relevant literature was remarkable; on the pre-adamite option alone he traced its origins back to La Peyrère, referenced its manifestation among fundamentalists such as Pember and Torrey, and spoke of its interpretive versatility. He himself felt that the theory had "vexing problems," but his review clearly substantiated its viability as an evangelical option.[45]

To these theological conservatives pre-adamism in its evolutionary guise has done much to keep alive the marriage of science and religion. For others it has played an even greater role, substantially shaping exegesis of the Genesis account. This is so, for example, in the case of E. K. Victor Pearce, whose book *Who Was Adam?* was specifically written to substantiate pre-adamic claims from a detailed reading of the text as much as from the field evidence of prehistoric anthropology and archaeology. With a clear commitment to biblical inerrancy, Pearce set out to square the claims of science with scripture in a manner worthy of the grand concordist schemes of the nineteenth-century harmonizing geologists Hugh Miller, Arnold Henry Guyot, and Benjamin Silliman. Pearce's claim is simply that the early chapters of Genesis house two distinct creation stories, the former focusing on the creation of the human species, the latter narrating the life story of an individual, Adam. As evidence for the existence of the former pre-adamites, Pearce painstakingly correlated specific biblical phrases with particular archaeological artifacts from prehistory, thereby showing how pre-adamism could be incorporated into the most literal reading of the Genesis chronicle.[46]

None of the foregoing critique should be taken to imply that evolutionary pre-adamism has yet won general approval from conservative Catholics or Protestants. Latter-day scientific creationists, for instance, who brook no tampering with the special creation of Adam from the dust of the ground, make the pre-adamite as unwelcome as those liberals for whom Adam has nothing but mythological significance. But for those who retain an investment in a first fully human being, Adam, yet who find persuasive the evidence for human evolution, the pre-adamites remain attractive. There are, however, other conservative groups, much farther to the political right, who have resurrected pre-adamism for altogether different, and much more sinister and abusive, purposes.

Seed Lines and Supremacy

The racialist pre-adamism of the late nineteenth century dramatically resurfaced, though with some modifications, in the mid-twentieth century among groups associated with what is known as the Christian Identity, or Kingdom Identity, movement. Rooted to one degree or another in nineteenth-century British Israelism, what Colin Kidd labels this "religion of race hatred" cultivates pre-adamite polygenism in its vigorous pursuit of racial purity.[47] For its part British Israelism was, and remains, a kind of biblical ethnology that posited that the Anglo-Saxons of Britain and North America were, in a fundamental respect, of Israelite origin, traceable to Ephraim, one of the ancient tribes of Israel. Fascinated with the identity of the so-called Ten Lost Tribes, a number of writers, since the late eighteenth century, have speculated on their destiny, some arguing that they might account for the native peoples of the New World.[48] The idea that the English nation was of Jewish origin gripped others as the eighteenth century wore on and flourished in works such as John Wilson's *Lectures on Our Israelitish Origin,* which first appeared in 1840 and then in many editions, and Edward Hine's *Forty-seven Identifications of the British Nation with the Lost Ten Tribes of Israel* (1874).[49]

The Christian Identity movement builds, to one degree or another, on this ethnocentric theology but prosecutes its case with explicit reliance on a version of pre-adamism that posits two distinct "seed lines"—one adamic the other pre-adamic—as the basis of its theology and cultural politics.[50] The story, for all its Byzantine exegetical maneuverings, is simple. The white races are adamic, the others pre-adamic. Adamic races are superior and possessed of some intrinsic spiritual and cultural capacities denied to others. It is the duty of the adamite people to preserve their seed line from being polluted through interracial mixing. In pursuit of this biblicist anthropology partisans have resorted to many of the claims made by figures such as Alexander Winchell, Charles Carroll, and Buckner H. Payne (Ariel)—such as the idea that Cain was the son of Eve's satanic union with a pre-adamite and that Cain perpetuated that corrupt bloodline by his own racial intermarriage. Significant, too, is the insistence that while adamic humans have a tripartite constitution—body, soul, and spirit—non-adamic races lack the latter component and are therefore spiritually inferior.

Something of the tactics and ethos of the sect may be gleaned from a brief inspection of some of its pamphlet literature. Its lineage in the ethno-theology of the turn of the century was perhaps nowhere more dramatically revealed than in the 1967 booklet *In the Image of God,* published by Destiny Publishers. One direct link was the writer's assertion that "modern Israel is found in the Anglo-Saxon race."[51] Another was the connection to the theological anthropology of turn-of-the-century white supremacists. In fact, this entire work was intended to re-present the findings of Carroll's 1900 *"The Negro a Beast"* or *"In the Image of God"* and to provide "the scientific evidence sustaining the argument concerning the true origin of the Negro as apart from the Adamic race."[52] Like its inspiration, it was a virulent attack on amalgamation. But now, at a time when Dr. Martin Luther King Jr. was championing civil rights, it became a tool in support of the politics of racial segregation. The introduction encapsulated the booklet's phobia about hybridity. The Genesis account revealed how "divinely-set barriers were crossed," readers were told, and how this "brought about a chain of biological and physiological developments upon the earth leading to dire physical and spiritual eventualities." Without missing a beat, the piece goes on: "One consequence today is that Negroes aspire to wrest a place for themselves in the white man's domain." Black aspiration galled the writer, and to justify the pressing need for segregation, readers needed to learn about "the true origin of races."[53] The remainder was a rehash of Carroll-style racist abhorrence in order to find anthropo-theological warrant for a separate pre-adamite origin for the black races, an attack on black humanity, a segregationist agenda, a preoccupation with dual seed lines, and a rejection of evolution because through it "the Negro has obtained his present unjustifiable, unnatural position 'in the family of man.'"[54]

One implication of the ethnohistorical scenario concocted by Christian Identity supremacists is the restriction of the Noachian deluge to a local region in the ancient Near East in order to explain the global distribution of pre-adamic races in the few thousand years since the Flood. Wesley Swift (1913–70), minister, one-time member of the Ku Klux Klan, and perhaps the key figure in the transmutation of British Israelism—or, better, Anglo-Israelism—into the Identity movement, took up this subject in a pamphlet version of a sermon delivered to his California congregation. In large part this effort stemmed from his passion to restrict the scope of biblical religion to the adamic race, to put the world right on the racial truths of "scriptures and of anthropology and archaeology," and to explain the continued existence of pre-adamite races to the present day.[55] In the published version of another sermon he told his audience that the tree of knowledge in the Garden of Eden was "a racial tree." "You can't touch that one," he went on. "You are not to mongrelize. You are to maintain a holy seed."[56]

These motifs characterize the entire movement's pathological fixation with breeding. Bertrand Comparet (1901–83), staged as a San Diego attorney, began one of his typical sermons with the assertion that "the Bible is the history of OUR people—the White people of the race that loosely we group under the term Anglo-Saxon," and within a few breaths he was railing against the "mongrelization" of "purebred Adamites." Discrimination against nonwhite races, his listeners were told, was entirely justified; after all, God "discriminated in favor of Adam and against all the pre-Adamic people."[57] Charles Lee Magne's booklet *The Negro and the World Crisis* continues the same offensively tedious refrain, presenting yet again the Christian Identity movement's standard vilification of black origins, anthropology, culture, and history. The purpose was to inject into his readers a sense of crisis that a day of racial reckoning was imminent, when adamic whites would have to relocate to ethnically pure geographical zones, prepare for racial war, and take steps to defend their homes "in the coming day of Red Revolution."[58] Everett Ramsey, another Identity Baptist pastor from Missouri, writing around 1995, began his account of *Racial Difference* with a repudiation of "race mixing" as "the sin for which God has & will kill." The defeat of the South more than a century earlier echoed down through the generations, for Ramsey introduced his harangue by claiming that before the Civil War, "our land was the land of Christian Adamic Israel," but changed immigration laws and "the false state gospel of 'equality of all people'" were now prostituting the nation. Again, dual seed lines were all important. "If . . . all people came from Adam," he presently asserted, then—horror of horrors—there would be no justification "to forbid interracial marriage, interracial reproduction and rich whites should have their assets stolen by the government to make the lives of all people equal; and, we should do all we can to get all peoples redeemed and into our churches."[59] The concluding pages seek to stir up readers by attributing various nasty diseases to race mixing, by appealing to followers to confront head-on all efforts at multiculturalism, and by attacking blood transfusions and organ transplants. Under his previous name, Everett Sileven, Ramsey is reported to have been jailed for contempt of court while serving as a minister in Nebraska for running an unapproved faith school in the state.

And so it continues. Pamphlets privately printed as recently as 2001 carried titles such as *Not of One Blood* and *False Biblical Teachings on the Origins of the Races and Interracial Marriages*.[60] Even with a smattering of genetics and blood chemistry thrown into the mix, they do little more than demonstrate the Identity movement supremacists' continued appeal to pre-adamism to satisfy its morbid lust for material that justifies talk of seed lines, vulgar excoriations of nonwhite peoples, and bilious attacks on interracial marriage.[61] And the flood continues with offerings such as *Biblical Anthropology: The Doctrine of Adam Man* (2002) and *Species of Men: A Polygenetic Hypothesis* (1999), whose titles amply convey their sentiments.[62]

The continuities with earlier versions of theological supremacism are obvious enough. But there are significant departures too. Chief among them is the endemic anti-Semitism that characterizes some versions of Christian Identity political theology. Grounded in the British Israelite ambivalence about the Jews, at once claiming affinity with Old Testament Judaism yet despising modern Jewry, advocates of Seedline Identity cast the Jews as "demon-seed."[63] This castigation of the Jews as "Satan's spawn" has been championed by Richard Butler (1918–2004), a student of Wesley Swift and former Lockheed aerospace engineer, who, after moving to Idaho in the 1970s, developed a political wing of the movement known as Aryan Nations, which catapulted him into public prominence. With violent language and the rhetoric of holy war, Butler emerged as the leading voice of neo-Nazi white supremacists during the mid-1980s, sustaining his aggressive anti-Semitism with a theology positing the Jews as non-adamic children of Satan.[64] As he expressed it in the movement's creedal formula: "The Bible is the family history of the White Race, the children of Yahweh placed here through the seedline of Adam . . . We believe there is a battle being fought this day between the children of darkness (today known as Jews) and the children of light (God), the Aryan race, the true Israel of the Bible."[65] This view represents a significant departure from earlier pre-adamist racism. As Leonard Rogoff has shown, American School anthropologists and even inflammatory writers such as Payne incorporated the Jews within the white race.[66] As for earlier pre-adamite writings on African inferiority, McCausland and others made it crystal clear than the black races fell within the arc of redemption and that there was no requirement "for his reception into the kingdom of God, that the blood of Adam should circulate in his veins."[67]

℘ The pre-adamite theory, in one incarnation or another, has continued to cast its shadow over the twentieth and into the twenty-first century. That it surfaces exclusively within the ranks of conservative religious communities reflects the fact that it is only for such groups that the idea of an historic Adam makes sense. But its versatility is remarkable. Both creationist and evolutionist religious believers have continued to find in the idea of pre-adamites a resource that enables them to retain their scientific and theological convictions. At the same time, the theory's capacity to be given a racial twist has made it extraordinarily attractive to radical supremacists, who have been able to exploit the idea of two lines, seed lines, of human descent for the grossest forms of racial abuse.

Just what the future of the pre-adamite will be remains to be seen. But there certainly still seems to be plenty of scope for keeping open the monogenist-polygenist debate. First there are the scientific controversies revolving around so-called Mitochondrial Eve—the name given by scientists to the East African female who is posited as the most recent common matrilineal ancestor of modern humans. Then the unearthing of the remains of *Homo floresiensis*—the hobbit-like creature whose fossil skeleton was excavated on the Indonesian island of Flores in 2003 and thought to be contemporaneous with modern humans—has been used to support the idea that humans may have emerged in different localities. The "Out of Africa" and "Multiregional" hypotheses, as these are called, continue to retain committed advocates. No doubt both adamites and pre-adamites stand ready to enter the fray if invited again onto the stage of controversy.

9 ✌ DIMENSIONS
Concluding Reflections

THE IDEA THAT ALL HUMAN BEINGS are descended from Adam is deeply embedded in the thought forms of Western culture and of the monotheistic religions more generally. In both sacred and secular consciousness the assumption of the universal descent of all peoples from a common stock is widely shared. These accounts—whether religiously inspired or scientifically sanctioned—are fundamental to the way in which the human species thinks of itself and to our understanding of the elemental bonds that unite all branches of the human race into a single family. Human genesis and human solidarity go together. Alongside this dominant picture, however, has been another story, one that dwells on human difference rather than resemblance, one that appeals to multiple origins rather than common descent, one that conjures up an inhabited world before Adam, a primeval universe whose legacy still lingers to the present day. That story has been our quarry in this book.

At base the concept of a pre-adamite world is profoundly simple. Adam was not the first inhabitant of the earth. But the theory's simplicity belies its versatility. In fact, it is multidimensional. Pre-adamism's early life was plagued by charges of heresy, infidelity, and skepticism. Both on the grounds of its biblical exegesis and its reliance on evidence from non-Western chronology and archaeology, the postulation of a pre-adamic cosmos was seen as an assault on the integrity of scripture. For the self-same reasons it has subsequently been read as the first move in the development of textual criticism. The spiritual fate of pre-adamism's most conspicuous early champion, Isaac La Peyrère, is encapsulated in the following piece of doggerel, composed as an epitaph and loosely translated from the French:

Here lies La Peyrère, first a good Israelite,
Then Hugenot, Catholic, Pre-adamite;
Four religions he tried, till, perplexed with so many,
At eighty he died, and went off without any.[1]

Whatever its heretical beginnings, however, the theory later came to reside among religious conservatives displaying nothing of the sense of irresolution advertised in this epigram. Now, rather than undermining biblical authority, the pre-adamites were co-opted as a means of preserving its integrity in the face of challenges to its infallibility from a variety of scientific disclosures. Some depicted the changing morphology and animal life of pre-adamic landscapes in response to the findings of historical geology; others speculated on the possibility of pre-adamic human races in order to accommodate biblical chronology to archaeological artifacts; still others postulated that the adamic family had evolved from pre-adamite hominids. In each case the motivation was to keep faith with both science and the Bible. Pre-adamism's migration from heresy to orthodoxy, from its role as a weapon of skepticism to a tool of fundamentalism, speaks to the contingency of theological labeling. The very idea that initially represented a secularizing trend in scientific endeavor by challenging the detailed accuracy of scriptural history and by liberating anthropological investigation from Mosaic strictures, has ended up being valorized as a reconciling tactic among conservative believers.

In each of these roles pre-adamism operates as a harmonizing strategy, as a means of keeping the claims of religion and science in tandem. As such, it discloses something about the general nature of concordist proposals. By working to preserve the peace between science and theology, it is not so much that pre-adamism acted as a conceptual bridge between two discrete spheres of knowledge and belief. Rather, it functioned as a kind of mold that sculpted both scientific commitment and theological conviction into a distinctive shape. Harmonizing schemes are not to be thought of as passively zipping together two disparate sets of beliefs. They are, rather, agents actively fashioning both scientific theory and religious doctrine into new forms. Scientifically, pre-adamism—depending on the precise version adopted—predisposed its advocates to certain explanatory alignments: they frequently found themselves committed to particular theses about linguistic origins, about the role of environmental modification in organic history, and about the fixity of type. By the same token it also had religious implications for the doctrines of the Fall, the image of God, the soul, and biblical hermeneutics.

ADAM'S ANCESTORS

Harmonizing strategies are thus rarely single-unit ideas; rather, they are conceptual systems—packages of ideas—that transform the very notions they seek to unite.

The changing fortunes and status of the pre-adamite over the centuries, moreover, are connected in significant ways with the secularization of scientific explanation more generally. Up to the middle of the nineteenth century at least, the language of Adam, adamites, and pre-adamites was perfectly good currency within scientific arenas. But as theological talk was more and more filtered out of general scientific discourse, the pre-adamite theory came to operate within more restricted spheres and in particular those communities concerned with accommodating theology and science to one another—that is, to internal theological dialogue or to works of religious apologetic. Again, the irony is plain: the very ideas that aided the secularization of science in the first instance became rehabilitated within the more restricted confines of the theological world.

Our analysis of pre-adamism, however, has revealed that its history cannot be conducted simply on the intellectual plane of a dialogue between science and religion. From its earliest days it has been implicated, in one way or another, in cultural politics. The reason, of course, is that the story of Adam and Eve itself is pregnant with sociopolitical significance. The nature of human identity, the relations between the sexes, the unity of the human family, the genesis of hatred, the emergence of crime—all these are incorporated into the Mosaic narrative. Any tampering with adamic genealogy disrupts this settlement. The supposition of an inhabited world before Adam immediately raised profound anxieties about a common human nature, the relationship between the adamic family and their co-adamite counterparts, and about who owns the biblical story. From their earliest days pre-adamite speculations were implicated in racial matters. But the trajectory has been entirely remarkable. Beginning as a theory to recapture the centrality of the Jewish experience in scriptural economy and global history, it has ended up as a weapon of flagrant anti-Semitism. La Peyrère cast Adam as the father of the Jews, with the pre-adamites as the gentile nations. Numerous other writers later followed his example. Latter-day white supremacists presume the adamic line to be Caucasian and find justification in their reading of Genesis for locating Jewish origins in some act of interracial sexual union with corrupt pre-adamites. The gross racism that came to monopolize certain segments of pre-adamite opinion thus marks a striking departure from its earlier incarnations in which, notably in the case of La Peyrère himself, the purpose was to sweep all nations into the benefits enjoyed by the chosen people—the adamic Jews. The genealogy of pre-adamism traces out a path from communalism to elitism, from humanitarianism to bigotry.

The intellectual history of this journey, of course, is not disconnected from geographical location in several different ways.[2] The seventeenth-century world of the pro-Semitic La Peyrère, with its millennial hopes for the coming of a Jewish Messiah, represented a markedly different space from that of the American South in the immediate ante- and postbellum years, where questions of the relations between black and white dominated consciousness. The eighteenth-century Scotland of Kames, with its enlightened taste for conjectural history, casts his polygenism is a different light from twenty-first-century midwestern neo-Nazi supremacists. In different spaces theories are differently appropriated, and texts are differently read. The fact that American southerners could not read the Bible without seeing slavery justified in its pages, while northern abolitionists could not read it without perceiving the need for emancipation on every page, points to the profound significance of what might be termed the geographies of reading. In a similar way the significance of Adam, and of necessity any possible pre-adamite races, were differently mobilized in different historical-geographical circumstances.

The social spaces that pre-adamite enthusiasts have occupied have also changed dramatically over the centuries. Surfacing initially in esoteric locations among writers engaged in metaphysical speculation, it achieved its first public airing among those renegade thinkers who congregated around the court of Queen Christina. Branded as heresy by the pope, it thereafter occupied the secret worlds of those reveling in skepticism, conjecture, and unbelief. Within a century or so it began to emerge more prominently into the republic of letters, finding supporters among the intellectual elite and among travelers and colonial planters, who found its polygenism attractive for different reasons. By the nineteenth century it was a prominent feature of the professionalizing science of anthropology on both sides of the Atlantic. In Britain its translation into a merciless justification for racial hierarchy suited it, at least in the eyes of some, to the requirements of an imperial nation; in the United States its potential to serve the interests of a slaveholding class made it attractive to those whose anticlerical stance brooked no patience with traditionalists who rested slavery on biblical foundations. As secularism tightened its grip on professional science, pre-adamism retreated to increasingly exclusive theological territory. Conservative believers, who all the while retained admiration for empirical science, found it a useful device to hold onto the historical centrality of Adam, even while allowing for the evolution of pre-adamite hominids as preparation for the first true man. As for racial pre-adamism, it moved increasingly to the fringes. Turn-of-the century supremacists, haunted by fears of hybridity in the aftermath of black emancipation, found in it a resource they could call upon to justify their disquiet over race mixing. Present-day extremists, marginalized on the dangerous fringes of society, preach a violent gospel of race hate and seed line purity grounded in the conviction that a corrupt pre-adamic bloodline continues to snake its malign way through human history.

Pre-adamism clearly has performed different cognitive roles. And it has served different social groups as well. Its multidimensional reach has been remarkable. Pre-adamism has been implicated in everything from hermeneutics and theology to archaeology and anthropology, from linguistics and evolutionary theory to slavery and race relations. Perhaps it is because it could fulfill such wide-ranging needs that it continues to lurk in the cultural memory of the West, periodically emerging to serve some scientific, theological, or political interest.

♔ During the winter of 1906–7 Jack London published a sequence of pieces in *Everybody's Magazine,* his famous evocation of life on the cusp of human emergence. *Before Adam,* as the resulting book was called, was an imaginative reconstruction of pre-adamic near-humans, with a primitive vocabulary of a mere thirty or forty words, struggling against the tyranny of a race of newcomers, the Fire People—fully human beings. In his portrayal London presents this new species of tool users, who overrun the semiterrestrial "Folk," as aggressive, relentless, callous conquerors. They are depicted in their imperial irresistibility as beating in the heads of the defeated Folk with rocks, murdering with the ruthless efficiency of arrow users, and overcoming their foes with clinical cunning and heartlessness. One of the Folk who engages in murder within his own tribe is portrayed in all "his tremendous atavistic tendencies" as foreshadowing "the coming of man."[3] Commenting on London's achievement, the anthropologist, ecologist, and science writer Loren Eiseley praised his capacity to evoke these pre-adamic creatures, "for all their speechless inadequacy," as "more appealing than those of the true men, the Fire People who destroyed them." To Eiseley it was plain where London's sympathies lay: They rested "with the underdog, the arrowless ones, the people who had not acquired the deadly killing nature of the true men." The pitiless horror that Eiseley, like London, observes in the humans of Adam's species seems all too plainly evident in those advocates of adamic racial superiority whose violent denigration of presumed pre-adamites has surfaced with ghastly vigor in the latter stages of the narrative I have sketched in this book.[4] Perhaps we still need a writer of Jack London's stature to champion the cause of those on the receiving end of the vilification meted out by these self-appointed heirs of adamic man.

NOTES

1. BEGINNINGS

1. See Bennett and Mandelbrote, *Garden, the Ark, the Tower, the Temple,* 16–20.

2. See Prest, *Garden of Eden.*

3. Scheuchzer's work is discussed in Rudwick, *Scenes from Deep Time,* 4–17. The quotation from Scheuchzer is cited on 8.

4. Quoted in Nicolson, *Power and Glory,* 104.

5. See Grove, *Green Imperialism;* Withers, "Geography, Enlightenment, and the Paradise Question," 67–92.

6. See Haber, *Age of the World;* Dean, "Age of the Earth Controversy," 435–56; and Rossi, *Dark Abyss of Time.*

7. Barr, "Luther and Biblical Chronology," 51–67.

8. Patrides, "Renaissance Estimates," 315–22. See also Patrides, *Premises and Motifs.*

9. See Barr, "Why the World Was Created in 4004 BC," 575–608; Leeman, "Was Bishop Ussher's Chronology Influenced by a Midrash?" 127–30; Gould, "Fall in the House of Ussher," 181–93; Brice, "Bishop Ussher, John Lightfoot and the Age of Creation," 18–24; Gorst, *Measuring Eternity,* chap. 2.

10. Woodward, "Medieval *Mappaemundi,*" 286–370.

11. See Mass, "Preadamites," 12:370–71; Popkin, *Isaac La Peyrére,* 27–28.

12. Harnack, "Origen," s.v.

13. Gregory of Nyssa, "On the Making of Man," 394. Given these convictions, it is not surprising that in later centuries Christian evolutionists—Catholics in particular—would look back to Gregory of Nyssa to find legitimacy within the Christian tradition for their evolutionary proposals. See Messenger, *Evolution and Theology;* Dorlodot, *Darwinism and Catholic Thought.*

14. Popkin, "Development of Religious Scepticism," 271–80; and "Pre-Adamite Theory in the Renaissance," 50–69.

15. Maimonides, *Guide of the Perplexed,* 516.

16. The discovery of human prehistory and deep geological time is charted in Rudwick, *Bursting the Limits of Time.*

17. Barr, "Pre-Scientific Chronology," 379–87.

18. Augustine, *City of God,* bk. 12, chap. 10; bk. 18, chap. 40. See the brief discussion in Popkin, *Isaac La Peyrére,* 27–28.

19. For this, of course, he was later censured by Andrew Dickson White in *History of the Warfare.*

20. On Scaliger, see Grafton's two-volume *Joseph Scaliger.* See also Grafton, *Defenders of the Text,* chap. 4: "Scaliger's Chronology: Philology, Astronomy, World History"; and "Dating History," 74–85. Problems arising from Chinese chronology are discussed in Van Kley, "Europe's 'Discovery' of China," 358–85; and Mungello, *Great Encounter.*

21. See Tavakoli-Targhi, "Contested Memories," 149–75.

22. See the discussion in Tavakoli-Targhi, "Orientalism's Genesis Amnesia," 1–14.

23. Eco, *Serendipities,* 64.

24. Montaigne, "An Apologie of *Raymond Sebond,*" *Montaigne's Essays,* vol. 2, bk. 2, chap. 12, 288.

25. Charron, *Of Wisdom;* Lanquet, *Epitome of Chronicles.*

26. Rossi, *Dark Abyss of Time.*

27. Kidd, *British Identities before Nationalism,* 17.

28. See Siraisi, "Vesalius and Human Diversity," 60–88.

29. See Wittkower, "Marvels of the East," *Journal* 159–97.

30. An extensive listing is provided in Woodward, "Medieval *Mappaemundi,*" 331; and in Friedman, *Monstrous Races,* 5–25.

31. Campbell, *Witness and the Other World,* 8, 55.

32. See the wide-ranging inquiry into this subject by Daston and Park, *Wonders and the Order of Nature.*

33. Ibid., 25.

34. Mason, *Deconstructing America,* 75. Mason notes that depictions of monstrous races in the ancient world followed the same principle. In both Pliny and Herodotus "the monstrous human races are a part of a system of roughly concentric circles with their centre in the region of Italy or Greece" (79).

35. Woodward, "Medieval *Mappaemundi,*" 332.

36. Jeffrey, "Medieval Monsters," 62.

37. Campbell, *Witness and the Other World,* 55.

38. See the discussion in Friedman, *Monstrous Races,* 178.

39. The following explanations are detailed in ibid.

40. Augustine, *City of God,* bk. 16, chap. 8.

41. Friedman, *Monstrous Races,* 88.

42. Campbell, *Witness and the Other World,* 10.

43. See Mason, *Deconstructing America,* 97–99.

44. See Greenblatt, *Marvelous Possessions.*

45. Campbell, *Witness and the Other World,* 180; Greenblatt, *Marvelous Possessions,* 75.

46. Raleigh, *Discovery,* 85.

47. Lafitau, *Moeurs des Sauvages Amériquains.*

48. On this subject the work of J. Brian Harley is recommended. See Harley, *Maps and the Columbian Encounter*; "Rereading the Maps of the Columbian Encounter," 522–42; and "Maps and the Invention of America," 8–12.

49. Lestringant, *Mapping the Renaissance World*, 64.

50. The quotations from Bodin and de Bry are to be found in Grafton, *New Worlds*, 130, 152.

51. Ryan, "Assimilating New Worlds," 519–38.

52. Grafton, *New Worlds*, 207.

53. Quoted in Huxley, "Aristotle, Las Casas and the American Indians," 58.

54. For various perspectives on this episode, see Hanke, *Aristotle and the American Indians;* and *All Mankind Is One;* Huxley, "Aristotle, Las Casas and the American Indians"; Padgen, *Fall of Natural Man.*

55. Quoted in Hanke, *All Mankind Is One*, 80.

56. Hanke, *Aristotle and the American Indians*, 54.

57. Grafton, *New Worlds*, 208.

58. These, and other accounts, are reviewed in Huddleston, *Origins of the American Indians;* Pagden, *Fall of Natural Man;* Kidd, *British Identities before Nationalism.*

59. Quoted in Slotkin, *Readings in Early Anthropology*, 182.

60. See Almond, *Adam and Eve*, 49.

61. On the skeptical reputations of Raleigh and Harriot, see Strathmann, "History of the World," 265–87; Lefranc, *Sir Walter Ralegh*, chap. 12; Jacquot, "Thomas Harriot's Reputation," 164–87.

62. Dove, *Confutation of Atheisme*, 4–5.

63. These extracts from Paracelsus and Bruno appear in Slotkin, *Readings in Early Anthropology*, 42–43. Bruno apparently had learned of Chinese thinking on the subject and of their theory that the human race had descended from three different protoplasts over thirty thousands years ago.

64. On this general theme, see Dick, *Plurality of Worlds.*

65. See Yates, *Giordano Bruno.* For a popular account of the whole episode, see White, *Pope and the Heretic.*

66. Quoted in Slotkin, *Readings in Early Anthropology*, 81.

67. Poole, "Seventeenth-Century Preadamism," 1–35.

2. HERESY

1. The standard biography is Popkin, *Isaac la Peyrère.* Popkin describes Renaissance pre-adamism as a "monumentally heretical doctrine."

2. See Grafton, *New Worlds*, 237.

3. I have discussed this subject in Livingstone, "Geographical Inquiry, Rational Religion and Moral Philosophy," 93–119.

4. I shall work from the English edition of the text, which was published by the Hakluyt Society in 1855. This translation was from the 1663 edition of *Relation du Groenland* printed in Paris by Thomas Jolly.

5. Popkin, *High Road to Pyrrhonism.*

6. In constructing the map for his Iceland volume, La Peyrère sought the advice of Pierre du Val, Sanson d'Abbeville's nephew and also geographer to the king of France.

7. La Peyrère, *Relation du Greonland,* 183.

8. Debenham, *Discovery and Exploration,* 111.

9. Hazard, *European Mind,* 27–28. Originally published as *La Crise de la con-science européenne* (Paris: Boivin, 1935).

10. On Worm, see Daniel, *Short History of Archaeology;* Trigger, *History of Ar-chaeological Thought.*

11. Worm's influence on La Peyrère is stressed in Schnapp, "Pre-adamites," 399–412.

12. La Peyrère *Relation du Groenland,* 193–95.

13. Burke, "Philosopher as Traveller," 130.

14. La Peyrère, *Relation du Groenland,* 217.

15. Popkin, *La Peyrère,* 11.

16. Hugo Grotius was the author of a treatise on the subject entitled *Dissertatio Altera de Origine Gentium Americanarum Adverses Obtrectatorem* (1643).

17. In *De Originibus Americanis Libri* Horn presented his own conjectures on the subject. See "Horn," *Biographie Universelle,* q.v.; and Grafton, *New Worlds,* 234–37.

18. La Peyrère, *Relation du Groenland,* 249.

19. La Peyrère, *Account of Iceland,* 2:435. La Peyrère, moreover, was of the opinion that the way in which northern environments retarded the growth of civilized cul-ture was more restricted to "the Vulgar sort" and that "People of Fashion, ought to be Exempted of this Rule, (less or more) in most Places" (435).

20. Ibid., 437.

21. See Skovgaard-Petersen, *Historiography at the Court of Christian IV.*

22. La Peyrère, *Account of Iceland,* 441–42.

23. The textual history of the work is discussed in some detail in Popkin, *La Pey-rère,* chap. 2.

24. See Poole, "Seventeenth-Century Preadamism," 6; also Bennett and Mandel-brote, *Garden, the Ark, the Tower, the Temple,* 195.

25. I have discussed various aspects of La Peyrère's thesis and its implications in a number of publications, in particular Livingstone, "Preadamites," 41–66; "Pread-amism," 25–34; *Preadamite Theory.* Specifically on La Peyrère, see also the now stan-dard biography by Popkin, *La Peyrère;* Rossi, *Dark Abyss of Time,* chap. 18: "History before Adam"; Gorst, *Measuring Eternity,* 43–48; Kidd, *Forging of Races,* 62–66.

26. La Peyrère, *Men before Adam,* 4, 19.

27. Grafton, *New Worlds,* 211.

28. La Peyrère, *Men before Adam,* preface.

29. La Peyrère, *Theological System,* 164.

30. La Peyrère, *Men before Adam,* 29–30.

31. See Popkin, "Biblical Criticism and Social Science," 339–60; and "Develop-ment of Religious Skepticism," 271–80.

32. Strauss, "Isaac de la Peyrère," in *Spinoza's Critique of Religion,* 64–65.

33. Smith, *Spinoza,* 57.

34. La Peyrère, *Men before Adam,* 60–61.

35. Grafton, *New Worlds,* 211.

36. The "rationalistic and scathing utilization of the polemic spirit of Judaism" in the cause of antichristian sentiments more generally during the period is discussed in Berti, "At the Roots of Unbelief," 555–75.

37. See the various discussions of this whole subject in Yardeni, "La Religion de La Peyrère," 245–59; Popkin, "Influence of La Peyrère's French Oriented Messianism," chap. 8 of *La Peyrère;* and "Jewish Messianism and Christian Millenarianism," 67–90. Such views have led several scholars to insist that Peyrère was a Marrano, but conclusive evidence for this claim is apparently lacking. See Grafton, "Vision of the Past and Future," 151–52; and Popkin, "Marrano Theology of Isaac La Peyrère," 97–126.

38. Although composed in an age fascinated by millennial eschatology, speculations about the Antichrist, and efforts to date the end of the world, La Peyrère's messianism seems to bear no trace of such influence.

39. La Peyrère, *Apologie de La Peyrère,* 21–22.

40. Grafton, *Defenders of the Text,* 205.

41. Gay, *Enlightenment.* See also Outram, *Enlightenment,* chap. 3.

42. Berti, "At the Roots of Unbelief," 562.

43. La Peyrère, *Apologie de La Peyrère.*

44. Popkin, *La Peyrère,* 15.

45. Kidd, *British Identities before Nationalism,* 16.

46. Ibid., 16.

47. White, *History of the Warfare,* 2:317.

48. Grafton, *Defenders of the Text,* chap. 8: "Isaac La Peyrère and the Old Testament."

49. Quoted in Grafton, *New Worlds,* 211; Rossi, *Dark Abyss of Time,* 148. See also Rubiés, "Hugh Grotius's Dissertation," 221–44.

50. See Rossi, *Dark Abyss of Time,* 145–47.

51. See Young, *Biblical Flood,* 52–53.

52. Rossi, *Dark Abyss of Time,* 148–50.

53. Stillingfleet, *Origines Sacrae,* 124. See also Kidd, *British Identities,* 40; Popkin, "Philosophy of Bishop Stillingfleet," 305; Carroll, *Common-Sense Philosophy.*

54. Stillingfleet, *Origines Sacrae,* xv.

55. Ibid., 127.

56. Ibid., 128.

57. Ibid., xiv.

58. Hale, *Primitive Origination of Mankind,* 185. On Hale's contribution to seventeenth-century debates about the spontaneous generation of human beings, see Goodrum, "Atomism, Atheism, and the Spontaneous Generation of Human Beings," 207–24.

59. Quoted in Rooden, "Conceptions of Judaism," 299–308.

60. Poole, "Seventeenth-Century Preadamism."

61. See also Poole, "Francis Lodwick's Creation," 245–63.

62. Salmon, *Works of Francis Lodwick*. See also Umberto Eco's account of Lodwick in *Search for the Perfect Language*, chap. 13.

63. Poole, "Divine and the Grammarian Theological Disputes," 273–300.

64. The anonymous manuscript is reproduced, along with commentary, in Poole, "Seventeenth-Century Preadamism."

65. Almond, *Adam and Eve*, 51.

66. The pre-adamism of these English radical sectarians, and the sources of the extracts I have quoted, are noted in Poole, "Seventeenth-Century Preadamism," 7–8.

67. See Kidd, *Forging of Races*, 88.

68. Blount, *Oracles of Reason*, 218.

69. Ibid., 8, 11.

70. Kidd, *British Identities before Nationalism*, 20.

71. Lecky, *History of the Rise and Influence of the Spirit of Rationalism*, 1:301.

72. Popkin, *La Peyrère*, 131; Dobbs, *First Volume of Universal History*. The work eventually extended to nine volumes.

73. Dobbs, *Concise View*, 228.

74. Ibid., 232.

75. Ibid., 238.

76. Ibid., 235.

77. Popkin, *La Peyrère*, 131. Dobbs's strange millennial politics are to be found in *Memoirs of Francis Dobbs*.

78. On Newton's eschatology, see Force, "Nature of Newton's 'Holy Alliance'"; Iliffe, "'Making a Shew,'" 55–88; Hutton, "More, Newton, and the Language of Biblical Prophecy," 39–43.

79. Dobbs, *Memoirs of Francis Dobbs*, 46. See the discussion of Dobbs's eschatology and its political implications in Orr, "From Orange Grove to Armageddon," 57–70.

80. Lammey, "Dobbs, Francis," q.v.

81. Biographical details are available in *Dictionary of National Biography*, s.v.; and in Sellers, "Lardner, Nathaniel," q.v.

82. Lardner, "Essay on the Mosaic Account," 11:227, 244–45.

83. *Co-adamitae*, 3.

84. Ibid.

85. Ibid., 8.

86. Ibid., 21.

87. Bennett and Mandelbrote, *Garden, the Ark, the Tower, the Temple*, 195.

88. Grafton, *Defenders of the Text*, 211, 212.

89. See Quennehen, "Lapeyrère," 243–55. She also contends that Chinese and other ancient chronologies were not the source of La Peyrère's pre-adamite speculations but were incorporated into them.

90. La Peyrère, *Men before Adam*, 208.

91. Lecky, *History of the Rise and Influence,* 1:107–8. Lecky pointed out that La Peyrère rejected the Mosaic authorship of the Pentateuch, radically reinterpreted the doctrine of original sin, and restricted the extent of the miraculous, thereby establishing the foundations of biblical criticism. White, *History of the Warfare,* 1:255.

92. Grafton, *Defenders of the Text,* 206.

93. Grafton, *New Worlds,* 241–42.

94. Popkin, "Biblical Criticism and Social Science."

95. See Poole, "Divine and Grammarian," 282–88; Eco, *Search for the Perfect Language,* chap. 11: "George Dalgarno."

96. Eco, *Serendipities,* 31. More generally on this theme, see chap. 2 of this work, "Languages in Paradise."

97. See Eco, *Search for the Perfect Language,* chap. 5: "The Monogenetic Hypothesis and the Mother Tongues."

98. Rossi, *Dark Abyss of Time,* 201.

99. Eco, *Search for the Perfect Language,* 89–90.

100. See, for example, Haddon, *History of Anthropology,* 52; McKee, "Isaac de la Peyrère," 456–85; Penniman, *Hundred Years of Anthropology,* 53; Comas, *Manual of Physical Anthropology,* 77, 80; Allen, *Legend of Noah,* 137; Gossett, *Race,* 15; Johnson, "Chronological Writing," 124–45; Hodgen, *Early Anthropology,* 272–76; Harris, *Rise of Anthropological Theory,* 89; Voget, *History of Ethnology,* 58–59; Stepan, *Idea of Race in Science,* 29; Grayson, *Establishment of Human Antiquity,* 140–42; Rossi, *Dark Abyss of Time,* 132–36; Trigger, *History of Archaeological Thought,* 112.

101. Sardar, Nandy, and Davies, *Barbaric Others,* 55. A more careful set of comments on La Peyrère's polygenesis and its racial significance is to be found in Bernasconi, "Who Invented the Concept of Race?" 11–36.

102. Myl, *De Origine Animalium.* More generally on this influence, see Browne, *Secular Ark,* 12–15.

103. Stillingfleet, *Origines Sacrae,* 130.

104. Popkin, *La Peyrère,* 1.

105. Malcolm, *Aspects of Hobbes,* chap. 12: "Hobbes, Ezra, and the Bible: The History of a Subversive Idea," 387.

106. Grafton, *New Worlds,* 237.

3. POLITY

1. [Al-Tajir], *Ancient Accounts of India and China,* 255.

2. Ridgley, *Body of Divinity,* 1:287.

3. Ibid., 288.

4. Here I work from the Dublin edition of this work, which first began to appear in 1744: *Universal History, from the Earliest Account of Time to the Present,* 1:98.

5. Ibid., 1:98–99.

6. Ibid., 1:99.

7. Ibid.

8. Carrithers, intro., *Spirit of Laws by Montesquieu,* 46–47. The importance of this

physiological context is clear in Montesquieu's "Essay on the Causes Affecting Minds and Characters," appended to this edition of *Spirit of Laws*. Paragraphs 9–15 deal with "How the condition of the fibres of the nervous system can affect minds and characters." More generally, see also Carrithers, "Enlightenment Science of Society," 232–70.

9. Montesquieu, *Spirit of Laws*, bk. 14, chap. 2, 316. In what follows I work from this edition.

10. Ibid., bk. 14, chap. 2, 316–17.

11. Ibid., bk. 14, chap. 2, 317.

12. Ibid., bk. 19, chap. 14, 427.

13. Ibid., bk. 14, chap. 10, 327.

14. Buffon, *Histoire naturelle*, 185.

15. See Gerbi, *Dispute of the New World*. Also Livingstone, "'Risen into Empire,'" 304–35.

16. Buffon, *Histoire naturelle*, 185.

17. Blumenbach, *De Generis Humani Varietate Nativa*, 2.23, 4.80. English translations in Slotkin, *Readings in Early Anthropology*, 189.

18. Goldsmith, *History of the Earth and Architecture*, 2:213.

19. Ibid., 2:222.

20. Ibid., 2:233.

21. Ibid., 2:239, 242.

22. Key biographies are Ross, *Lord Kames;* Lehmann, *Henry Home.* See also the review of Kames's standpoint in Kidd, *Forging of Races*, 95–100.

23. Henry Home of Kames, *Sketches of the History of Man.* In what follows I work from the posthumously published three-volume edition that appeared in Glasgow in 1817.

24. Ibid., 1:32, 27.

25. Ibid., 1:13.

26. Ibid., 1:11.

27. Ibid., 1:14.

28. See the discussion of Kames in Stanton, *Leopard's Spots*, 15–16; Stocking, *Race, Culture, and Evolution*, 44–45; Wokler, "Apes and Races in the Scottish Enlightenment."

29. Stocking, "Scotland as the Model of Mankind," 85.

30. Wood, "Science of Man," 204.

31. Kames, *Sketches of the History of Man*, 1:12.

32. Ibid., 2:237, 239.

33. For discussion of this point, see Sebastiani, "Race and National Characters in Eighteenth-Century Scotland," 1–14.

34. Kames, *Sketches of the History of Man*, 1:41.

35. Ibid.

36. See Beattie, *Elements of Moral Science*, 2:56–57. See the discussion in Kidd, *Forging of Races*, 100.

37. Hughes, *Natural History of Barbados*, bk. 1, 9.

38. Ibid., bk. 1, 10.

39. Ibid., bk. 1, 14.

40. Atkins, *Navy-Surgeon*, 23, 24. Atkins's commitment to co-adamism is noted in Curtin, *Image of Africa*, 41.

41. Atkins, *Voyage to Guinea, Brasil, and the West-Indies*, 39.

42. White, *Account of the Regular Gradation in Man*, 125.

43. Ibid., 133.

44. Ibid., 126.

45. Ibid., 136.

46. Ibid., 55.

47. King, "Dissertation Concerning the Creation of Man," 3:74.

48. Ibid., 76.

49. Ibid., 77.

50. Ibid., 79.

51. Ibid., 82.

52. Ibid., 85, 86, 83.

53. Ibid., 88, 93.

54. Ibid., 100.

55. Ibid., 118.

56. Ibid., 164. King's eschatological interests more generally surfaced in his *Remarks on the Signs of the Times*, which appeared in 1798.

57. King, "Creation of Man," 169.

58. Montesquieu, *Spirit of Laws*, bk. 14, chap. 2, 319.

59. Ibid., bk. 14, chap. 2, 320.

60. Ibid.

61. Ibid., bk. 14, chap. 2, 319.

62. See Poe, "What Did Russians Mean," 585–608; Rubiés, "Oriental Despotism and European Orientalism," 109–80.

63. Goldsmith, *History of the Earth*, 2:214, 221, 225, 227, 228.

64. Ibid., 2:231.

65. Kames, *Sketches of the History of Man*, 1:34.

66. Ibid., 1:36.

67. See Thomson, "Issues at Stake," 1–20.

68. Godwyn, *Negro's and Indians Advocate*, 15, 14, 18.

69. Ibid., preface.

70. Ibid., 3.

71. Ibid., 15.

72. Ibid., 18–19.

73. Ibid., 18.

74. Clarkson, *Essay on the Slavery and Commerce*, 190.

75. Ibid., 213–14.

76. Ibid., 214.

77. Ibid., 256.

78. Brogan, "Clarkson, Thomas," q.v.

79. Ramsay, *Essay on the Treatment*, 175.

80. Ibid., 180.

81. Ibid., 200–201.

82. Adair, *History of the American Indians,* 11.

83. Ibid., 12.

84. Hatley, "Adair, James," q.v.

85. White, *Regular Gradation in Man,* iii.

86. Ibid., 137.

87. Ibid., 42, 67.

88. Ibid., 135.

89. Ibid., 117.

90. Ibid., 145.

91. Morgan, "Long, Edward," q.v.

92. Long, *History of Jamaica,* 2:352.

93. Ibid., 2:353.

94. Ibid., 2:354.

95. Goldsmith, *History of the Earth,* 2:212.

96. Long, *History of Jamaica,* 2:370.

97. Ibid., 2:369.

98. Ibid., 2:370.

99. Ibid., 2:364.

100. Ibid., 2:374–75.

101. Ibid., 2:377.

102. Long, *Candid Reflections.* See the discussion in Wheeler, *Complexion of Race.*

103. Montagu, *Edward Tyson.*

104. Tyson, *Orang-Outang,* 55.

105. Ibid., 92.

106. Tyson, *Phocaena.*

107. Tyson, *Orang-Outang,* 54.

108. Long explicitly refers to Tyson's work in *History of Jamaica,* 2:359, though he cites him in order to confirm the humanlike behavior of the orang.

109. See Meijer, *Race and Aesthetics;* Wyhe, "Authority of Human Nature," 17–42.

110. On the Scottish concern to map human nature over time and place, see Garrett, "Anthropology," 79–93.

111. Monboddo, *Origin and Progress of Language,* 1:293.

112. See the discussion in Wokler, "Apes and Races in the Scottish Enlightenment," 145–68.

113. Monboddo, *Origin and Progress of Language,* 1:12.

114. Ibid., 1:i.

115. Ibid., 1:v.

116. Ibid., 1:vii–viii.

117. Ibid., 1:270.

118. Ibid., 1:290.

119. Ibid., 1:301, v.

120. Ibid., 1:579.

121. Hammett, "Burnett, James, Lord Monboddo," q.v.; Monboddo, *Origin and Progress of Language,* 1:144.

122. Monboddo, *Origin and Progress of Language,* 1:270.

123. Smith, *Essay* (1788 ed.), 136–37.

124. Greene, *American Science in the Age of Jefferson,* 323.

125. Greene, "American Debate on the Negro's Place in Nature," 384–96; Gossett, *Race,* 39–40.

126. Stanton, *Leopard's Spots,* 4.

127. Harris, *Rise of Anthropological Theory,* 86.

128. Jordan, intro. to Smith, *Essay* (1810 ed.), 1.

129. Smith, *Essay* (1810 ed.), 21.

130. For a general discussion, see Livingstone, "Human Acclimatization," 359–94.

131. Smith, *Essay* (1810 ed.), 40.

132. Ibid., 78.

133. Ibid., 84.

134. Jordan, intro., xxviii.

135. Smith, *Essay* (1810 ed.), 107.

136. Ibid., 129.

137. Ibid., 204.

138. Ibid., 185.

139. Noll, *Princeton and the Republic,* 121.

140. Smith, *Essay* (1810 ed.), 149.

141. Ibid., 7.

142. Ibid., 8.

143. Ibid., 149.

144. Noll, "Rise and Long Life of the Protestant Enlightenment," 100.

145. Smith, *Essay* (1878 ed.), 48.

146. Fiering, *Moral Philosophy,* 300.

147. See Adair, "'That Politics May Be Reduced to a Science'"; Branson, "James Madison," 235–50.

148. It is worth noting that Thomas Reid, a central figure within Scottish Common Sense philosophy, felt the need to resort to the writings of naturalists and travelers in developing his views on human nature. See Wood, "Science of Man," 210. This approach, moreover, was entirely in keeping with his suspicion that it might not be too long before the philosophy of the human mind would have its own Newton. See the discussion in Hatfield, "Remaking the Science of Mind," 184–231.

4. APOLOGETICS

1. King, "Dissertation," 69–70.

2. Ibid., 70.

3. Ibid., 70–71.

4. Ibid., 103.

5. Ibid., 121–22.

6. Rudwick, *Scenes from Deep Time.*

7. "Pre-Adamite World," 155.

8. Harris, *Pre-Adamite Earth,* 105.

9. Ibid., 271.

10. Ibid., 146, 147.

11. Chalmers, *Evidence and Authority;* this pamphlet appeared in many subsequent editions.

12. See Chalmers, *Series of Discourses.*

13. Buckland, *Vindicae Geologicae;* and *Geology and Mineralogy Considered with Reference to Natural Theology.* For further details, see Rupke, *Great Chain of History.*

14. Harris, *Man Primeval,* 1.

15. Ibid., 24.

16. Ibid., 26–27.

17. Ibid., 30.

18. Ibid., 185.

19. R. T. B., "Pre-Adamite Earth," 56, 55.

20. S. A. L., "Pre-Adamite World," 532–33.

21. Morris, *Science and the Bible; and Present Conflict.*

22. Morris, *Work Days of God,* 6.

23. Ibid., 71.

24. Advocates of this interpretation of the Genesis creation narrative are discussed in detail in Numbers, *Creationists.*

25. In arguing for literal days, Morris reflected on Hugh Miller's day-age theory—which assigned to each Genesis day a geological epoch—only to reject it.

26. Thomas, *Elpis Israel,* 10, 11.

27. Ibid., 11–12.

28. Spencer, "Christmas [*later* Noel-Fearn], Henry," q.v.

29. Christmas, *Echoes of the Universe,* v.

30. Ibid., 74.

31. Ibid., 85.

32. Ibid., 89.

33. Ibid., 92.

34. Ibid., 95.

35. See the discussions in Van Riper, *Men among the Mammoths;* Grayson, *Establishment of Human Antiquity.*

36. She notes, for example, visits to St. Acheu, near Amiens, where Boucher de Perthes unearthed in the drift deposits what he believed were stone implements from antediluvian times, and to the Duke of Buccleuch's coal pits near Dalkeith.

37. [Duncan], *Pre-Adamite Man,* 39, 40.

38. Ibid., preface.

39. Ibid., 3, 52.

40. Rudwick, *Scenes from Deep Time,* 136.

41. [Duncan], *Pre-Adamite Man,* 27.

42. Gould, "Pre-Adamite in a Nutshell," 76.

43. [Duncan], *Pre-Adamite Man,* 130.

44. Ibid., 149.

45. Gould, "Pre-Adamite in a Nutshell," 76.

46. These extracts from reviews are to be found in [Duncan], *Pre-Adamite Man,* iii, vii.

47. Snobelen, "Of Stones, Men and Angels," 87, 84.

48. Ibid., 98, 99.

49. I have noted this trajectory elsewhere. See Livingstone, "Preadamites," 41–66; and "Preadamism," 25–34.

50. Gall, "On Improved Monographic Projections," 148; and "Use of Cylindrical Projections," 119–23.

51. Peters and Gall are both discussed in Crampton, "Cartography's Defining Moment," 16–32. I am grateful to Dr. Crampton for sharing with me a chapter on this subject from his book *Mapping.*

52. Gall, *Primeval Man Unveiled,* 27.

53. Ibid., 27.

54. Ibid., 50.

55. Ibid., 176.

56. Ibid., 180, v–vi.

57. Ibid., 177.

58. Ibid., 182.

59. Pember, *Earth's Earliest Ages,* 74, 73.

60. Ibid., preface.

61. Ibid., 72.

62. So, for example, Custance, *Time and Eternity,* vol. 6, pt. 6, chap. 2.

63. This diversity, and the nonracial versions of pre-adamism defended by Duncan and Gall in particular, are noted, for example, in Allen, "Religious Heterodoxy and Nationalist Tradition," 2–34; and "Identity and Destiny," 163–214.

64. Good, *Book of Nature,* 2:92.

65. Ibid., 2:94. His reasoning here was based on the claim that "it was in his one hundred and thirtieth year, that Seth was given to him [Adam] in the place of Abel."

66. Ibid., 102.

67. Ibid., 103.

68. Van Amringe, *Investigation of the Theories.*

69. Agassiz, "Geographical Distribution," 181–204; and Agassiz, "Diversity of Origin," 110–45.

70. Agassiz, "Geographical Distribution," 193.

71. See Lurie, "Louis Agassiz," 227–42; and Winsor, "Louis Agassiz," 89–117.

72. Agassiz, "Geographical Distribution," 181.

73. Ibid., 184–85.

74. Ibid., 189.

75. Agassiz, "Diversity of Origin," 128.

76. Ibid., 143.

77. Caldwell, *Thoughts on the Original Unity;* Morton, *Crania Americana.*

78. Agassiz, "Diversity of Origin," 119.

79. Moore, "Unity of the Human Races," 207. In this work Moore also reviewed the work of Prichard, Smyth, Bachman, and Van Amringe.

80. Ibid., 214.

81. Ibid., 230.

82. Ibid., 231.

83. F[rothingam], "Men before Adam," 80.

84. Ibid., 82.

85. Ibid.

86. Ibid., 85.

87. Ibid., 93.

88. Smith, *Natural History of the Human Species,* 167.

89. Ibid., 173.

90. Ibid., 176.

91. Kneeland's own polygenist perspective is clear in his article "Hybrid Races of Animals and Men," 535–39. Here he concluded that "the *genus* HOMO consists of *several species*" (539).

92. Kneeland, intro., in Smith, *Natural History of Human Species,* 60.

93. Ibid., 67.

94. Ibid., 69, 71, 72.

95. On account of his orientalist scholarship, Lane features prominently in Said's *Orientalism.*

96. Lane, *Genesis of the Earth and of Man,* xxii. The Huxley-Wilberforce encounter has been subject to much revision. Readers are referred to Lucas, "Wilberforce and Huxley," 313–30; Gilley, "Huxley-Wilberforce Debate," 325–40; Jensen, "Return to the Wilberforce-Huxley Debate," 161–79.

97. Lane, *Genesis of Earth and Man,* 137.

98. Ibid., 142–43.

99. Ibid., 64, 69, 128.

100. Gliddon, "Monogenists and the Polygenists," 415.

101. Lane, *Genesis of Earth and Man,* xviii.

102. Ibid., 200, 201.

103. Ibid., 202.

104. Ibid., 209.

105. Ibid., 267, 228.

106. Ibid., 253.

107. Woodward, review of Huddleston, "McCausland, Dominick," q.v.

108. McCausland, *Adam and the Adamite,* 4.

109. Ibid., 3.

110. Ibid., 160.

111. Ibid., 90.

112. Ibid., 102.

113. Ibid., 115.

114. Ibid., 183.

115. Ibid., 182–83.

116. "Was Adam the First Man," 583.

117. Ibid., 582.

118. Wayne, "After Babel," 402.

119. Campbell, "Unity of the Human Race," 74.

120. Ibid., 100.

121. Nemo, *Man,* 1.

122. Ibid., 51.

123. Ibid., 55.

124. Ibid., 47.

125. Ibid., 110.

126. Ibid., 115.

5. ANTHROPOLOGY

1. Randolph, *Eulis.*

2. The most complete biography is by John Patrick Deveney, *Paschal Beverly Randolph.*

3. Randolph, *Pre-Adamite Man.* The work was originally published under the pseudonym Griffith Lee as *Pre-Adamite Man: The Story of the Human Race, from 35,000 to 100,000 Years Ago!* (New York: Sinclair Toussey, 1863).

4. Weiser, review of Lee, *Pre-Adamite Man,* 222.

5. Randolph, *Pre-Adamite Man,* 67.

6. Ibid., 68, 162.

7. Ibid., 71.

8. Ibid., 253.

9. Ibid., 79, 76.

10. Ibid., 84.

11. Ibid., 75–76.

12. Much later, in 1949, the same stance was adopted by Anne Terry White, whose book *Men before Adam* was used as a rhetorical tool to attack the creationism of William Jennings Bryan and to present in popular form the standard anthropological evidence for human evolution. The work was published by the Scientific Book Club, was thoroughly rationalist in spirit, and contained no references to the biblical narrative.

13. On the origins of the Ethnological Society, see Stocking, *Victorian Anthropology.*

14. The story of the formation of the Anthropological Society of London has frequently been told. See Burrow, "Evolution and Anthropology," 137–54; Stocking,

"What's in a Name," 369–90; Rainger, "Race, Politics and Science," 51–70; Stocking, *Victorian Anthropology,* 247–54.

15. Crawfurd's interactions with Murchison feature in Stafford, *Scientist of Empire.*

16. Crawfurd, "On the Classification of the Races," 355–56.

17. Crawfurd, "On the Effects of Commixture," 76–92.

18. Crawfurd's virulently anti-Darwinian sentiments with respect to ethnology are apparent in "On the Theory of the Origin of Species," 27–38.

19. Crawfurd, "Classification," 365, 371.

20. See Alter, *Darwin and the Linguistic Image,* 39.

21. Latham, *Man and His Migrations,* 78.

22. Darwin, *On the Origin of Species,* 422.

23. Crawfurd, "On the Aryan or Indo-Germanic Theory," 285–86. See also his paper "On Language as a Test," 1–8.

24. Crawfurd, "Classification," 361.

25. Crawfurd, "On the Connexion between Ethnology and Physical Geography," 4, 8. I have discussed Crawfurd's climatic thinking in Livingstone, "Race, Space and Moral Climatology," 159–80.

26. Poole, "Ethnology of Egypt," 263.

27. Ibid., 264.

28. Farrar, "Traditions, Real and Fictitious," 298.

29. Farrar, "Fixity of Type," 399.

30. Vance, "Farrar, Frederic William."

31. Farrar, "Language and Ethnology," 204.

32. Bendysche, "History of Anthropology," 335–420.

33. Philalethes, "Peyrerius," 109.

34. Ibid., 116. The image of a feud between progressive science and theological bigotry was not uncommonly projected among anthropologists. See, for example, the anonymous review of "Antiquity of Man," 1136–52.

35. Philalethes, "Distinction between Man and Animals," 163.

36. Charnock, "Science of Language," 194.

37. Ibid., 212.

38. Harris, "Plurality of Races," 175.

39. Ibid., 177.

40. Moore was the author of works on medical subjects, theology, and Scottish archaeology and a work seeking to establish connections between Buddhism and Judaism through an analysis of rock records in India; he also composed poetry and hymns.

41. Hunt, "On the Application of the Principle," 320–21.

42. Ibid., 329.

43. "Dr. Moore and His First Man," 108.

44. Wake, "Relation of Man," 367.

45. Ibid., 370.

46. Wake, "Adamites," 373, 370.

47. See Blake, review of Huxley et al., *Edinburgh Review* (April 1863): 541–69.

48. Blake, comment following Wake, "Adamites," 375.

49. Prichard, "On the Cosmogony of Moses," 46 (1815): 285–92; 47 (1816): 110–17, 258–63; 48 (1816): 111–17.

50. Prichard, *Researches,* 1st ed., 178, 41.

51. Prichard, *Natural History of Man,* 39, 40.

52. See Stocking, "From Chronology to Ethnology," lxxx.

53. Prichard, *Researches,* 3rd ed., 1:473.

54. Prichard, *Researches,* 1st ed., 233.

55. See Augstein, *James Cowles Prichard's Anthropology,* 131–33. See also Stocking, *Victorian Anthropology,* 48–53. Stocking claims that Prichard later drew back from the heterodoxy of making Adam black, though this abandonment of primitive blackness had probably less to do with social niceties as with changes in his attitude to the theories of Jacob Bryant, who regarded early culture bearers as descendants of Ham. Stocking, "From Chronology to Ethnology," liv.

56. Prichard, "On the Extinction of Human Races," 166–70. See the discussion in Augstein, *Prichard's Anthropology,* 145.

57. Kidd, *Forging of Races,* 132–34.

58. Kennedy, "On the Probable Origin of the American Indians," 227.

59. Ibid., 229.

60. FitzRoy, "Outline Sketch," 11.

61. Dunn, "On the Physiological and Psychological Evidence," 191–92.

62. Ibid., 201.

63. Ibid., 199.

64. Dunn, "Some Observations on the Psychological Differences," 11. Dunn read this paper to the society in 1863, having already delivered it at the 1862 Cambridge meeting of the British Association. See also Dunn, "Some Observations on the Tegumentary Differences," 59–71.

65. Dunn, "Some Observations on the Psychological Differences," 19.

66. See Osborne, *Nature.*

67. Quatrefages, *Human Species,* 31, 33. See also his essay "Histoire naturelle de l'homme," 807–33.

68. Peschel, *Races of Man,* 33–34.

69. Wagner, "Creation of Man," 227.

70. Ibid., 230. While it is not clear just who Wagner had in mind, German polygenists who maintained that humans were of multiple, autochthonous, origins included Johann Georg Justus Ballenstedt, a scientifically literate Lutheran minister, who argued his case in *Die Vorwelt und die Mitwelt* (1824), and the Giessen professor of surgery Ferdinand August Ritgen, who voiced the same views during the 1830s. I am most grateful to Nicolaas Rupke for drawing these figures to my attention. See Rupke, "Neither Creation nor Evolution," 143–72.

71. See Stocking, *Victorian,* 67.

72. Reddie, "Slavery," 280.

73. Ibid., 282.

74. Ibid., 286.

75. Ibid.

76. Ibid., 287.

77. Ibid., 289.

78. Ibid., 290.

79. Ibid., 289.

80. Ibid., 290.

81. Alexander, "Bible," 101–20. Reprinted in Noll, *Princeton Theology,* 103, 100, 104.

82. Ibid., 104.

83. See the discussion in Numbers, "Charles Hodge," 77–102. On Green, see Numbers, "'Most Important Biblical Discovery,'" 257–76.

84. [Hodge], "Unity of Mankind," 103–49; and "Examination of Some Reasonings," 435–64.

85. Brock, "Common Origin of the Human Species," 115–43.

86. Noyce, *Outlines of Creation,* 327–38.

87. Miller, "Unity of the Human Races," 392, 396. Later Miller seems to have rejected this view when he asserted: "I do not see how we are to avoid the conclusion that this Caucasian type was the type of Adamic man." Miller, *Testimony of the Rocks,* 251–52.

88. Duns, *Science and Christian Thought,* 207. A few scattered biographical fragments are available in Watt, *New College Edinburgh,* 55–57, 248–49.

89. Ibid., 301.

90. Ibid.

91. Ibid., 298.

92. Ibid., 292.

93. Duns, *Biblical Natural Science,* 537. Duns's final reference was probably to Randolph's *Pre-Adamite Man.*

94. Astore, "Gentle Skeptics?" 40–76. The following paragraphs on Catholic attitudes to polygenism draw on Astore's analysis.

95. Wiseman, *Twelve Lectures,* 99.

96. Ibid., 77.

97. Ibid., 124.

98. Ibid., 125.

99. Ibid., 138, 139.

100. Ibid., 142.

101. Reusch, *Nature and the Bible,* 2:200, 217, 225.

102. Ibid., 2:235.

103. Walworth, *Gentle Skeptic,* 332.

104. Brownson, *American Republic,* 351. On Brownson more generally, see Gilhooley, *Contradiction and Dilemma.*

105. Brownson, "Faith and Reason," 137.

106. Reusch, *Nature and the Bible,* 209–10.

107. Ibid., 232.

108. Brownson, "Faith and the Sciences." Available online at http://orestes brownson.com/index.php?id=98.

109. See Helmstadter, "Condescending Harmony," 167–95.

110. Smith, *On the Relation*, 387.

111. Ibid., 389.

112. Ibid., 390.

113. Ibid., 391.

114. Ibid., 391–92.

115. Ibid., 391.

116. Ibid., 393.

117. Thompson, *Photographic Views of Egypt*.

118. Thompson, *Question of Races;* and *Teachings of the New Testament*.

119. Thompson, *Man in Genesis and Geology*, 106, 107.

120. Ibid., 98.

121. Ibid., 101, 103. One reviewer of this volume presented readers with the central question: "If it could be proved that man is historically developed from an ape, or that human beings existed on the earth before Adam, would the Mosaic narrative of the creation be impeached?" In this writer's opinion Thompson had approached the question with a "candid and liberal spirit" given "the instability of scientific theories." Review of *Man in Genesis and Geology*, 786, 787.

122. Ponton, *Beginning*, 505–6.

123. Ponton spelled out eleven defects of Darwin's theory in *Beginning*, chap. 31.

124. See Foucault, *Order of Things;* Gillespie, *Charles Darwin*.

6. ANCESTORS

1. Greene, *Death of Adam*.

2. Robinson, *Death of Adam*.

3. See Gruber, *Conscience in Conflict;* O'Leary, *Roman Catholicism and Modern Science,* chap. 4.

4. I have examined Murphy's views in Livingstone, "Darwin in Belfast," 387–408.

5. Mivart, *On the Genesis of Species,* 300.

6. Ibid., 303.

7. See Gilley and Loades, "Thomas Henry Huxley," 285–308. See also Desmond, *Huxley,* 25.

8. Desmond, *Huxley,* 340.

9. O'Leary, *Roman Catholicism and Modern Science,* 93.

10. See Artigas, Glick, and Martínez, *Negotiating Darwin,* chap. 7.

11. Stocking, "Persistence of Polygenist Thought," 46.

12. Biographical details are available in Davenport, "Alexander Winchell," 185–201; and Yoder, "Winchell, Alexander," 14:438–40. Aspects of his thought are also discussed in Haller, *Outcasts from Evolution;* Livingstone, "History of Science and the History of Geography," 271–302; and *Darwin's Forgotten Defenders,* 85–92.

13. Winchell's style of natural theology is revealed in Winchell, *Theologico-Geology; Creation;* and *Reconciliation of Science and Religion,* 156–77. On different conceptions of design in the period, see Bowler, "Darwinism and the Argument from Design," 29–43; and Livingstone, "Idea of Design," 329–57.

14. Winchell, *Sketches of Creation,* 34.

15. Winchell, *Doctrine of Evolution,* 8.

16. Ibid., 95, 97.

17. Ibid., 98.

18. Ibid., 8. Cope's Lamarckianism is discussed in Bowler, "Edward Drinker Cope," 249–65.

19. Winchell, *Reconciliation of Science and Religion,* v.

20. Winchell, *Doctrine of Evolution,* 17.

21. Winchell, "Huxley and Evolution," 295.

22. Winchell's letter is reproduced *in extenso* in Alberstadt, "Alexander Winchell's Preadamites," 97–112.

23. Cited in ibid., 108.

24. Cited in ibid.

25. It was reported widely in newspapers at the time, including the *Nashville Banner, Chattanooga Commercial, Memphis Appeal, Knoxville Chronicle,* and *Rochester Democrat and Chronicle.* See Davenport, "Scientific Interests," 500–521.

26. Conkin, *Gone with the Ivy,* 51.

27. Davenport, "Scientific Interests," 517.

28. Cited in Alberstadt, "Alexander Winchell's Preadamites," 109.

29. Winchell, "Religious Ideas," 5–26; and "Religious Nature of Savages," 357–78.

30. Winchell, *Sketches of Creation,* 350.

31. Cited in Alberstadt, "Alexander Winchell's Preadamites," 104.

32. Winchell, *Adamites and Preadamites;* and "Preadamite," 8: q.v.

33. Winchell, *Preadamites.*

34. Winchell, "Preadamite," 486.

35. Ibid., 484.

36. Cited in Alberstadt, "Alexander Winchell's Preadamites," 110.

37. Winchell, *Adamites and Preadamites,* 6; Winchell, "Preadamite," 484. In his later 1880 volume he also identified the hints of pre-adamism detectable in the works of figures such as Bory de Saint-Vincent and Bernard Hombron. While he did not specify his sources, it seems likely that he had in mind Bory de Saint-Vincent, *L'Homme;* and Hombron, *Aventures.*

38. Winchell, *Preadamites,* 457.

39. Alberstadt, "Alexander Winchell's Preadamites," 107.

40. Winchell, *Preadamites,* 155.

41. Winchell, *Adamites and Preadamites* 26–27.

42. Ibid., 19.

43. Ibid., 20.

44. Winchell, *Preadamites,* v.

45. Ibid., 286.

46. See Brooke, "Natural Theology," 221–86; Baxter, "Brewster," 45–50.

47. Thus, Chalmers, *Series of Discourses,* discourse 4; Brewster, *More Worlds than One.*

48. Winchell, *Preadamites,* 290.

49. Winchell, *Adamites and Preadamites,* 21.

50. Ibid., 42–43.

51. Winchell, *Preadamites,* 412–13.

52. Ibid., 7.

53. The idea of black originality in the pre-adamite thinking of Winchell and others has been identified as a significant backdrop to comparable debates on race and identity among black American Muslims. See Allen, "Religious Heterodoxy and Nationalist Tradition," 2–34; and "Identity and Destiny," 163–214.

54. Winchell, *Preadamites,* 161.

55. See Bucke, *History of American Methodism,* 2:189–90. Whedon regularly solicited articles from Winchell on scientific matters for the serial. See ibid., 385.

56. [Whedon], review of McCausland, *Adam and the Adamite,* 153–55.

57. Winchell, *Preadamites,* 287.

58. [Whedon], review of Winchell, *Adamites and Preadamites,* 567; and review of Fontaine, *How the World Was Peopled,* 521–23.

59. Winchell, *Preadamites,* 288.

60. Editorial footnotes to Winchell, "Preadamites."

61. "Professor Winchell's 'Preadamites,'" 86–90.

62. Colman, "Pre-Adamites," 902, 891.

63. Haeckel, *History of Creation,* 2:304.

64. See Clements, "Study of the Old Testament," 3:109–41. See also Chadwick, *Secularization of the European Mind;* Frei, *Eclipse of Biblical Narrative;* Rogerson, *Old Testament Criticism.*

65. Brief biographical details are available in "Shields, Charles Woodruff," in Wilson and Fiske, *Appleton's Cyclopaedia,* s.v.; and Harper, "Shields, Charles Woodruff," s.v.

66. Shields, *Philosophia Ultima,* 1:69, 370. Shields had already outlined the same options in *Religion and Science.*

67. Shields, *Scientific Evidences,* 124.

68. Ibid., 2:197.

69. Ibid., 1:169.

70. Ibid., 1:163, 170.

71. Ibid., 2:343.

72. Shields, *Scientific Evidences,* 118.

73. The story surrounding the appointment is discussed in Gundlach, "Evolution Question at Princeton."

74. I have discussed something of Macloskie's project in Livingstone, *Darwin's Forgotten Defenders;* and Livingstone and Wells, *Ulster-American Religion.*

75. For example, Macloskie, "Scientific Speculation," 617–25; "Concessions to Sci-

ence," 220–28; "Theistic Evolution," 1–22; "Outlook of Science and Faith," 597–615; and "Mosaism and Darwinism," 425–41.

76. Macloskie, "Preliminary Talks on Science and Faith," 27.

77. Macloskie, "Theistic Evolution," 4, 5.

78. Macloskie, "Origin of New Species," 262.

79. Ibid., 267.

80. Ibid., 273.

81. See the discussion in Livingstone and Noll, "B. B. Warfield," 283–304.

82. Warfield, review of *Philosophia Ultima,* 541–42.

83. Warfield, "On the Antiquity," 1–25, reprinted in *Biblical and Theological Studies,* 256.

84. Popkin, *Peyrère,* 30.

85. Warfield, "On the Antiquity," 256.

86. Warfield, "Calvin's Doctrine," 208, 209. Warfield's interpretation of Calvin on this point has been rejected by Murray, "Calvin's Doctrine of Creation," 21–43.

87. Warfield, review of Orr, *God's Image in Man,* 555–58.

88. Short, *Bible and Modern Research,* 57.

89. Short, *Modern Discovery,* 114.

90. In some ways, though without explicit reference to Adam, this arrangement was similar to that of the Cambridge prehistorian Miles Burkitt during the 1920s and 1930s, who wrote for the modernist wing of the Anglican Church and considered that the generation of the soul—humanization—was likely to have taken place during the Upper Paleolithic, a period marked by the disappearance of Neanderthal man. See Burkitt, "Some Reflections," 347–57. I am grateful to Pamela J. Smith for drawing this to my attention. See Smith, "Splendid Idiosyncrasy."

91. Short, *Modern Discovery,* 114.

92. Ibid., 49–53, 65, 101, 115.

93. See Brundell, "Catholic Church Politics," 81–95.

94. See Artigas, Glick, and Martínez, *Negotiating Darwin.* The authors discuss several controversial cases at the time.

95. Zahm, *Evolution and Dogma,* 354.

96. He made approving reference to a pamphlet by a Jesuit, François Dierckx, on ape men and Adam's precursors. Dierckx, *L'Homme-Singe.* A Spanish translation appeared the following year. This work was critical of Leroy's claim that Adam's body was the product of evolutionary forces whereas Eve was immediately formed by direct divine creation.

97. Zahm, *Evolution and Dogma,* 355.

98. Ibid., 357.

99. Appleby, "Exposing Darwin's 'Hidden Agenda,'" 173–207.

100. Quoted in Garrigan, "Preadamites," 11:702.

101. Johnson, "Problem of Prehistoric Man," 211.

102. Quoted in Lattey, "Encyclical 'Humani Generis,'" 279.

103. Johnson, *Anthropology and the Fall,* 51.

104. Garrigan, "Preadamites," 702.

105. Joly, *Man before Metals,* 186.

106. Mass, "Preadamites," 12:370–71. See also Amann, "Préadamites," 6:2793–2800.

107. See De Bont, "Rome and Theistic Evolutionism," 457–78.

108. Dorlodot, *Darwinism and Catholic Thought,* 109–10.

109. O'Leary, *Roman Catholicism and Modern Science,* 126–28.

110. Gross, "Problem of Origins," 130. Trans. from *Revue des sciences religieuses* for January 1933.

111. See Messenger, *Evolution and Theology,* 277, 279.

112. Messenger, intro. in ibid., 1.

113. Bernard Ramm, for example, later commended the volume as an exemplary piece of theological probing, "a masterpiece of erudition" as he called it, that evangelical Protestantism would do well to emulate. Ramm, *Christian View,* 225.

114. Rhodes, "Problem of Man's Origin," 9. Other reviewers are discussed in O'Leary, *Roman Catholicism and Modern Science,* 136–37.

115. Messenger, "Evolution and Theology To-Day," 194–195. In this volume Messenger also wrote the following articles: "Outline of Embryology, in the Light of Modern Science" (221–32); "A Short History of Embryology" (233–42); and "The Embryology of St. Thomas Aquinas" (243–58). The volume also included the following pieces by Henry de Dorlodot: "A Vindication of the Mediate Animation Theory" (259–83); "An Objection from Moral Theology: The Question of Abortion and the Mediate Animation Theory" (301–12); "A Formal Answer to Objections against the Mediate Animation Theory" (313–26).

116. In his mind, therefore, advocates of the immediate animation theory, such as W. McGarry, generally opposed human evolution, while defenders of the mediate theory welcomed it.

7. BLOODLINES

1. Kitson, "'Candid Reflections,'" 11–25. See also Wheeler, *Complexion of Race.*

2. Levy, "How the Dismal Science Got Its Name," 5.

3. Peart and Levy, *"Vanity of the Philosopher,'"* 5.

4. Quoted in Levy, "How the Dismal Science Got Its Name," 22.

5. See chap. 3 and the bibliographical details provided there.

6. Quoted in Stocking, *Victorian Anthropology,* (249.

7. Hunt, "On the Application of Natural Selection," 320–40.

8. Hunt, "On Ethno-Climatology," 50–79. I have discussed the issue of acclimatization in Livingstone, "Human Acclimatization," 359–94.

9. Knox, *Races of Men,* 2. The second edition was entitled *The Races of Men: A Philosophical Enquiry into the Influence of Race over the Destinies of Nations.*

10. Biographical details are available in Lonsdale, *Sketch of the Life and Writings.* Various aspects of Knox's thinking are discussed in Rehbock, *Philosophical Naturalists,* chap. 2: "Robert Knox: Idealism Imported"; and Richards, "'Moral Anatomy' of Robert Knox," 373–436.

11. Emerson, *Emerson's Complete Works,* vol. 6: *Conduct of Life,* 21.

12. Hunt, *Negro's Place in Nature,* 7.

13. "Hunt on the Negro," li.

14. Knox, *Races of Men,* 6.

15. "Plurality of the Human Race," 121.

16. Ibid., 123.

17. Ibid., 130.

18. Stepan, *Idea of Race in Science,* 44.

19. Nott et al., *Types of Mankind,* 45.

20. Augstein, *James Cowles Prichard's Anthropology,* 233.

21. See Nelson, "'Men before Adam,'" 161–81.

22. Gliddon, "Monogenists and the Polygenists ," 410, 411, 440.

23. Cited in Stanton, *Leopard's Spots,* 176.

24. On these figures, see Stanton, *Leopard's Spots;* Gould, *Mismeasure of Man,* 50–68; Bieder, *Science Encounters the Indian,* 55–103; Horsman, *Josiah Nott of Mobile.*

25. Morton, *Crania Americana;* Nott, "Mulatto," 252–56; Gliddon, *Ancient Egypt.*

26. Gould, "Morton's Ranking of Races," 503–9.

27. Meigs, "Cranial Characteristics," 203–352.

28. I have discussed this case in Livingstone, "Moral Discourse of Climate," 413–34.

29. Meigs, "Cranial Characteristics," 299.

30. On the role of art in earlier projections of racial hierarchy, see Bindman, *Ape to Apollo;* Higgins, "Art, Genius, and Racial Theory," 17–40.

31. For a general analysis of the rhetorical role of scientific illustrations, see Daston and Galison, "Image of Objectivity," 81–128. For the use of art in anthropology, see Cowling, *Artist as Anthropologist.* The later use of photography for similar purposes is discussed in Green, "Veins of Resemblance," 3–16.

32. Gliddon, "Monogenists and the Polygenists," 547, 548.

33. Gliddon, "On the Ethnographic Tableau," in 637.

34. Gliddon, "On the Geographical Distribution," 650.

35. Morton, *Crania Americana,* 82, 54.

36. Morton, "Hybridity in Animals," 39–50, 203–12.

37. Nott, *Two Lectures,* 40–41.

38. Gliddon, "Monogenists and the Polygenists," 443.

39. Frederickson, *Black Image in the White Mind,* 78. Chap. 3 deals with "Science, Polygenesis, and the Proslavery Argument." See also Frederickson, *Racism.*

40. Bachman, *Doctrine of the Unity of the Human Race.* See the discussion in Stanton, *Leopard's Spots,* 123–36; and especially Stephens, *Science, Race, and Religion.*

41. Bachman, review of *Types of Mankind,* 627–59.

42. Smyth, *Unity of the Human Races,* 110–11, xv.

43. Ibid., xvii, xix.

44. Cabell, *Testimony of Modern Science,* 22.

45. Ibid., 158.

46. [Hodge], "Unity of Mankind," 112. See the discussion in Stewart, "Mediating the Center."

47. [Hodge], "Unity of Mankind," 132.

48. Bachman, review of *Types of Mankind,* 657.

49. Smyth, *Unity of the Human Races,* 333–34.

50. Ibid., 337.

51. See Smith, *In His Image.*

52. See the discussions in Carwardine, *Evangelicals and Politics;* Genovese, *"Slavery Ordained of God";* Genovese and Fox-Genovese, *Mind of the Master Class;* Stout, *Upon the Altar of the Nation;* Noll, *Civil War as a Theological Crisis.*

53. Armstrong, *Christian Doctrine of Slavery.*

54. [Howe], "Nott's Lectures," 487.

55. Dabney, *Defence of Virginia.*

56. Dabney, "New South."

57. Noll, "Bible and Slavery," 64.

58. The connections between pro-slavery and antievolution among the southern Presbyterians are explored in Hampton, "'Handmaid' or 'Assailant.'"

59. Junkin, *Integrity of Our National Union;* Stringfellow, *Scriptural and Statistical Views.*

60. Haynes, *Noah's Curse.*

61. There is an extensive literature on southern honor. See in particular Wyatt-Brown, *Southern Honor;* and Patterson, *Slavery and Social Death.*

62. Holifield, *Theology in America,* chap. 25: "The Dilemma of Slavery."

63. Kidd provides an overall survey of black counter-theologies vis-à-vis racial issues in chap. 8 of *Forging of Races.*

64. Douglass, *Claims of the Negro,* 12.

65. Ibid., 16, 17, 12.

66. Ibid., 32.

67. "Chapters on Ethnology." See the discussion in Noll, *Civil War,* 69.

68. Delany, *Principia of Ethnology.*

69. Williams, *History of the Negro in America,* 2:416–17.

70. Tanner, "Descent of the Negro," 513–28; reprinted as *Descent of the Negro,* 23.

71. See, for example, Haynes, *Black;* Brent, "Ancient Glory of the Hamitic Race," 272–75.

72. Thompson, "Negro," 509.

73. Anderson, "Black Responses to Darwinism," 252. Anderson here discusses most of the figures I have surveyed in this paragraph. See also Little, "African Methodist Episcopal Church Media," 1–14.

74. Something of the way in which Winchell's negative evaluation of Africans allowed for other groups—notably Hawaiians—to be placed higher up the scale and much more positively represented is discussed in Desmond, "Picturing Hawai'i," 459–501.

75. Winchell, *Preadamites,* 192.

76. Winchell, *Adamites and Preadamites,* 22, 23.

77. Winchell, *Preadamites,* 187.

78. See Stepan, *Idea of Race,* 78–81.

79. Huxley, "On the Methods and Results," 248.

80. Winchell, *Preadamites,* 253, 255.

81. Ibid., 81.

82. Ibid., 81, 82.

83. Ibid., 83, 85.

84. Ibid., 83, 80, 81.

85. Ibid., 285, 237, 252.

86. Stokes, "Someone's in the Garden," 722.

87. Dailey, "Sex, Segregation, and the Sacred," 125.

88. Kidd, *Forging of Races,* 147.

89. Cartwright, "Report on the Diseases," 691–715.

90. Cartwright, "Unity of the Human Race," 129, 130.

91. Ibid., 131.

92. Ibid., 134.

93. Frederickson, *Black Image,* 88.

94. Ariel's pamphlet has been discussed by several writers. See Frederickson, *Black Image,* 188–89; Stokes, "Someone's in the Garden," 722–23; Haynes, *Noah's Curse,* 112–13; Kidd, *Forging of Races,* 149–50.

95. Ariel, *Negro,* 10.

96. Ibid., 13, 14.

97. Ibid., 20, 21, 22.

98. Ibid., 27.

99. Ibid., 30.

100. Ibid., 46.

101. Ibid., 48.

102. Young, *Negro,* 15, 35.

103. Ibid., 4.

104. Ibid., 11.

105. Ibid., 20.

106. "Dr. Blackie's Letter," appended to ibid., 38 (the statement is italicized in the original).

107. Ibid., 40.

108. Ibid., 45.

109. Quoted in Stokes, "Someone's in the Garden," 723. On the issue of the genuineness of his authorship, see 739.

110. M.S., *Adamic Race,* 13.

111. Ibid., 69.

112. Ibid., 25, 26.

113. Ibid., 35.

114. Ibid., 54.

115. Ibid., 69.

116. Stokes, "Someone's in the Garden," 724.

117. Prospero, *Caliban,* excerpted in Kvam, Schering, and Ziegler, *Eve and Adam,* 491–95.

118. Minister, *Nachesh,* 486–91.

119. Stokes, "Someone's in the Garden," 724–26.

120. Lester, *Pre-Adamites,* 22.

121. Ibid., 25.

122. Ibid., 26.

123. Hasskarl, *"Missing Link,"* 9.

124. Ibid., 19, 41.

125. Bolding, *"What of the Negro Race,"* 2.

126. Ibid., 12, 27.

127. Ibid., 7, 36–37.

128. Carroll, *"Negro a Beast,"* 148, 39.

129. Ibid., 22, 48.

130. Ibid., 63.

131. One horrified reviewer, Edward Atkinson, described it as "the most sacrilegious book ever issued from the press in this country" and lamented that it "was securing a very wide circulation among the poor whites of the Cotton States." Atkinson, "Negro a Beast," 202.

132. Carroll, *"Negro a Beast,"* 66.

133. Carroll, *Tempter of Eve.*

134. Carroll, *"Negro a Beast,"* 120.

135. Harvey, *Freedom's Coming,* 43.

136. Caucasian, *Anthropology for the People,* 97, 213.

137. Ibid., 291, 267.

138. Ibid., 301, 327–28.

8. SHADOWS

1. See Marsden, *Fundamentalism and American Culture,* 47–48.

2. See Numbers, *Creationists,* 53.

3. Torrey, *Difficulties and Alleged Errors,* 31, 36.

4. Biographical details are available in Fleming, *Memories of a Scientific Life;* Süsskind, "Fleming, John Ambrose," q.v. His accomplishments in electrical engineering are recorded in MacGregor-Morris, "Sir Ambrose Fleming (Jubilee of the Valve)," 134–44.

5. Numbers, *Creationists,* 164.

6. These aims and objectives appeared on the inside cover of each issue of the institute's *Transactions.*

7. See also Fleming, "Evolution and Revelation," 11–40.

8. Fleming, *Evolution or Creation,* 21.

9. Ibid., 71. Fleming had been attacking this viewpoint at least since 1928. In a piece published in *Nineteenth Century and After* he had written of the "difficulties in

the acceptance of this spontaneous development of man from an ape-like progenitor." See *Creation or Evolution*, 68. On Bishop Barnes and science, see Bowler, "Evolution and the Eucharist," 453–67; and *Reconciling Science and Religion*, 260–70.

10. Fleming, *Evolution or Creation*, 79.

11. Fleming, "Modern Anthropology," 15–42.

12. "Evolution and Our Faith"; also reported in "Age of Man and the Earth."

13. Fleming, *Origin of Mankind*, 149.

14. Ibid., 75.

15. Ibid., 115.

16. Ibid., 116.

17. Ibid., 143–44.

18. Keith, *Darwinism and What It Implies*, 34.

19. Keith, *Darwinism and Its Critics*, 24, 20–21.

20. Lang, "Human Origin and Christian Doctrine," 168–70.

21. Review of *Origin of Mankind*, 186.

22. Berkhof, *Systematic Theology*, 188–90.

23. Fisher, *Origins Solution*, 99, 105.

24. Ibid., 28.

25. Ibid., 194.

26. Ibid., 376, 378.

27. See the discussion in O'Leary, *Roman Catholicism and Modern Science*, 149–59.

28. An introduction to some of the immediate commentaries at the time in English, French, German, Spanish, and Italian is provided in Weigel, "Gleanings," 520–49.

29. For example, Asensio, "De Persona Adae," 464–526; Marazuela, "Poligenismo y Evoluzionismo," 347–72; Havet, "L'Encyclique," 114–27; Picard, "La science expérimentale," 65–89; Bataini, "Monogénisme et polygénisme," 363–69; Denis, *Les origines du monde;* Vandebroek and Renwart, "'Humani Generis,'" 3–20; Cordero, "Evolucionismo," 465–75, 477–79; Labourdette, *Le péché originel;* Colombo, "Transformismo," 17–43; Ceuppens, "Le polygénisme," 20–32. These works are a small sample of a larger body of Catholic literature on the subject. A more comprehensive listing is to be found in Rahner, *Theological Investigations*, 1:229–30.

30. Bea, "Die Enzyklika 'Humani Generis,'" 36–56.

31. See Weigel, "Gleanings," 544. Even those who gave a scientific assessment of the question of the unity of the species still reviewed the early pre-adamite theory of La Peyrère. See Carles, "Polygénisme ou monogénisme," 84–100.

32. O'Brien, *God and Evolution*, 120.

33. Ewing, "Human Evolution," 294.

34. Lienart, "Evolution Harmonizes with Christian Faith," 307.

35. O'Leary, *Roman Catholicism*, 154.

36. Rahner, "Theological Reflexions," 244.

37. Ibid., 294–95.

38. Rahner, *Hominization*, 93–94.

39. Rahner, "Theological Reflexions," 295. See also Mascall's Bampton Lectures for 1956, *Christian Theology and Natural Science*, 254–89, on "Man's Origin and Ancestry." Rahner's polygenetic inclinations are also evident in his essay "Evolution and Original Sin," 61–73, in which he considers that it would be "better if the magisterium refrained from censuring polygenism" (64).

40. Kasujja, *Polygenism.*

41. Brunner, *Christian Doctrine*, 50, 81

42. Kidner, *Genesis*, 28–29.

43. Stott, *Understanding the Bible*, 5.

44. Berry, *Adam and the Ape*, 44. Berry specifically uses Kidner's federal theology as a means of preserving human spiritual unity without committing to physical descent of all humankind from Adam.

45. Ramm, *Christian View*, 222.

46. Pearce, *Who Was Adam?*

47. Kidd, *Forging of Races*, 218.

48. See, for example, the discussion of John Adair's 1775 work in chap. 3.

49. See Kidd, *Forging of Races*, 203–18.

50. The movement is charted in Barkun, *Religion and the Racist Right.*

51. *In the Image of God*, viii.

52. Ibid., v.

53. Ibid., vi, vii.

54. Ibid., 21.

55. Swift, *Were All the People*, 2.

56. Swift, *God, Man, Nations & the Races*, excerpted in Kvam, Schering, and Ziegler, *Eve and Adam*, 498.

57. Comparet, *By Divine Appointment*, 1, 2, 8. Numerous sermons by Comparet are available online with titles such as "Adam Was Not the First Man," "Christianity Discriminates," "God's Immigration Laws," and "The Cain-Satanic Seedline"; see www.churchoftrueisrael.com/comparet/.

58. Magne, *Negro and the World Crisis*, 99.

59. Ramsey, *Racial Difference*, 3, 1, 35.

60. Weisman, *Not of One Blood;* Allen, *False Biblical Teachings.*

61. This paragraph, and other passages in which moral language finds its way into my narrative, are composed in full acknowledgment of Philip Morgan's telling observation that the "historians' greatest moral obligation" is "to enter sympathetically into a milieu different from their own, bring it to life, and make sense of it." Yet, as he himself acknowledges, "historians cannot escape making moral judgments." Morgan, "Morality and Slavery," 397, 393. In my opinion the mind-set depicted here constitutes one of those circumstances in which robust language is justified. Morgan's observations are drawn from his contribution to a discussion on "Historians and Moral Judgments" convened by Nicholas Canny. In this exchange Hermann W. von der Dunk, reflecting on Nazi crimes, notes: "The study of such episodes . . . raises the question of whether the classical directive for the historian, to explain events so that people can better understand them, does not collide with the more fundamen-

tal moral imperative which forbids us to ignore or belittle the suffering of innocent victims." Von der Dunk, "German Historians and the Crucial Dilemma," 382. Chris Bayly, too, reflecting on his own work, confesses that it "was very difficult to write" about empire in the late 1930s and early 1940s in India and Southeast Asia "without feeling that Empire *at this time and in these places* had become irremediably morally bankrupt." Bayly, "Moral Judgment," 389–90. In writing here about pre-adamist racism over the past century, I can only echo these judgments.

62. Gayman, *Biblical Anthropology;* Allen, *Species of Men.*

63. Kidd, *Forging of Races,* 222.

64. See Kvam, Schering, and Ziegler, "Preadamite Theory," 483–502; Barkun, *Religion and the Racist Right;* Aho, *Politics of Righteousness;* Swain, *New White Nationalism in America.*

65. Butler, "Who, What, Why, When, Where," 500.

66. Rogoff, "Is the Jew White," 195–230.

67. McCausland, *Adam and the Adamite,* 300.

9. DIMENSIONS

1. This epitaph appeared in "Epigram, Part II," 719.

2. More generally on the location of knowledge, see Livingstone, *Putting Science in Its Place*; and Withers, *Placing the Enlightenment.*

3. London, *Before Adam,* 76.

4. Eiseley, "Epilogue," 105.

BIBLIOGRAPHY

Adair, Douglass. "'That Politics May Be Reduced to a Science': David Hume, James Madison, and the Tenth Federalist." In *Fame and the Founding Fathers: Essays,* ed. Trevor Colbourn. New York: Norton, for the Institute of Early American History and Culture at Williamsburg, Va., 1974.

Adair, James. *The History of the American Indians; Particularly Those Nations Adjoining to the Mississippi, East and West Florida, Georgia, South and North Carolina, and Virginia.* London: printed for Edward and Charles Dilly, 1775.

Agassiz, Louis. "The Diversity of Origin of the Human Races." *Christian Examiner and Religious Miscellany* 49 (1850): 110–45.

———. "Geographical Distribution of Animals." *Christian Examiner and Religious Miscellany* 48 (1850): 181–204.

"Age of Man and the Earth." *Times of London,* January 15, 1935.

Aho, James A. *The Politics of Righteousness: Idaho Christian Patriotism.* Seattle: University of Washington Press, 1990.

Alberstadt, Leonard. "Alexander Winchell's Preadamites—A Case for Dismissal from the Vanderbilt University." *Earth Sciences History* 13 (1994): 97–112.

Alexander, Archibald. "The Bible, A Key to the Phenomena of the Natural World." *Biblical Repertory and Princeton Review* 1 (1829): 101–20.

Allen, D. C. *The Legend of Noah: Renaissance Rationalism in Art, Science, and Letters.* Urbana: University of Illinois Press, 1963.

Allen, Ernest, Jr. "Identity and Destiny: The Formative Views of the Moorish Science Temple and the Nation of Islam." In *Muslims on the Americanization Path?* ed. Yvonne Yazbeck Haddad and John L. Esposito, 163–214. New York: Oxford University Press, 2000.

———. "Religious Heterodoxy and Nationalist Tradition: The Continuing Evolution of the Nation of Islam." *Black Scholar* 26 (Fall–Winter 1996): 2–34.

Allen, Thomas Coley. *False Biblical Teachings on the Origins of the Races and Interracial Marriages.* Franklington, N.C.: T. C. Allen, 2001.

———. *Species of Men: A Polygenetic Hypothesis.* Franklinton, N.C.: T. C. Allen, 1999.

Almond, Philip C. *Adam and Eve in Seventeenth Century Thought.* Cambridge: Cambridge University Press, 1999.

[al-Tajir, Sulayman]. *Ancient Accounts of India and China, by Two Mohammedan Travellers. Who Went to Those Parts in the 9th Century; Translated from the Arabic, by the Late Learned Eusebius Renaudot. With Notes, Illustrations, and Inquiries by the Same Hand.* London, 1733.

Alter, Stephen G. *Darwin and the Linguistic Image: Language, Race, and Natural Theology in the Nineteenth Century.* Baltimore: Johns Hopkins University Press, 1999.

Amann, E. "Préadamites." In *Dictionnaire de théologie Catholique contenant l'exposé des doctrines de la théologie Catholique. Leurs preuves et leur histoire,* 6:2793–2800. Paris: Librairie Letouzey, 1935.

Anderson, Eric D. "Black Responses to Darwinism, 1859–1915." In *Disseminating Darwinism: The Role of Place, Race, Religion, and Gender,* ed. Ronald L. Numbers and John Stenhouse, 247–66. New York: Cambridge University Press, 1999.

"The Antiquity of Man." *Anthropological Review* 7 (1869): 1136–52.

Appleby, R. Scott. "Exposing Darwin's 'Hidden Agenda': Roman Catholic Responses to Evolution, 1875–1925." In *Disseminating Darwinism: The Role of Place, Race, Religion, and Gender,* ed. Ronald L. Numbers and John Stenhouse, 173–207. New York: Cambridge University Press, 1999.

Ariel [Buckner H. Payne]. *The Negro: What Is His Ethnological Status? Is He the Progeny of Ham? Is He a Descendant of Adam and Eve? Has He a Soul? Or Is He a Beast in God's Nomenclature? What Is His Status as Fixed by God in Creation? What Is His Relation to the White Race?* Cincinnati: Publisher for the Proprietor, 1867.

Armstrong, George D. *The Christian Doctrine of Slavery.* New York: Charles Scribner's, 1857.

Artigas, Mariano, Thomas F. Glick, and Rafael A. Martínez. *Negotiating Darwin: The Vatican Confronts Evolution, 1877–1902.* Baltimore: Johns Hopkins University Press, 2006.

Asensio, F. "De Persona Adae et de Peccato Originali Originante Secundum Genesim." *Gregorianum* 29 (1948): 464–526.

Astore, William J. "Gentle Skeptics? American Catholic Encounters with Polygenism, Geology, and Evolutionary Theories from 1845 to 1875." *Catholic Historical Review* 82 (1996): 40–76.

Atkins, John. *The Navy-Surgeon: Or, a Practical System of Surgery. Illustrated with Observations on Such Remarkable Cases as Have Occurred to the Author's Practice in the Service of the Royal Navy.* London: printed for Caesar Ward and Richard Chandler, 1734.

———. *A Voyage to Guinea, Brasil, and the West-Indies; In His Majesty's Ships, the Swallow and Weymouth.* London: printed for Caesar Ward and Richard Chandler, 1735.

Atkinson, Edward. "The Negro a Beast." *North American Review* 181 (1905): 202.

Augstein, H. F. *James Cowles Prichard's Anthropology: Remaking the Science of Man in Early Nineteenth-Century Britain.* Clio Medica, 52. Amsterdam: Editions Rodopi, 1999.

Bachman, John. *The Doctrine of the Unity of the Human Race Examined on the Principles of Science.* Charleston: Canning, 1850.

————. Review of *Types of Mankind*. *Charleston Medical Journal and Review* 9 (1854): 627–59.

Baldwin, Samuel Davies. *Dominion; or, the Unity and Trinity of the Human Race; with the Divine Political Constitution of the World, and the Divine Rights of Shem, Ham, and Japheth*. Nashville: Stevenson and F. A. Owen, 1858.

Barkun, Michael. *Religion and the Racist Right*, rev. ed. Chapel Hill: University of North Carolina Press, 1997.

Barr, James. "Luther and Biblical Chronology." *Bulletin of the John Rylands University Library* 72 (1990): 51–67.

————. "Pre-Scientific Chronology: The Bible and the Origin of the World." *Proceedings of the American Philosophical Society* 143 (1999): 379–87.

————. "Why the World Was Created in 4004 B.C.: Archbishop Ussher and Biblical Chronology." *Bulletin of the John Rylands University Library of Manchester* 67 (1985): 575–608.

Bataini, J. "Monogénisme et polygénisme. Une explication hybride." *Divus Thomas (Piac.)* 30 (1953): 363–69.

Baxter, Paul. "Brewster, Evangelism and the Disruption of the Church of Scotland." In *"Martyr of Science": Sir David Brewster, 1781–1868*, ed. A. D. Morrison-Low and J. R. R. Christie, 45–50. Edinburgh: Royal Scottish Museum, 1984.

Bayly, Christopher A. "Moral Judgment: Empire, Nation and History." *European Review* 14 (2006): 385–91.

Bea, Augustin. "Die Enzyklika 'Humani Generis': Ihre Grundgedanken und Ihre Bedeutung." *Scholastik* 26 (1951): 36–56.

Beattie, James. *Elements of Moral Science*. Edinburgh: printed by Mundell, Doig, and Stevenson, 1807.

Bendysche, Thomas. "The History of Anthropology." *Memoirs Read before the Anthropological Society of London* 1 (1863–64): 335–420.

Bennett, Jim, and Scott Mandelbrote. *The Garden, the Ark, the Tower, the Temple: Biblical Metaphors of Knowledge in Early Modern Europe*. Oxford: Museum of the History of Science, 1998.

Berkhof, Louis. *Systematic Theology*. 1939. Reprint. London: Banner of Truth, 1971.

Bernasconi, Robert. "Who Invented the Concept of Race? Kant's Role in the Enlightenment Construction of Race." In *Race*, ed. Robert Bernasconi, 11–36. Oxford: Blackwell, 2001.

Berry, R. J. *Adam and the Ape: A Christian Approach to the Theory of Evolution*. London: Falcon, 1975.

Berti, Silvia. "At the Roots of Unbelief." *Journal of the History of Ideas* 56 (1995): 555–75.

Bieder, Robert E. *Science Encounters the Indian, 1820–1880: The Early Years of American Ethnology*. Norman: University of Oklahoma Press, 1986.

Bindman, David. *Ape to Apollo: Aesthetics and the Idea of Race in the 18th Century*. Ithaca: Cornell University Press, 2002.

Blake, Carter. Review of articles by Huxley, Owen, Webb, and Gratiolet. *Edinburgh Review* (April 1863): 541–69.

Blount, Charles. *The Oracles of Reason*. London, 1693.

Blumenbach, Johann Friedrich. *De Generis Humani Varietate Nativa*. 3rd ed. Göttingen: Vandenhoek and Ruprecht, 1795.

Bolding, B. J. *"What of the Negro Race?" Bolding vs. Hasskarl*. Chambersburg: Democratic News, ca. 1898.

Bory de Saint-Vincent, J.B.G. *L'Homme—Homo. Essai zoologique sur le genre humaine*. Paris, 1827.

Boucher de Perthes, Jacques. *Antiquités celtiques et antédiluviennes*. Paris, 1847.

Bowler, Peter J. "Darwinism and the Argument from Design: Suggestions for a Reevaluation." *Journal of the History of Ideas* 10 (1977): 29–43.

———. "Edward Drinker Cope and the Changing Structure of Evolutionary Theory." *Isis* 68 (1977): 249–65.

———. "Evolution and the Eucharist: Bishop E. W. Barnes on Science and Religion in the 1920s and 1930s." *British Journal for the History of Science* 31 (1988): 453–67.

———. *Reconciling Science and Religion: The Debate in Early Twentieth-Century Britain*. Chicago: University of Chicago Press, 2001.

Branson, Roy. "James Madison and the Scottish Enlightenment." *Journal of the History of Ideas* 40 (1979): 235–50.

Brent, George Wilson. "The Ancient Glory of the Hamitic Race." *AME Church Review* 12 (October 1895): 272–75.

Brewster, David. *More Worlds than One*. London: Murray, 1854.

Brice, William R. "Bishop Ussher, John Lightfoot and the Age of Creation." *Journal of Geological Education* 30 (1982): 18–24.

Brock, William. "The Common Origin of the Human Species." In *Lectures to Young Men; Delivered before the Young Men's Christian Association, in Exeter Hall, from November 21, 1848, to February 6, 1849*, 115–43. London: Jones, 1849.

Brogan, Hugh. "Clarkson, Thomas (1760–1846)." *Oxford Dictionary of National Biography*. Oxford: Oxford University Press, 2004–6.

Brooke, John Hedley. "Natural Theology and the Plurality of Worlds: Observations on the Brewster-Whewell Debate." *Annals of Science* 34 (1977): 221–86.

Browne, Janet. *The Secular Ark: Studies in the History of Biogeography*. New Haven: Yale University Press, 1983.

Brownson, Orestes. *The American Republic: Its Constitution, Tendencies, and Destiny*. New York: P. O'Shea, 1866.

———. "Faith and Reason." *Brownson's Quarterly Review* (April 1863): 129–60.

———. "Faith and the Sciences." *Catholic World* (December 1873). Available online at http://orestesbrownson.com/index.php?id=98.

Brundell, Barry. "Catholic Church Politics and Evolution Theory, 1894–1902." *British Journal for the History of Science* 34 (2001): 81–95.

Brunner, Emil. *The Christian Doctrine of Creation and Redemption*. Vol. 2: *Dogmatics*. Trans. Olive Wyon. London: Lutterworth, 1955.

Bucke, Emory Stevens, ed. *The History of American Methodism. In Three Volumes*. New York: Abingdon Press, 1964.

Buckland, William. *Geology and Mineralogy Considered with Reference to Natural Theology.* 2 vols. London: Pickering, 1836.

———. *Vindicae Geologicae: Or the Connexion of Geology with Religion.* Oxford: Oxford University Press, 1820.

Burke, Peter. "The Philosopher as Traveller: Bernier's Orient." In *Voyages and Visions: Towards a Cultural History of Travel,* ed. Jaś Elsner and Joan-Pau Rubiés, 124–37. London: Reaktion Books, 1999.

Burkitt, M. C. "Some Reflections on Man and Nature in the Light of Past and Recent Prehistoric Research." *Scientia* (December 1934): 347–57.

Burrow, John W. "Evolution and Anthropology in the 1860s: The Anthropological Society of London, 1863–1871." *Victorian Studies* 7 (1963): 137–54.

Butler, Richard G. "Who, What, Why, When, Where: Aryan Nations." Reprinted in *Eve and Adam: Jewish, Christian, and Muslim Readings on Genesis and Gender,* ed. Kristan E. Kvam, Linda S. Schering, and Valerie H. Ziegler, 500. Bloomington: Indiana University Press, 1999.

Cabell, J. L. *The Testimony of Modern Science to the Unity of Mankind; Being a Summary of the Conclusion Announced by the Highest Authorities in the Several Departments of Physiology, Zoölogy, and Comparative Philology in Favor of the Specific Unity and Common Origin of All the Varieties of Man.* New York: Robert Carter & Bros., 1859.

Caldwell, Charles. *Thoughts on the Original Unity of the Human Race.* New York: E. Bliss, 1830.

Campbell, John. "The Unity of the Human Race, Considered from an American Standpoint." *British and Foreign Evangelical Review* 29 (1880): 74–101.

Campbell, Mary B. *The Witness and the Other World: Exotic European Travel Writing, 400–1600.* Ithaca: Cornell University Press, 1988.

Carles, J. "Polygénisme ou monogénisme. Le problème de l'unité de l'espèce humaine." *Archives de philosophie* 17 (1954): 84–100.

Carrithers, David W. "The Enlightenment Science of Society." In *Inventing Human Science: Eighteenth-Century Domains,* ed. Christopher Fox, Roy Porter, and Robert Wokler, 232–70. Berkeley: University of California Press, 1995.

———. Introduction to *The Spirit of Laws by Montesquieu. A Compendium of the First English Edition,* ed., with an introduction, notes, and appendixes, David Wallace Carrithers, 1–88. Berkeley: University of California Press, 1977.

Carroll, Charles. *"The Negro a Beast" or "In the Image of God."* St. Louis: American Book and Bible House, 1900.

———. *The Tempter of Eve, or the Criminality of Man's Social, Political, and Religious Equality with the Negro, and the Amalgamation to Which These Crimes Inevitably Lead.* St. Louis: Adamic Press, 1902.

Carroll, Robert T. *The Common-Sense Philosophy of Religion of Bishop Edward Stillingfleet.* The Hague: Martinus Nijhoff, 1975.

Cartwright, Samuel A. "Report on the Diseases and Physical Peculiarities of the Negro Race." *New Orleans Medical and Surgical Journal* (May 1851): 691–715.

———. "Unity of the Human Race Disproved by the Hebrew Bible." *De Bow's Review* 29 (1860): 129–36.

Carwardine, Richard J. *Evangelicals and Politics in Antebellum America.* New Haven: Yale University Press, 1993.

Caucasian [William H. Campbell]. *Anthropology for the People: A Refutation of the Theory of the Adamic Origin of All Races.* Richmond, Va.: Everett Waddey, 1891.

Ceuppens, F. "Le Polygénisme et la Bible." *Angelicum* 24 (1947): 20–32.

Chadwick, Owen. *The Secularization of the European Mind in the Nineteenth Century.* Cambridge: Cambridge University Press, 1975.

Chalmers, Thomas. *The Evidence and Authority of the Christian Revelation.* Edinburgh: Blackwood, 1814.

———. *A Series of Discourses on the Christian Revelation, Viewed in Connection with the Modern Astronomy.* Glasgow: J. Smith, 1817.

"Chapters on Ethnology." *Christian Recorder,* February 23 and March 2, 1861.

Charnock, Richard Stephen. "The Science of Language." *Anthropological Review* 1 (1863): 193–215.

Charron, Pierre. *Of Wisdom.* London, 1612.

Christmas, Henry. *Echoes of the Universe: From the World of Matter and the World of Spirit.* Philadelphia: A. Hart, 1850.

Clarkson, Thomas. *An Essay on the Slavery and Commerce of the Human Species, Particularly the African.* London: J. Phillips, 1786.

Clements, R. E. "The Study of the Old Testament." In *Nineteenth Century Religious Thought in the West,* ed. Ninian Smart, John Clayton, Steven Katz, and Patrick Sherry, 3:109–41. 3 vols. Cambridge: Cambridge University Press, 1985.

Co-adamitae: or, an Essay to Prove the Two Following Paradoxes, viz. I. That There Were Other Men Created at the Same Time with Adam. II. That the Angels Did Not Fall. London: J. Wilford, 1732.

Colman, Henry. "Pre-adamites." *Methodist Review* 7 (1891): 891–902.

Colombo, G. "Transformismo Antropologico e Teologia." *Scuola Cattolica* 47 (1949): 17–43.

Comas, Juan. *Manual of Physical Anthropology.* Springfield, Ill.: Charles C. Thomas, 1960.

Comparet, Bertrand L. *By Divine Appointment.* San Diego, Calif.: Your Heritage, n.d.

Conkin, Paul K. *Gone with the Ivy: A Biography of Vanderbilt University.* Knoxville: University of Tennessee Press, 1985.

Cordero, M. García. "Evolucionismo, Poligenismo y Exegesis Biblica." *Ciencia* 78 (1951): 465–75, 477–79.

Cowling, Mary. *The Artist as Anthropologist: The Representation of Type and Character in Victorian Art.* Cambridge: Cambridge University Press, 1989.

Crampton, Jeremy W. "Cartography's Defining Moment: The Peters Projection Controversy 1974–1990." *Cartographica* 31, no. 4 (1994): 16–32.

———. *Mapping: A Critical Introduction to G.I.S. and Cartography.* Oxford: Blackwell, in press.

Crawfurd, John. "On Language as a Test of the Races of Man." *Transactions of the Ethnological Society of London,* n.s. 3 (1865): 1–8.

———. "On the Aryan or Indo-Germanic Theory." *Transactions of the Ethnological Society of London,* n.s. 1 (1861): 268–86.

———. "On the Classification of the Races of Man." *Transactions of the Ethnological Society of London,* n.s. 1 (1861): 354–78.

———. "On the Connexion between Ethnology and Physical Geography." *Transactions of the Ethnological Society of London,* n.s. 2 (1863): 4–23.

———. "On the Effects of Commixture, Locality, Climate, and Food on the Races of Man." *Transactions of the Ethnological Society of London,* n.s. 1 (1861): 76–92.

———. "On the Theory of the Origin of Species by Natural Selection in the Struggle for Life." *Transactions of the Ethnological Society of London,* n.s. 7 (1869): 27–38.

Curtin, Philip D. *The Image of Africa: British Ideas and Action, 1780–1850.* London: Macmillan, 1965.

Custance, Arthur. *Time and Eternity.* Vol. 6. Grand Rapids, Mich.: Zondervan, 1977.

Dabney, Robert L. *A Defence of Virginia: (and through Her, of the South) in Recent and Pending Contests against the Sectional Party.* New York: E. J. Hale, 1867.

———. "The New South." In *Discussions by Robert L. Dabney,* ed. C. R. Vaughan. Vol. 4. Mexico, Mo.: Crescent Book House, 1897.

Dagg, J. L. *The Elements of Moral Science.* New York: Sheldon, 1860.

Dailey, Jane. "Sex, Segregation, and the Sacred after *Brown.*" *Journal of American History* 91 (2004): 119–44.

Daniel, Glyn. *A Short History of Archaeology.* London: Thames and Hudson, 1981.

Darwin, Charles. *Essays and Reviews.* London: John W. Parker and Son, 1860.

———. *On the Origin of Species.* London: John Murray, 1859.

Daston, Lorraine, and Peter Galison. "The Image of Objectivity." *Representations* 40 (1992): 81–128.

Daston, Lorraine, and Katharine Park. *Wonders and the Order of Nature, 1150–1750.* New York: Zone Books, 1998.

Davenport, F. Garvin. "Alexander Winchell: Michigan Scientist and Educator." *Michigan History* 35 (1951): 185–201.

———. "Scientific Interests in Kentucky and Tennessee, 1870–1890." *Journal of Southern History* 14 (1948): 500–521.

Dean, Dennis R. "The Age of the Earth Controversy: Beginnings to Hutton." *Annals of Science* 38 (1981): 435–56.

Debenham, Frank. *Discovery and Exploration: An Atlas-History.* New York: Doubleday, 1960.

De Bont, Raf. "Rome and Theistic Evolutionism: The Hidden Strategies behind the 'Dorlodot Affair,' 1920–1926." *Annals of Science* 62 (2005): 457–78.

Delany, Martin R. *Principia of Ethnology: The Origin of Races and Color, with an Archaeological Compendium of Ethiopian and Egyptian Civilization from Years of Careful Examination and Enquiry.* Philadelphia: Harper & Bros., 1879.

Denis, Paul. *Les Origines du monde et de l'humanité.* Liege: Pensée Catholique, 1950.

Desmond, Adrian. *Huxley: Evolution's High Priest.* London: Michael Joseph, 1997.

———. *Huxley: The Devil's Disciple*. London: Michael Joseph, 1994.

Desmond, Jane C. "Picturing Hawai'i: The 'Ideal' Native and the Origins of Tourism, 1880–1915." *Positions* 7 (1999): 459–501.

Deveney, John Patrick. *Paschal Beverly Randolph: A Nineteenth-Century Black American Spiritualist, Rosicrucian, and Sex Magician*. SUNY Series in Western Esoteric Traditions. New York: State University of New York Press, 1996.

Dick, Steven J. *Plurality of Worlds: The Origins of the Extraterrestrial Life Debate from Democritus to Kant*. Cambridge: Cambridge University Press, 1982.

Dierckx, François. *L'Homme-Singe et les précurseurs d'Adam en face de la science et de la théologie*. Brussels: Société de Librairie, 1894.

Dobbs, Francis. *A Concise View, from History and Prophecy, of the Great Predictions in the Sacred Writings, That Have Been Fulfilled; Also of Those That Are Now Fulfilling, and That Remain to Be Accomplished*. London: printed for the author, 1800.

———. *First Volume of Universal History, Commencing with the Creation and Ending 536 Years before the Christian Era*. In *Letters to His Son*. London: printed for the author, 1787.

———. *Memoirs of Francis Dobbs, Esq. Also Genuine Reports of His Speeches in Parliament on the Subject of an Union, and His Prediction of the Second Coming of the Messiah; With Extracts from His Poem on the Millennium*. Dublin: Jones, 1800.

Dorlodot, Henry de. *Darwinism and Catholic Thought*. Trans. E. C. Messenger. London: Burns, Oates and Co., 1922.

———. "A Formal Answer to Objections against the Mediate Animation Theory." In *Evolution and Theology*, ed. E. C. Messenger, 313–26. London: Burns, Oates and Washburne, 1931.

———. "An Objection from Moral Theology: The Question of Abortion and the Mediate Animation Theory." In *Evolution and Theology*, ed. E. C. Messenger, 301–12. London: Burns, Oates and Washburne, 1931.

———. "A Vindication of the Mediate Animation Theory." In *Evolution and Theology*, ed. E. C. Messenger, 259–83. London: Burns, Oates and Washburne, 1931.

Douglass, Frederick. *The Claims of the Negro, Ethnologically Considered. An Address, before the Literary Societies of Western Reserve College, at Commencement, July 12, 1854*. Rochester: Lee, Mann & Co., 1854.

Dove, John. *A Confutation of Atheisme*. London: Henry Rockett, 1605.

"Dr. Moore and His First Man." *Anthropological Review* 5 (1867): 105–10.

[Duncan, Isabelle]. *Pre-Adamite Man; or, the Story of Our Old Planet and Its Inhabitants, Told by Scripture and Science*. 3rd ed. London: Saunders, Otley, and Co., 1860.

Dunn, Robert. "On the Physiological and Psychological Evidence in Support of the Unity of the Human Species." *Transactions of the Ethnological Society of London*, n.s. 1 (1861): 186–202.

———. "Some Observations on the Psychological Differences Which Exist among the Typical Races of Man." *Transactions of the Ethnological Society of London*, n.s. 3 (1865): 9–25.

———. "Some Observations on the Tegumentary Differences Which Exist among the Races of Man." *Transactions of the Ethnological Society of London*, n.s. 1 (1861): 59–71.

Duns, John. *Biblical Natural Science, Being the Explanation of the All the References in Holy Scripture to Geology, Botany, Zoology, and Physical Geography.* Edinburgh: William Mackenzie, 1863–66.

———. *Science and Christian Thought.* London: Religious Tract Society, [ca. 1866].

Eco, Umberto. *Serendipities: Language and Lunacy.* Trans. William Weaver. London: Weidenfeld and Nicolson, 1999.

———. *The Search for the Perfect Language.* Trans. James Fentress. London: Fontana, 1997.

Eiseley, Loren. "Epilogue. Jack London, Evolutionist." In Jack London, *Before Adam*, 105–11. New York: Bantam Books, 1970.

Emerson, Ralph Waldo. *Emerson's Complete Works.* Vol. 6: *The Conduct of Life.* London: George Routledge and Sons, 1883.

"The Epigram, Part II." *Southern Literary Messenger* 14 (December 1848): 718–20.

"Evolution and Our Faith." *Daily Telegraph,* January 15, 1935.

F[rothingam], N. L. "Men before Adam." *Christian Examiner and Religious Miscellany* 50 (1851): 79–96.

Farrar, Frederic W. "Fixity of Type." *Transactions of the Ethnological Society of London*, n.s. 3 (1865): 394–99.

———. "Language and Ethnology." *Transactions of the Ethnological Society of London*, n.s. 4 (1866): 196–204.

———. "Traditions, Real and Fictitious." *Transactions of the Ethnological Society of London*, n.s. 3 (1865): 298–307.

Fiering, Norman. *Moral Philosophy at Seventeenth-Century Harvard: A Discipline in Transition.* Chapel Hill: University of North Carolina Press, 1981.

Fisher, Dick. *The Origins Solution: An Answer in the Creation-Evolution Debate.* Lima, Ohio: Fairway Press, 1996.

FitzRoy, Adm. [Robert]. "Outline Sketch of the Principal Varieties and Early Migrations of the Human Race." *Transactions of the Ethnological Society of London*, n.s. 1 (1861): 1–11.

Fleming, Ambrose. "Evolution and Revelation." *Journal of the Transactions of the Victoria Institute* 59 (1927): 11–40.

———. *Evolution or Creation?* 2nd ed. London: Marshall, Morgan and Scott, 1938.

———. *Memories of a Scientific Life.* London: Marshall, Morgan and Scott, 1934.

———. "Modern Anthropology *versus* Biblical Statements on Human Origin." *Journal of the Transactions of the Victoria Institute* 67 (1935): 15–42.

———. *The Origin of Mankind Viewed from the Standpoint of Revelation and Research.* London: Marshall, Morgan and Scott, 1935.

Force, James E. "The Nature of Newton's 'Holy Alliance' between Science and Religion: From the Scientific Revolution to Newton (and Back Again)." In *Rethinking the Scientific Revolution,* ed. Margaret Osler, 247–70. Cambridge: Cambridge University Press, 2000.

Foucault, Michel. *The Order of Things: An Archaeology of the Human Sciences*. London: Tavistock Publications, 1970.

Frederickson, George M. *The Black Image in the White Mind: The Debate on Afro-American Character and Destiny, 1817–1914*. New York: Harper and Row, 1972.

———. *Racism: A Short History*. Princeton: Princeton University Press, 2002.

Frei, Hans W. *The Eclipse of Biblical Narrative: A Study of Eighteenth and Nineteenth Century Hermeneutics*. New Haven: Yale University Press, 1974.

Friedman, John Block. *The Monstrous Races in Medieval Art and Thought*. Cambridge: Harvard University Press, 1981.

Gall, James. "On Improved Monographic Projections of the World." *Report of the Twenty-fifth Meeting of the British Association for the Advancement of Science*. London: John Murray, 1856.

———. *Primeval Man Unveiled: or, the Anthropology of the Bible*. 2nd ed. London: Hamilton, Adams and Co., 1880.

———. *The Stars and the Angels; or, The Natural History of the Universe and Its Inhabitants*. London: Hamilton, Adams, 1858.

———. "Use of Cylindrical Projections for Geographical, Astronomical, and Scientific Purposes." *Scottish Geographical Magazine* 1 (1885): 119–23.

Garrett, Aaron. "Anthropology: The 'Original' of Human Nature." In *The Cambridge Companion to the Scottish Enlightenment*, ed. Alexander Broadie, 79–93. Cambridge: Cambridge University Press, 2003.

Garrigan, O. W. "Preadamites." In *New Catholic Encyclopedia*, 11: 702. New York: McGraw-Hill, 1967.

Gay, Peter. *The Enlightenment: An Interpretation. The Rise of Modern Paganism*. 1966. Reprint. New York: Norton, 1995.

Gayman, Dan. *Biblical Anthropology: The Doctrine of Adam Man*. Schell City, Mo.: Watchman Outreach Ministries, 2002.

Genovese, Eugene D. *"Slavery Ordained of God": The Southern Slaveholders' View of Biblical History and Modern Politics*. Gettysburg, Pa.: Gettsyburg College, 1985.

Genovese, Eugene D., and Elizabeth Fox-Genovese. *The Mind of the Master Class: History and Faith in the Southern Slaveholders' Worldview*. New York: Cambridge University Press, 2005.

Gerbi, Antonello. *The Dispute of the New World: The History of a Polemic, 1750–1900*. Rev. and enl. ed. Trans. Jeremy Moyle. Pittsburgh: University of Pittsburgh Press, 1973.

Gilhooley, Leonard *Contradiction and Dilemma: Orestes Brownson and the American Ideal*. New York: Fordham University Press, 1972.

Gillespie, Neal C. *Charles Darwin and the Problem of Creation*. Chicago: University of Chicago Press, 1979.

Gilley, Sheridan. "The Huxley-Wilberforce Debate: A Reconstruction." In *Religion and Humanism*, ed. Keith Robbins, 325–40. Oxford: Blackwell, 1981.

Gilley, Sheridan, and Ann Loades. "Thomas Henry Huxley: The War between Science and Religion." *Journal of Religion* 61 (1981): 285–308.

Gliddon, George R. "The Monogenists and the Polygenists: Being an Exposition of the Doctrines of Schools Professing to Sustain Dogmatically the Unity or the Diversity of Human Races; with an Inquiry into the Antiquity of Mankind upon Earth, Viewed Chronologically, Historically, and Palaeontologically." In *Indigenous Races of the Earth; or, New Chapters of Ethnological Inquiry*, ed. J. C. Nott and George R. Gliddon. Philadelphia: Trübner, J. B. Lippincott, 1857.

———. *Ancient Egypt: A Series of Chapters on Early Egyptian History, Archaeology, and Other Subjects Connected with Hieroglyphical Literature*. New York: J. Winchester, 1843.

Godwyn, Morgan. *The Negro's and Indians Advocate, Suing for Their Admission into the Church: or, A Persuasive to the Instructing and Baptizing of the Negro's and Indians in Our Plantations*. London: printed for the author, 1680.

Goldsmith, Oliver. *An History of the Earth and Architecture, in Eight Volumes*. Vol. 2. London: J. Nourse, 1779.

Good, John Mason. *The Book of Nature*. 2nd ed. London: Longman, Rees, Orme, Brown, and Green, 1828.

Goodrum, Matthew R. "Atomism, Atheism, and the Spontaneous Generation of Human Beings: The Debate over a Natural Origin of the First Humans in Seventeenth-Century Britain." *Journal of the History of Ideas* 63 (2002): 207–24.

Gorst, Martin. *Measuring Eternity: The Search for the Beginning of Time*. New York: Broadway Books, 2001.

Gossett, Thomas F. *Race: The History of an Idea in America*. Dallas: Southern Methodist University Press. 1963.

Gould, Stephen Jay. "Fall in the House of Ussher." In *Eight Little Piggies: Reflections in Natural History*, 181–93. London: Penguin, 1993.

———. *The Mismeasure of Man*. Harmondsworth: Penguin, 1984.

———. "Morton's Ranking of Races by Cranial Capacity: Unconscious Manipulation of Data May Be a Scientific Norm." *Science* 200, May 5, 1978, 503–9.

———. "The Pre-Adamite in a Nutshell." *Natural History* 108 (1999): 24–27.

Grafton, Anthony. "Dating History: The Renaissance and the Reformation of Chronology." *Daedalus* (Spring 2003): 74–85.

———. *Defenders of the Text: The Traditions of Scholarship in an Age of Science, 1450–1800*. Cambridge: Harvard University Press, 1991.

———. *Joseph Scaliger: A Study in the History of Classical Scholarship*. Vol. 1: *Textual Criticism and Exegesis*. Oxford: Clarendon, 1983. Vol. 2: *Historical Chronology*. Oxford: Clarendon Press, 1993.

———. *New Worlds, Ancient Texts: The Power of Tradition and the Shock of Discovery*. Cambridge: Belknap Press of Harvard University Press, 1992.

———. "A Vision of the Past and Future." *Times Literary Supplement*, February 12–18, 1988, 151–52.

Grayson, Donald K. *The Establishment of Human Antiquity*. New York: Academic Press, 1983.

Green, David. "Veins of Resemblance: Photography and Eugenics." *Oxford Art Journal* 7, no. 2 (1985): 3–16.

Greenblatt, Stephen. *Marvelous Possessions: The Wonder of the New World*. Chicago: University of Chicago Press, 1991.

Greene, John C. "The American Debate on the Negro's Place in Nature, 1780–1815." *Journal of the History of Ideas* 15 (1954): 384–96.

———. *American Science in the Age of Jefferson*. Ames: Iowa State University Press, 1984.

———. *The Death of Adam: Evolution and Its Impact on Western Consciousness*. Ames: Iowa State University Press, 1959.

Gregory of Nyssa. "On the Making of Man." In *Nicene and Post-Nicene Fathers*. Vol. 5: *Gregory of Nyssa*, ed. Philip Schaff and Henry Wace. 1892. Reprint. Grand Rapids, Mich.: Eerdmans, 1976.

Gross, Abbé J. "The Problem of Origins in Recent Theology." In *Theology and Evolution (A Sequel to Evolution and Theology)*, ed. E. C. Messenger, 124–45. London: Sand & Co., 1949.

Grove, Richard H. *Green Imperialism: Colonial Expansion, Tropical Island Edens and the Origins of Environmentalism, 1600–1860*. Cambridge: Cambridge University Press, 1995.

Gruber, Jacob W. A. *A Conscience in Conflict: The Life of St. George Jackson Mivart*. Westport, Conn.: Greenwood Press, 1960.

Gundlach, Bradley John. "The Evolution Question at Princeton, 1845–1929." Ph.D. diss., University of Rochester (N.Y.), 1995.

Haber, Francis C. *The Age of the World: Moses to Darwin*. Baltimore: Johns Hopkins Press, 1959.

Haddon, A. C. *History of Anthropology*. London: Watts and Co., 1910.

Haeckel, Ernst. *The History of Creation: or the Development of the Earth and Its Inhabitants by the Action of Natural Causes. A Popular Exposition of the Doctrine of Evolution in General, and of That of Darwin, Goethe, and Lamarck in Particular*. Trans. and rev. E. Ray Lankester. 2 vols. London: Kegan Paul, Trench and Co., 1883.

Hale, Matthew. *The Primitive Origination of Mankind*. London, 1677.

Haller, John S., Jr. *Outcasts from Evolution. Scientific Attitudes of Racial Inferiority, 1859–1900*. Urbana: University of Illinois Press 1971.

Hammett, Iain Maxwell. "Burnett, James, Lord Monboddo." *Oxford Dictionary of National Biography*. Oxford: Oxford University Press, 2004–6.

Hampton, Monte Harrell. "'Handmaid' or 'Assailant': Debating Science and Scripture in the Culture of the Lost Cause." Ph.D. diss., University of North Carolina at Chapel Hill, 2004.

Hanke, Lewis. *All Mankind Is One: A Study of the Disputation between Bartolomé de las Casas and Juan Ginés de Sepúlveda in 1550 on the Intellectual and Religious Capacity of American Indians*. Dekalb: Northern Illinois University Press, 1974.

———. *Aristotle and the American Indians: A Study in Race Prejudice in the Modern World*. London: Hollis and Carter, 1959.

Harley, J. B. *Maps and the Columbian Encounter*. Milwaukee: Golda Meir Library, 1990.

———. "Maps and the Invention of America." *Map Collector* 58 (1992): 8–12.

———. "Rereading the Maps of the Columbian Encounter." *Annals of the Association of American Geographers* 82 (1992): 522–42.

Harnack, Adolf. "Origen." *Encyclopaedia Britannica,* 14th ed. London: Encyclopaedia Britannica, 1929.

Harper, George McLean. "Shields, Charles Woodruff." *Dictionary of American Biography.* New York: Scribner's, 1935.

Harris, George. "The Plurality of Races, and the Distinctive Character of the Adamite Species." *Anthropological Review* 5 (1867): 175–87.

Harris, John. *Man Primeval: or, the Constitution and Primitive Condition of the Human Being. A Contribution to Theological Science.* Boston: Gould and Lincoln, 1852.

———. *The Pre-Adamite Earth: Contributions to Theological Science.* Rev. and enl. ed. London: Ward, 1850.

Harris, Marvin. *The Rise of Anthropological Theory: A History of Theories of Culture.* New York: Thomas Y. Crowell, 1968.

Harvey, Paul. *Freedom's Coming: Religious Culture and the Shaping of the South from the Civil War through the Civil Rights Era.* Chapel Hill: University of North Carolina Press, 2005.

Hasskarl, G. G. H. *"The Missing Link"; or, The Negro's Ethnological Status. Is He a Descendant of Adam and Eve? Is He the Progeny of Ham? Has He a Soul? What Is His Relation to the White Race? Is He a Subject of the Church, of the State, Which?* Chambersburg, Pa.: Democratic News, 1898.

Hatfield, Gary. "Remaking the Science of Mind: Psychology as Natural Science." In *Inventing Human Science: Eighteenth-Century Domains,* ed. Christopher Fox, Roy Porter, and Robert Wokler, 184–231. Berkeley: University of California Press, 1996.

Hatley, Tom. "Adair, James (fl. 1736–1775)." *Oxford Dictionary of National Biography.* Oxford: Oxford University Press, 2004.

Havet, J. "L'Encyclique 'Humani generis' et le polygénisme." *Revue diocesaine de namur* 6 (1951): 114–27.

Haynes, Joseph E. *The Black: or the Natural History of the Hamitic Race.* Raleigh, N.C.: Edwards and Broughton, 1894.

Haynes, Stephen R. *Noah's Curse: The Biblical Justification of American Slavery.* Oxford: Oxford University Press, 2002.

Hazard, Paul. *The European Mind: 1680–1715.* Cleveland: Meridian Books, 1969.

Helmstadter, Richard. "Condescending Harmony: John Pye Smith's Mosaic Geology." In *Science and Dissent in England, 1688–1945,* ed. Paul Wood, 167–95. London: Ashgate, 2004.

Higgins, David. "Art, Genius, and Racial Theory in the Early Nineteenth Century: Benjamin Robert Haydon." *History Workshop Journal* 58 (2004): 17–40.

[Hodge, Charles]. "Examination of Some Reasonings against the Unity of Mankind." *Biblical Repertory and Princeton Review* 34 (1862): 435–64.

———. "The Unity of Mankind." *Biblical Repertory and Princeton Review* 31 (1859): 103–49.

Hodgen, Margaret T. *Early Anthropology in the Sixteenth and Seventeenth Centuries*. Philadelphia: University of Pennsylvania Press, 1964.

Holifield, E. Brooks. *Theology in America: Christian Thought from the Age of the Puritans to the Civil War*. New Haven: Yale University Press, 2003.

Hombron, Bernard. *Aventures les plus curieuses de voyageurs*. 2 vols. Paris: Belin-Leprieur, Morizot, 1847.

"Horn." *Biographie universelle, ancienne et moderne*. Vol. 19. Paris, 1817.

Horsman, Reginald. *Josiah Nott of Mobile. Southerner, Physician, and Racial Theorist*. Baton Rouge: Louisiana State University Press, 1987.

[Howe, George]. "Nott's Lectures." *Southern Presbyterian Review* 3 (1850): 426–90.

Huddleston, L. E. *Origins of the American Indians: European Concepts, 1492–1729*. Austin: University of Texas Press, 1967.

Hughes, Griffith. *The Natural History of Barbados. In Ten Books*. London, 1750.

Hunt, James. *The Negro's Place in Nature: A Paper Read before the Anthropological Society*. New York: Van Evrie, Horton and Co., 1864.

———. "On Ethno-Climatology; or the Acclimatization of Man." *Transactions of the Ethnological Society of London*, n.s. 2 (1863): 50–79.

———. "On the Application of the Principle of Natural Selection to Anthropology, in Reply to Views Advocated by Some of Mr. Darwin's Disciples." *Anthropological Review* 4 (1866): 320–40.

"Hunt on the Negro." *Journal of the Anthropological Society of London* 2 (1864): xv–lvi.

Hutton, Sarah. "More, Newton, and the Language of Biblical Prophecy." In *The Books of Nature and Scripture: Recent Essays on Natural Philosophy, Theology, and Biblical Criticism in the Netherlands of Spinoza's Time and the British Isles of Newton's Time*, ed. James E. Force and Richard H. Popkin, 39–43. Dordrecht: Kluwer, 1994.

Huxley, G. L. "Aristotle, Las Casas and the American Indians." *Proceedings of the Royal Irish Academy* 80C (1980): 57–68.

Huxley, Thomas Henry. *Man's Place in Nature and Other Anthropological Essays*. London: Macmillan, 1894.

Iliffe, Rob. "'Making a Shew': Apocalyptic Hermeneutics and the Sociology of Christian Idolatry in the Work of Isaac Newton and Henry More." In *The Books of Nature and Scripture: Recent Essays on Natural Philosophy, Theology, and Biblical Criticism in the Netherlands of Spinoza's Time and the British Isles of Newton's Time*, ed. James E. Force and Richard H. Popkin, 55–88. Dordrecht: Kluwer, 1994.

In the Image of God. Merrimac, Mass.: Destiny Publishers, 1967.

Jacquot, Jean. "Thomas Harriot's Reputation for Impiety." *Notes and Records of the Royal Society* 9 (1952): 164–87.

Jeffrey, David Lyle. "Medieval Monsters." In *Manlike Monsters on Trial*, ed. Marjorie M. Halpin and Michael M. Ames, 47–62. Vancouver: University of British Columbia Press, 1980.

Jensen, J. Vernon. "Return to the Wilberforce-Huxley Debate." *British Journal for the History of Science* 21 (1988): 161–79.

Johnson, Humphrey J. T. *Anthropology and the Fall.* Oxford; Blackwell, 1923.

———. "The Problem of Prehistoric Man." *Tablet,* August 12, 1939, 211.

Johnson, James. "Chronological Writing: Its Concept and Development." *History and Theory* 2 (1962): 124–45.

Joly, N. *Man before Metals.* 1883 Reprint. New York: Appleton, 1889.

Jordan, Winthrop. Introduction to Samuel Stanhope Smith, *Essay on the Causes of the Variety of Complexion and Figure in the Human Species.* Cambridge: Belknap Press of Harvard University Press, 1965.

Junkin, George. *The Integrity of our National Union, vs. Abolitionism: An Argument from the Bible, in Proof of the Position That Believing Masters Ought to Be Honored and Obeyed by Their Own Servants, and Tolerated in, Not Excommunicated from, the Church of God: Being Part of a Speech Delivered before the Synod of Cincinnati, on the Subject of Slavery, September 19th and 20th, 1843.* Cincinnati: printed by R. P. Donogh, 1843.

Kames, Henry Home, Lord. *Sketches of the History of Man.* 2 vols. Edinburgh: W. Strahan and T. Cadell, 1774.

Kasujja, Augustine. *Polygenism and the Theology of Original Sin Today. Eastern African Contribution to the Solution of the Scientific Problem: The Impact of Polygenism in Modern Theology.* Rome: Urbaniana University Press, 1986.

Keith, Arthur. *Darwinism and Its Critics.* London: Watts and Co., 1935.

———. *Darwinism and What It Implies.* London: Watts and Co., 1928.

Kennedy, James. "On the Probable Origin of the American Indians, with Particular Reference to That of the Caribs." *Journal of the Ethnological Society of London* 4 (1856): 226–67.

Kidd, Colin. *British Identities before Nationalism: Ethnicity and Nationhood in the Atlantic World, 1600–1800.* Cambridge: Cambridge University Press, 1999.

———. *The Forging of Races: Race and Scripture in the Protestant Atlantic World, 1600–2000.* Cambridge: Cambridge University Press, 2006.

Kidner, Derek. *Genesis. An Introduction and Commentary.* London: Tyndale, 1967.

King, Edward. "Dissertation Concerning the Creation of Man." *Morsels of Criticism, Tending to Illustrate Some Few Passages in the Holy Scriptures upon Philosophical Principles and an Enlarged View of Things,* 3:69–169. 2nd ed. London: J. Davis, 1800.

Kitson, Peter. "'Candid Reflections': The Race Issue in the Discourse of Slavery and Abolition." In *Discourses of Slavery and Abolition,* ed. Markman Ellis and Brycchan Carey, 11–25. London: Palgrave, 2003.

Kneeland, Samuel, Jr. "The Hybrid Races of Animals and Men." *DeBow's Review* 19 (1855): 535–39.

Knox, Robert. *The Races of Men: A Fragment.* London: Henry Renshaw, 1850.

Kvam, Kristan E., Linda S. Schering, and Valerie H. Ziegler. "The Preadamite Theory and the Christian Identity Movement: Race, Hierarchy, and Genesis 1–3 at the Turn of the Millennium." In *Eve and Adam: Jewish, Christian, and Muslim Readings on Genesis and Gender,* ed. Kvam, Schering, and Ziegler, 483–502. Bloomington: Indiana University Press, 1999.

Labourdette, M. M. *Le péché originel et les origines de l'homme.* Paris: Alsatia, 1953.

Lafitau, Joseph-François. *Moeurs des sauvages amériquains comparées aux moeurs des premiers temps.* Paris: Saugrain l'Aîné; C. E. Hochereau, 1724.

Lammey, David. "Dobbs, Francis." *Oxford Dictionary of National Biography.* Oxford: Oxford University Press, 2004.

Lane, Edward William. *The Genesis of the Earth and of Man: Or the History of Creation, and the Antiquity and Races of Mankind, Considered on Biblical and Other Grounds.* Ed. Reginald Stuart Poole. 2nd ed. London: Williams and Norgate, 1860.

Lang, W. D. "Human Origin and Christian Doctrine." *Nature* 136 (1935): 168–70.

Lanquet, Thomas. *Epitome of Chronicles.* London: Marshe, 1559.

La Peyrère, Isaac. *An Account of Iseland Sent to Monsieur de la Mothe de Vayer,* English translation of 1644 document in *A Collection of Voyages and Travels, Some Now First Printed from Original Manuscripts. Others Translated Out of Foreign Languages, and Now First Publish'd in English.* 4 vols. 2:431–46. London: Printed for Awnsham and John Churchill, 1704.

———. *Apologie de la Peyrère.* Paris: L. Billaine, 1663.

———. *Men before Adam. Or a Discourse upon the Twelfth, Thirteenth, and Fourteenth Verses of the Fifth Chapter of the Epistle of the Apostle Paul to the Romans. By Which Are Prov'd, That Men Were Created before Adam.* London, 1656.

———. *Relation du Greonland.* In *A Collection of Documents on Spitzbergen and Greenland, Comprising a Translation from F. Martens' Voyage to Spitzbergen: A Translation from Isaac de la Peyrère's Histoire de Groenland . . . ,* ed. Adam White. London: Hakluyt Society, 1855.

———. *A Theological System upon That Presupposition That Men Were before Adam.* London, 1655.

Lardner, Nathaniel. *The Works of Nathaniel Lardner, D.D.* 11 vols. London, 1788.

Latham, R. G. *Man and His Migrations.* London: John Van Voorst, 1851.

Lattey, C. "The Encyclical 'Humani Generis.'" *Scripture* 4 (1951): 278–79.

Lecky, W. E. H. *History of the Rise and Influence of the Spirit of Rationalism in Europe.* New York: D. Appleton, 1879.

Leeman, Saul. "Was Bishop Ussher's Chronology Influenced by a Midrash?" *Semeia* 8 (1977): 127–30.

Lefranc, Pierre. *Sir Walter Ralegh. Ecrivain, l'oeuvre et les idées.* Quebec: Les Presses de l'Université Laval, 1968.

Lehmann, William C. *Henry Home, Lord Kames, and the Scottish Enlightenment: A Study in National Character and in the History of Ideas.* The Hague: Martinus Nijhoff, 1971.

Lester, A. Hoyle. *The Pre-Adamites, or Who Tempted Eve? Scripture and Science in Unison as Respects the Antiquity of Man.* Philadelphia: J. B. Lippincott, 1875.

Lestringant, Frank. *Mapping the Renaissance World: The Geographical Imagination in the Age of Discovery.* Oxford: Polity Press, 1994.

Levy, David M. "How the Dismal Science Got Its Name: Debating Racial Quackery." *Journal of the History of Economic Thought* 23 (2001): 5–35.

Liénart, Cardinal Achille. "Evolution Harmonizes with Christian Faith." In John O'Brien, *God and Evolution: The Bearing of Evolution upon the Christian Faith,* 297–309. 2nd ed. Notre Dame: University of Notre Dame Press, 1961.

Little, Lawrence S. "The African Methodist Episcopal Church Media and Racial Discourse, 1880–1900." *North Star: A Journal of African American Religious History* 2, no. 1 (1998): 1–14.

Livingstone, David N. "Darwin in Belfast: The Evolution Debate." In *Nature in Ireland: A Scientific and Cultural History,* ed. John W. Foster, 387–408. Dublin: Lilliput Press, 1997.

———. *Darwin's Forgotten Defenders: The Encounter between Evangelical Theology and Evolutionary Thought.* Edinburgh: Scottish Academic Press, 1987.

———. "Geographical Inquiry, Rational Religion and Moral Philosophy: Enlightenment Discourses on the Human Condition." In *Geography and Enlightenment,* ed. David N. Livingstone and Charles W. J. Withers, 93–119. Chicago: University of Chicago Press, 1999.

———. "The History of Science and the History of Geography: Interactions and Implications." *History of Science* 22 (1984): 271–302.

———. "Human Acclimatization: Perspectives on a Contested Field of Inquiry in Science, Medicine and Geography." *History of Science* 25 (1987): 359–94.

———. "The Idea of Design: The Vicissitudes of a Key Concept in the Princeton Response to Darwin." *Scottish Journal of Theology* 37 (1984): 329–57.

———. "The Moral Discourse of Climate: Historical Considerations on Race, Place, and Virtue." *Journal of Historical Geography* 17 (1991): 413–34.

———. "Preadamism: The History of a Harmonizing Strategy." *Fides et Historia* 22 (1990): 25–34.

———. "Preadamites: The History of an Idea from Heresy to Orthodoxy." *Scottish Journal of Theology* 40 (1987): 41–66.

———. *The Preadamite Theory and the Marriage of Science and Religion.* Philadelphia: American Philosophical Society, 1992.

———. *Putting Science in Its Place: Geographies of Scientific Knowledge.* Chicago: University of Chicago Press, 2003.

———. "Race, Space and Moral Climatology: Notes toward a Genealogy." *Journal of Historical Geography* 28 (2002): 159–80.

———. "'Risen into Empire': Moral Geographies of the American Republic." In *Geography and Revolution,* ed. Charles W. J. Withers and David N. Livingstone, 304–35. Chicago: University of Chicago Press, 2005.

Livingstone, David N., and Mark A. Noll. "B. B. Warfield (1851–1921): A Biblical Inerrantist as Evolutionist." *Isis* 91 (2000): 283–304.

Livingstone, David N., and Ronald A. Wells. *Ulster-American Religion: Episodes in the History of a Cultural Connection.* Notre Dame, Ind.: University of Notre Dame Press, 1999.

London, Jack. *Before Adam.* New York: Bantam Books, 1970.

Long, Edward. *Candid Reflections upon the Judgment Latterly Awarded by the Court of the King's Bench on What Is Commonly Called the Negroe-Cause.* London, 1772.

—. *The History of Jamaica. Or, General Survey of the Antient and Modern State of That Island: With Reflections on Its Situation, Settlements, Inhabitants, Climate, Products, Commerce, Laws, and Government.* London: T. Lowndes, 1774.

Lonsdale, Henry. *A Sketch of the Life and Writings of Robert Knox, the Anatomist.* London: Macmillan, 1870.

Lucas, J. R. "Wilberforce and Huxley: A Legendary Encounter." *Historical Journal* 22 (1979): 313–30.

Lurie, Edward. "Louis Agassiz and the Races of Man." *Isis* 45 (1954): 227–42.

MacGregor-Morris, J. T. "Sir Ambrose Fleming (Jubilee of the Valve)." *Notes and Records of the Royal Society of London* 11, no. 2 (1955): 134–44.

Macloskie, George. "Concessions to Science." *Presbyterian Review* 10 (1889): 220–28.

—. "Mosaism and Darwinism." *Princeton Theological Review* 2 (1904): 425–41.

—. "Preliminary Talks on Science and Faith." Notebook. Macloskie Papers, CO498, carton 3, Firestone Library, Princeton University.

—. "The Origin of New Species and of Man." *Bibliotheca Sacra* 60 (1903): 261–75.

—. "The Outlook of Science and Faith." *Princeton Theological Review* 1 (1903): 597–615.

—. "Scientific Speculation." *Presbyterian Review* 8 (1887): 617–25.

—. "Theistic Evolution." *Presbyterian and Reformed Review* 33 (January 1898): 1–22.

Magne, Charles Lee. *The Negro and the World Crisis.* Harrison, Ark.: Kingdom Identity Ministeries, n.d.

Maimonides, Moses. *The Guide of the Perplexed.* Trans. Shlomo Pines. Chicago: University of Chicago Press, 1963.

Malcolm, Noel. *Aspects of Hobbes.* Oxford: Clarendon Press, 2002.

Marazuela, T. Ayuso. "Poligenismo y Evoluzionismo a la Luz de la Biblia y de la Teología." *Arbor* 9 (1951): 347–72.

Marsden, George M. *Fundamentalism and American Culture: The Shaping of Twentieth-Century Evangelicalism, 1870–1925* (New York: Oxford University Press, 1980.

Marvin, Enoch Mather. *The Work of Christ, or, the Atonement Considered in Its Influence upon the Intelligent Universe.* St. Louis: P. M. Pinckard, 1867.

Mascall, E. L. *Christian Theology and Natural Science.* London: Longmans, Green, 1956.

Mason, Peter. *Deconstructing America: Representations of the Other.* London: Routledge, 1990.

Mass, A. J. "Preadamites." In *The Catholic Encyclopedia: An International Work of Reference on the Constitution, Doctrine, Discipline, and History of the Catholic Church,* 12:370–71. New York: Encyclopedia Press, 1913.

McCausland, Dominick. *Adam and the Adamite; or, the Harmony of Scripture and Ethnology* 5th ed. 1864. Reprint. London: Richard Bentley, 1882.

————. *The Builders of Babel*. London, 1871.

McKee, D. R. "Isaac de la Peyrère, A Precursor of Eighteenth-Century Critical Deists." *Publications of the Modern Language Association* 56 (1944): 456–85.

Meigs, J. Aitken "The Cranial Characteristics of the Races of Men." In *Indigenous Races of the Earth; or, New Chapters of Ethnological Inquiry*, ed. J. C. Nott and George R. Gliddon, 203–352. Philadelphia: Trübner, J. B. Lippincott, 1857.

Meijer, Miriam Claude. *Race and Aesthetics in the Anthropology of Petrus Camper (1720–1789)*. Amsterdam: Rodopi, 1999.

Messenger, E. C. "The Embryology of St. Thomas Aquinas." In Messenger, *Theology and Evolution*, 243–58.

————. "Evolution and Theology To-Day: A Re-Examination of the Problems." In Messenger, *Theology and Evolution*, 172–216.

————. "Outline of Embryology, in the Light of Modern Science." In Messenger, *Theology and Evolution*, 221–32.

————. "A Short History of Embryology." In Messenger, *Theology and Evolution*, 233–42.

————. *Evolution and Theology*. London: Burns, Oates and Washburne, 1931.

————. ed. *Theology and Evolution (A Sequel to Evolution and Theology)*. London: Sands and Co., 1949.

Miller, Hugh. *Testimony of the Rocks; or, Geology in Its Bearing on the Two Theologies, Natural and Revealed*. Edinburgh: Thomas Constable, 1857.

————. "Unity of the Human Races." 1850. *Essays: Historical and Biographical, Political and Social, Literary and Scientific*, 387–97. 4th ed. Edinburgh: William P. Nimmo, 1870.

Minister [D. G. Phillips]. *Nachesh: What Is It? Or, An Answer to the Question, "Who and What Is the Negro?" Drawn from Revelation*. Augusta, Ga.: Jas. L. Gow, 1868.

Mivart, St. George Jackson. *On the Genesis of Species*. New York: Appleton, 1871.

Monboddo, James Burnet, Lord. *Of the Origin and Progress of Language*. Edinburgh: J. Balfour, 1774.

Montagu, M. F. Ashley. *Edward Tyson, M.D., F.R.S., 1650–1708, and the Rise of Human and Comparative Anatomy in England: A Study in the History of Science*, Memoir 20. Philadelphia: American Philosophical Society, 1943.

Montaigne, Michel de. *Essays*. Trans. John Florio. 3 vols. London: Dent, 1965.

Montesquieu, M. de Secondat, Baron de. *The Spirit of Laws*. Trans. from the French. London: printed for J. Nourse and P. Vaillant, 1750.

Moore, W. H. "Unity of the Human Races." *British and Foreign Evangelical Review* 1 (1852): 207–32.

Morgan, Kenneth. "Long, Edward (1734–1813)." *Oxford Dictionary of National Biography*. Oxford: Oxford University Press, 2004.

Morgan, Philip D. "Morality and Slavery." *European Review* 14 (2006): 393–99.

Morris, H. W. *Present Conflict of Science with the Christian Religion, or, Modern Scepticism Met on Its Own Ground*. Philadelphia: Ziegler, 1885.

————. *Work Days of God or Science and the Bible*. 2nd ed. 1890. Reprint. London: Pickering and Inglis, ca. 1924.

———. *Science and the Bible; or, the Mosaic Creation and Modern Discoveries.* Philadelphia: Ziegler and McCurdy, 1871.

Morton, Samuel George. *Crania Americana; or, A Comparative View of the Skulls of Various Aboriginal Nations of North and South America, to Which Is Added an Essay on the Varieties of the Human Species.* London: J. Dobson, Simpkin, Marshall and Co., 1839.

———. "Hybridity in Animals, Considered in Reference to the Question of the Unity of the Human Species." *American Journal of Science,* 2nd ser., 3 (1847): 39–50, 203–12.

M.S. *The Adamic Race: Reply to "Ariel," Drs. Young and Blackie, on the Negro.* New York: Russell Bros., 1868.

Mungello, D. E. *The Great Encounter of China and the West, 1500–1800.* New York: Rowman and Littlefield, 1999.

Murray, John. "Calvin's Doctrine of Creation." *Westminster Theological Journal* 17 (1954): 21–43.

Nelson, G. Blair. "'Men before Adam!': American Debates over the Unity and Antiquity of Humanity." In *When Science and Christianity Meet,* ed. David C. Lindberg and Ronald L. Numbers, 161–81. Chicago: University of Chicago Press, 2003.

Nemo [W. Moore]. *Man: Palaeolithic, Neolithic and Several Other Races, Not Inconsistent with Scripture.* Dublin: Hodges, Foster, 1876.

Nicolson, Adam. *Power and Glory: Jacobean England and the Making of the King James Bible.* London: HarperCollins, 2003.

Noll, Mark A. "The Rise and Long Life of the Protestant Enlightenment in America." In *Knowledge and Belief in America. Enlightenment Traditions and Modern Religious Thought,* ed. William M. Shea and Peter A. Huff, 88–124. New York: Cambridge University Press, 1995.

———. *The Civil War as a Theological Crisis.* Chapel Hill: University of North Carolina Press, 2006.

———. "The Bible and Slavery." In *Religion and the American Civil War,* ed. Randall M. Miller, Harry S. Stout, and Charles Reagan Wilson, 43–73. New York: Oxford University Press, 1998.

———. *Princeton and the Republic, 1768–1822: The Search for a Christian Enlightenment in the Era of Samuel Stanhope Smith.* Princeton: Princeton University Press, 1989.

———. ed. *The Princeton Theology: Scripture, Science, and Theological Method from Archibald Alexander to Benjamin Warfield.* Grand Rapids, Mich.: Baker, 1983.

Nott, Josiah Clark. "The Mulatto a Hybrid—Probable Extermination of the Two Races if the Whites and Blacks Are Allowed to Intermarry." *American Journal of the Medical Sciences* 6 (1843): 252–56.

———. *Two Lectures on the Natural History of the Caucasian and Negro Races.* Mobile, Ala.: Dade and Thompson, 1844.

Nott, Josiah Clark, and George R. Gliddon. *Types of Mankind; or, Ethnological Researches, Based upon the Ancient Monuments, Paintings, Sculptures, and Crania of*

Races, and upon Their Natural, Geographical, Philological, and Biblical History. Philadelphia: Lippincott, Gambo, 1854.

Noyce, Elisha. *Outlines of Creation.* London: Ward and Lock, 1858.

Numbers, Ronald L. "Charles Hodge and the Beauties and Deformities of Science." In *Charles Hodge Revisited: A Critical Appraisal of His Life and Work,* ed. John W. Stewart and James H. Moorhead, 77–102. Grand Rapids: Eerdmans, 2002.

———. *The Creationists: From Scientific Creationism to Intelligent Design.* Exp. ed. Cambridge: Harvard University Press, 2006.

———. "'The Most Important Biblical Discovery of Our Time': William Henry Green and the Demise of Ussher's Chronology." *Church History* 69 (2000): 257–76.

O'Brien, John A. *God and Evolution: The Bearing of Evolution upon the Christian Faith.* 2nd ed. Notre Dame: University of Notre Dame Press, 1961.

O'Leary, Don. *Roman Catholicism and Modern Science: A History.* New York: Continuum, 2006.

Orr, James. *God's Image in Man and Its Defacement in the Light of Modern Denials.* London: Hodder and Stoughton, 1905.

Orr, Philip. "From Orange Grove to Armageddon." In *A Man Stepped Out for Death: Thomas Russell and County Down,* ed. Brian S. Turner, 57–70. Newtownards: Colourpoint Books, 2003.

Osborne, Michael A. *Nature, the Exotic, and the Science of French Colonialism.* Bloomington: Indiana University Press, 1994.

Outram, Dorinda. *The Enlightenment.* Cambridge: Cambridge University Press, 1995.

Padgen, Anthony. *The Fall of Natural Man: The American Indian and the Origins of Comparative Ethnology.* New York: Cambridge University Press, 1982.

Patrides, C. A. *Premises and Motifs in Renaissance Thought and Literature.* Princeton: Princeton University Press, 1982.

———. "Renaissance Estimates of the Year of Creation." *Huntington Library Quarterly* 26 (1963): 315–22.

Patterson, Orlando. *Slavery and Social Death: A Comparative Study.* Cambridge: Harvard University Press, 1982.

Pearce, R. K. Victor. *Who Was Adam?* Exeter: Paternoster, 1969.

Peart, Sandra J., and David M. Levy, *The "Vanity of the Philosopher": From Equality to Hierarchy in Postclassical Economics.* Ann Arbor: University of Michigan Press, 2005.

Pember, G. H. *Earth's Earliest Ages and Their Connection with Modern Spiritualism and Theosophy.* 1876. Reprint. Glasgow: Pickering and Inglis, n.d.

Penniman, T. K. *A Hundred Years of Anthropology,* 2nd ed. London: Duckworth, 1952.

Peschel, Oscar. *The Races of Man, and Their Geographical Distribution.* 1874 (in German). New York: Appleton, 1906.

Philalethes. "The Distinction between Man and Animals." *Anthropological Review* 2 (1864): 153–63.

———. "Peyrerius and Theological Criticism." *Anthropological Review* 2 (1864): 109–16.

Picard, Guy. "La Science expérimentale est-elle favorable au polygénisme?" *Sciences Ecclesiastiques* 4 (1951): 65–89.

"The Plurality of the Human Race." *Anthropological Review* 3 (1865): 120–32.

Poe, Marshall. "What Did Russians Mean When They Called Themselves 'Slaves of the Tsar'?" *Slavic Review* 57 (1998): 585–608.

Ponton, Mungo. *The Beginning: Its When and Its How.* London: Longmans, Green, 1871.

Poole, Reginald Stuart. "The Ethnology of Egypt." *Transactions of the Ethnological Society of London,* n.s. 3 (1863): 260–64.

Poole, William. "The Divine and the Grammarian Theological Disputes in the 17th Century Universal Language Movement." *Historiographia Linguistica* 30 (2003): 273–300.

———. "Francis Lodwick's Creation: Theology and Natural Philosophy in the Early Royal Society." *Journal of the History of Ideas* 66 (2005): 245–63.

———. "Seventeenth-Century Preadamism, and an Anonymous English Preadamist." *Seventeenth Century* 19 (2004): 1–35.

Popkin, Richard H. "Biblical Criticism and Social Science." *Boston Studies in the Philosophy of Science* 14 (1974): 339–60.

———. "The Development of Religious Scepticism and the Influence of Isaac la Peyrère's Pre-Adamism and Bible Criticism." In *Classical Influences on European Culture, A.D. 1500–1700,* ed. R. R. Bolgar, 271–80. Cambridge: Cambridge University Press, 1976.

———. *The High Road to Pyrrhonism.* San Diego: Austin Hill Press, 1980.

———. *Isaac La Peyrère (1596–1676): His Life, Work and Influence.* Leiden: E. J. Brill, 1987.

———. "Jewish Messianism and Christian Millenarianism." In *Culture and Politics from Puritanism to the Enlightenment,* ed. Perez Zagorin, 67–90. Berkeley: University of California Press, 1980.

———. "The Marrano Theology of Isaac La Peyrère." *Studi Internazionali di Filosofia* 5 (1973): 97–126.

———. "The Philosophy of Bishop Stillingfleet." *Journal of the History of Philosophy* 9 (1971): 303–19.

———. "The Pre-Adamite Theory in the Renaissance." In *Philosophy and Humanism: Renaissance Essays in Honor of Paul Oskar Kristeller,* ed. Edward P. Mahoney, 50–69. Leiden: E. J. Brill, 1976.

"The Pre-Adamite World." *Sharpe's London Magazine* 12 (1850): 155.

Prest, John. *The Garden of Eden: The Botanic Garden and the Recreation of Paradise.* New Haven: Yale University Press, 1981.

Prichard, James Cowles. "On the Cosmogony of Moses." *Philosophical Magazine* 46 (1815): 285–92; 47 (1816): 110–17, 258–63; 48 (1816): 111–17.

———. "On the Extinction of Human Races." *Edinburgh New Philosophical Journal* 28 (1839–40): 166–70.

———. *Researches into the Physical History of Man.* London: John and Arthur Arch, 1813.

———. *Researches into the Physical History of Mankind.* 3rd ed. 5 vols. London: Sherwood, Gilbert, Piper, 1836–47.

———. *The Natural History of Man; Comprising Inquiries into the Modifying Influence of Physical and Moral Agencies on the Different Tribes of the Human Family.* 3rd ed. London: Hippolyte Billiere, 1848.

"Professor Winchell's 'Preadamites.'" *Appleton's Journal: A Magazine of General Literature* 9 (July 1880): 86–90.

Prospero. *Caliban: A Sequel to "Ariel."* New York, 1868.

Quatrefages, A. de. "Histoire naturelle de l'homme. Unité de l'espèce humaine." *Revue de Deux Mondes* 30 (1860): 807–33.

———. *The Human Species.* 2nd ed. London: Kegan Paul, 1879.

Quennehen, Elisabeth. "Lapeyrère, la Chine et la chronologie biblique." *La Lettre Clandestine* 9 (2000): 243–55.

R. T. B. "The Pre-Adamite Earth." *Southern Quarterly Review* 5, no. 9 (1852): 48–71.

Rahner, Karl. "Evolution and Original Sin." In *Consilium.* Vol. 26: *The Evolving World and Theology,* ed. Johannes Metz, 61–73. New York: Paulist Press, 1967.

———. *Hominization: The Evolutionary Origin of Man as a Theological Problem.* New York: Herder and Herder, 1965.

———. *Theological Investigations.* Vol. 1: *God, Christ, Mary and Grace.* 1954 (in German). London: Darton, Longman and Todd, 1961.

———. "Theological Reflexions on Monogenism." In *Theological Investigations.* Vol. 1: *God, Christ, Mary and Grace,* 229–96. 1954 (in German). London: Darton, Longman and Todd, 1961.

Rainger, Ronald. "Race, Politics and Science: The Anthropological Society of London in the 1860s." *Victorian Studies* 22 (1978): 51–70.

Raleigh, Sir Walter. *The Discovery of the Large, Rich and Beautiful Empire of Guiana.* London: Hakluyt Society, 1848.

Ramm, Bernard. *The Christian View of Science and Scripture.* 1954. Reprint. Exeter: Paternoster, 1971.

Ramsay, James. *Essay on the Treatment and Conversion of African Slaves in the British Sugar Colonies.* Dublin: printed for T. Walker, C. Jenkin, R. Marchbank, L. White, R. Burton, and P. Byrne, 1784.

Ramsey, Everett. *Racial Difference, More than Skin Deep.* Harrison, Ark.: Kingdom Identity Ministries, n.d.

Randolph, Paschal Beverly. *Eulis: Affectional Alchemy. The History of Love: Its Wondrous Magic, Chemistry, Rules, Laws, Moods, Modes and Rationale. Being the Third Revelation of Soul and Sex.* Toledo: Randolph Publishing Co., 1874.

———. *Pre-Adamite Man: Demonstrating the Existence of the Human Race upon This Earth 100,000 Thousand Years Ago!* Toledo: Randolph Publishing Co., 1888.

Reddie, James. "Slavery." *Anthropological Review* 2 (1864): 280–93.

Rehbock, Philip F. *The Philosophical Naturalists: Themes in Early Nineteenth-Century British Biology.* Madison: University of Wisconsin Press, 1983.

Reusch, Franz. Heinrich. *Nature and the Bible: Lectures on the Mosaic History of Creation in Its Relation to Natural Science.* Trans. Kathleen Lyttelton. Edinburgh: T. and T. Clark, 1886.

Review of *Man in Genesis and Geology. New Englander and Yale Review* 28 (1869): 785–89.

Review of *The Origin of Mankind Viewed from the Standpoint of Revelation and Research* by Sir Ambrose Fleming. *Christianity Today* 6, no. 6 (January 1936): 186.

Rhodes, P. G. M. "The Problem of Man's Origin." In *Theology and Evolution,* ed. E. C. Messenger, 3–9. London: Burns, Oates and Washburne, 1931.

Richards, Evelleen. "The 'Moral Anatomy' of Robert Knox: The Interplay between Biological and Social Thought in Victorian Scientific Naturalism." *Journal of the History of Biology* 22 (1989): 373–436.

Ridgley, Thomas. *A Body of Divinity: Wherein the Doctrines of the Christian Religion Are Examined and Defended Being the Substance of Several Lectures on the Assembly's Larger Catechism.* 2 vols. London, 1731.

Robinson, Marilyn. *The Death of Adam: Essays on Modern Thought.* New York: Picador, 1998.

Rogerson, John W. *Old Testament Criticism in the Nineteenth Century: England and Germany.* London: SPCK, 1984.

Rogoff, Leonard. "Is the Jew White? The Racial Place of the Southern Jew." *American Jewish History* 85 (1997): 195–230.

Rooden, Peter van. "Conceptions of Judaism as a Religion in the Seventeenth-Century Dutch Republic." In *The Church and the Jews,* ed. Diana Wood, 299–308. *Studies in Church History* 29. Oxford: Blackwell, 1992.

Ross, Ian Simpson. *Lord Kames and the Scotland of His Day.* Oxford: Clarendon Press, 1972.

Rossi, Paolo. *The Dark Abyss of Time: The History of the Earth and the History of Nations from Hooke to Vico.* Trans. Lydia G. Cochrane. Chicago: University of Chicago Press, 1984.

Rubiés, Joan-Pau. "Hugh Grotius's Dissertation on the Origin of the American Peoples and the Use of Comparative Methods." *Journal of the History of Ideas* 52 (1991): 221–44.

———. "Oriental Despotism and European Orientalism: Botero to Montesquieu." *Journal of Early Modern History* 9 (2005): 109–80.

Rudwick, Martin J. S. *Bursting the Limits of Time: The Reconstruction of Geohistory in the Age of Revolution.* Chicago: University of Chicago Press, 2005.

———. *Scenes from Deep Time: Early Pictorial Representations of the Prehistoric World.* Chicago: University of Chicago Press, 1992.

Rupke, Nicolaas. *The Great Chain of History: William Buckland and the English School of Geology.* Oxford: Clarendon, 1983.

———. "Neither Creation nor Evolution: The Third Way in Mid-Nineteenth Century Thinking about the Origin of Species." *Annals of the History and Philosophy of Biology* 10 (2005): 143–72.

Ryan, Michael T. "Assimilating New Worlds in the Sixteenth and Seventeenth Centuries." *Comparative Studies in Society and History* 23 (1981): 519–38.

S. A. L. "The Pre-Adamite World." *Southern Literary Messenger* 21, no. 9 (1855): 522–33.

Said, Edward. *Orientalism: Western Conceptions of the Orient.* London: Routledge and Kegan Paul, 1978.

Salmon, Vivian. *The Works of Francis Lodwick: A Study of His Writings in the Intellectual Context of the Seventeenth Century.* London: Longman, 1972.

Sardar, Zia, Ashis Nandy, and Merryl Wyn Davies. *Barbaric Others: A Manifesto on Western Racism.* London: Pluto Press, 1993.

Schnapp, Alain. "The Pre-Adamites: An Abortive Attempt to Invent Pre-History in the Seventeenth Century?" In *History of Scholarship: A Selection of Papers from the Seminar on the History of Scholarship Held Annually at the Warburg Institute,* ed. Christopher Ligota and Jean-Louis Quantin, 399–412. Oxford: Oxford University Press, 2006.

Sebastiani, Silvia. "Race and National Characters in Eighteenth-Century Scotland: The Polygenetic Discourse of Kames and Pinkerton." *Cromohs* 8 (2003): 1–14.

Sellers, Ian. "Lardner, Nathaniel." In *The New International Dictionary of the Christian Church,* ed. J. D. Douglas. Exeter: Paternoster, 1974.

"Shields, Charles Woodruff." In *Appleton's Cyclopaedia of American Biography,* ed. James Grant Wilson and John Fiske. New York: Appleton, 1888.

Shields, Charles Woodruff. *Philosophia Ultima, or Science of the Sciences.* 3 vols. London: Samson Low, Marston, Searle, and Rivington, 1889–1905.

———. *Religion and Science in Their Relation to Philosophy: An Essay on the Present State of the Sciences.* New York: Scribner, Armstrong and Co., 1875.

———. *The Scientific Evidences of Revealed Religion.* New York: Bishop Paddock Lectures, 1900.

Short, A. Rendle. *The Bible and Modern Research.* London: Marshall, Morgan and Scott, n.d.

———. *Modern Discovery and the Bible.* 1942. Reprint. London: Inter-Varsity, 1961.

Siraisi, Nancy G. "Vesalius and Human Diversity in *De Humani Corporis Fabrica.*" *Journal of the Warburg and Courtauld Institutes* 57 (1994): 60–88.

Skovgaard-Petersen, Karen. *Historiography at the Court of Christian IV (1588–1648): Studies in the Latin Histories of Denmark by Johannes Pontanus and Johannes Meursius.* Copenhagen: Museum Tusculanum Press, 2002.

Sloan, James A. *The Great Question Answered; or, Is Slavery a Sin in Itself (Per Se)? Answered according to the Teaching of the Scripture.* Memphis: Hutton, Gallaway, 1857.

Slotkin, J. S., ed. *Readings in Early Anthropology.* London: Methuen, 1965.

Smith, Lt. Col. Charles Hamilton. *The Natural History of the Human Species, Its Typical Forms, Primæval Distribution, Filiations, and Migrations.* Boston: Gould and Lincoln, 1851.

Smith, H. Sheldon. *In His Image, but . . . : Racism in Southern Religion, 1780–1910.* Durham: Duke University Press, 1972.

Smith, John Pye. *On the Relation between the Holy Scriptures and Some Parts of Geological Science.* 2nd ed. London: Jackson and Walford, 1840.

Smith, Pamela J. "A Splendid Idiosyncrasy: Prehistory at Cambridge, 1915–50." Ph.D. diss., University of Cambridge, 2004.

Smith, Samuel Stanhope. *An Essay on the Causes of the Variety of Complexion and Figure in the Human Species. To Which Are Added, Strictures on Lord Kames's Discourse on the Original Diversity of Mankind.* Philadelphia: Robert Aitken, 1788.

———. *An Essay on the Causes of the Variety of Complexion and Figure in the Human Species.* 1810. Reprint. Cambridge: Belknap Press of Harvard University Press, 1965.

Smith, Steven B. *Spinoza, Liberalism, and the Question of Jewish Identity.* New Haven: Yale University Press, 1997.

Smyth, Thomas. *The Unity of the Human Races Proved to Be the Doctrine of Scripture, Reason and Science with a Review of the Present Position and Theory of Professor Agassiz.* New York: Putnam, 1850.

Snobelen, Stephen David. "Of Stones, Men and Angels: The Competing Myth of Isabelle Duncan's *Pre-Adamite Man.*" *Studies in History and Philosophy of Biological and Biomedical Sciences* 32 (2001): 59–104.

Spencer, H. J. "Christmas [*later* Noel-Fearn], Henry (1811–1868)." *Oxford Dictionary of National Biography.* Oxford: Oxford University Press, 2004.

Stafford, Robert A. *Scientist of Empire: Sir Roderick Murchison, Scientific Exploration and Victorian Imperialism.* Cambridge: Cambridge University Press, 1989.

Stanton, William. *The Leopard's Spots: Scientific Attitudes toward Race in America, 1815–59.* Chicago: University of Chicago Press, 1960.

Stepan, Nancy. *The Idea of Race in Science: Great Britain, 1800–1960.* London: Macmillan, 1982.

Stephens, Lester. *Science, Race, and Religion in the American South: John Bachman and the Charleston Circle of Naturalists, 1815–1895.* Chapel Hill: University of North Carolina Press, 2000.

Stewart, John W. "Mediating the Center: Charles Hodge on American Science, Language, Literature, and Politics." *Studies in Reformed Theology and History* 3, no. 1 (1995).

Stillingfleet, Edward. *Origines Sacrae, or a Rational Account of the Grounds of Natural and Revealed Religion.* 1662. Reprint. Oxford: Clarendon Press, 1797.

Stocking, George W., Jr. "From Chronology to Ethnology: James Cowles Prichard and British Anthropology, 1800–1850." In *Researches into the Physical History of Man,* by James Cowles Prichard, ed. George W. Stocking Jr. Chicago: University of Chicago Press, 1973.

———. "The Persistence of Polygenist Thought in Post-Darwinian Anthropology." *Race, Culture, and Evolution: Essays in the History of Anthropology,* 42–68. Chicago: University of Chicago Press, 1982.

———. *Race, Culture, and Evolution: Essays in the History of Anthropology.* Chicago: University of Chicago Press, 1982.

———. "Scotland as the Model of Mankind: Lord Kames' Philosophical View of

Civilization." In *Toward a Science of Man: Essays in the History of Anthropology,* ed. Timothy H. H. Thoresen, 65–89. The Hague: Mouton, 1975.

———. *Victorian Anthropology.* New York: Free Press, 1987.

———. "What's in a Name? The Origins of the Royal Anthropological Institute, 1837–1871." *Man: The Journal of the Royal Anthropological Institute* 6 (1971): 369–90.

Stokes, Mason. "Someone's in the Garden with Eve: Race, Religion, and the American Fall." *American Quarterly* 50 (1998): 718–43.

Stott, John R. W. *Understanding the Bible: The Story of the Old Testament.* London: Scripture Union, 1978.

Stout, Harry S. *Upon the Altar of the Nation: A Moral History of the American Civil War.* New York: Viking, 2006.

Strathmann, Ernest A. "The History of the World and Ralegh's Scepticism." *Huntington Library Quarterly* 3 (1940): 265–87.

Strauss, Leo. *Spinoza's Critique of Religion.* New York: Schocken Books, 1982.

Stringfellow, Thornton. *Scriptural and Statistical Views in Favor of Slavery.* Richmond: J. W. Randolph, 1856.

Süsskind, Charles. "Fleming, John Ambrose." In *Dictionary of Scientific Biography,* ed. Charles Coulston Gillispie. New York: Charles Scribner's Sons, 1970–80.

Swain, Carol M. *The New White Nationalism in America: Its Challenge to Integration.* Cambridge: Cambridge University Press, 2002.

Swift, Wesley A. *God, Man, Nations & the Races.* Reprinted in *Eve and Adam: Jewish, Christian, and Muslim Readings on Genesis and Gender,* ed. Kristan E. Kvam, Linda S. Schering, and Valerie H. Ziegler, 496–99. Bloomington: Indiana University Press, 1999.

———. *Were All the People of the Earth Drowned in the Flood?* Harrison, Ark.: Kingdom Identity Ministries, n.d.

Tanner, Benjamin Tucker. "The Descent of the Negro." *AME Church Review* 15 (July 1898): 513–28. Reprinted as *The Descent of the Negro: Reply to Rev. Drs. J. H. Vincent, J. M. Freeman and J. L. Hurlbut.* Philadelphia: AME, 1898.

Tavakoli-Targhi, Mohamad. "Contested Memories: Narrative Structures and Allegorical Meanings of Iran's Pre-Islamic History." *Iranian Studies* 29 (1996): 149–75.

———. "Orientalism's Genesis Amnesia." *Comparative Studies of South Asia, Africa and the Middle East* 15, no. 1 (1996): 1–14.

Thomas, John. *Elpis Israel: A Book for the Times: Being an Exposition of the Kingdom of God, with Reference to "The Time of the End" and "Age to Come."* London: by the author, 1849.

Thompson, Joseph P. *Man in Genesis and Geology: Or, the Biblical Account of Creation, Tested by Scientific Theories of His Origin and Antiquity.* New York: Samuel R. Wells, 1870.

———. *Photographic Views of Egypt: Past and Present.* Glasgow: W. Collins, 1854.

———. *The Question of Races in the United States.* Glasgow: Robert Anderson, 1874.

———. *Teachings of the New Testament on Slavery*. New York: Joseph H. Ladd, 1856.

Thompson, W. O. "The Negro: The Racial Inferiority Argument in the Light of Science and History." *Voice of the Negro* 3 (1906): 507–13.

Thomson, Ann. "Issues at Stake in Eighteenth-Century Racial Classification." *Cromohs* 8 (2003): 1–20.

Torrey, R. A. *Difficulties and Alleged Errors and Contradictions in the Bible*. London: James Nisbet, n.d. [ca. 1907].

Trigger, Bruce G. *A History of Archaeological Thought*. Cambridge: Cambridge University Press, 1989.

Tyson, Edward. *Orang-Outang, sive Homo Sylvestris: or, the Anatomy of a Pygmie Compared with That of a Monkey, an Ape, and a Man. To Which Is Added, a Philological Essay Concerning the Pygmies, the Cynocephali, the Satyrs and Sphinges of the Ancients. Wherein It Will Appear That They Are All Either Apes or Monkeys, and Not Men*. London: printed for Thomas Bennet, 1699.

———. *Phocaena, or, the Anatomy of a Porpess, Dissected at Gresham College: With a Preliminary Discourse Concerning Anatomy, and a Natural History of Animals*. London: printed for Benj. Tooke, 1680.

An Universal History, from the Earliest Account of Time to the Present: Compiled from Original Authors; and Illustrated with Maps, Cuts, Notes, Chronological and Other Tables. Vol. 1. Dublin: printed by Edward Bate, 1744.

Van Amringe, William Frederick. *An Investigation of the Theories of the Natural History of Man, by Lawrence, Prichard, and Others Founded upon Human Analogies; and an Outline of a New Natural History of Man Founded upon History, Anatomy, Physiology, and Human Analogies*. New York: Baker and Scribner, 1848.

Vance, Norman. "Farrar, Frederic William (1831–1903)." *Oxford Dictionary of National Biography*. Oxford: Oxford University Press, 2004.

Vandebroek, G., and L. Renwart. "'Humani Generis' et les sciences naturelles." *Nouvelle revue théologique* 73 (1951): 3–20.

Van der Myl, Abraham. *De Origine Animalium et Migratione Populorum*. Geneva: n.p., 1667.

Van Kley, Edwin J. "Europe's 'Discovery' of China and the Writing of World History." *American Historical Review* 76 (1971): 358–85.

Van Riper, A. Bowdoin. *Men among the Mammoths: Victorian Science and the Discovery of Human Prehistory*. Chicago: University of Chicago Press, 1993.

Van Wyhe, John. "The Authority of Human Nature: The *Schuädellehre* of Franz Joseph Gall." *British Journal for the History of Science* 35 (2002): 17–42.

Voget, Fred W. *A History of Ethnology*. New York: Holt, Rinehart and Winston, 1975.

Von der Dunk, Hermann W. "German Historians and the Crucial Dilemma." *European Review* 14 (2006): 373–84.

Wagner, Rudolph. "Creation of Man, and Substance of the Mind." *Anthropological Review* 1 (1863): 227–32.

Wake, C. Staniland. "The Adamites." *Journal of the Anthropological Institute of Great Britain and Ireland* 1 (1872): 363–76.

Wake, Charles S. "The Relation of Man to the Inferior Forms of Animal Life." *Anthropological Review* 1 (1863): 365–73.

Walworth, Clarence A., ed. *The Gentle Skeptic; or, Essays and Conversations of a Country Justice on the Authenticity and Truthfulness of the Old Testament Records.* Ed. [or, rather, written by] Rev. C. Walworth. New York: D. Appleton, 1863.

Warfield, B. B. "On the Antiquity and the Unity of the Human Race." *Princeton Theological Review* 9 (1911): 1–25.

———. "Calvin's Doctrine of the Creation." *Princeton Theological Review* 13 (1915): 190–255.

———. Review of *God's Image in Man* by James Orr. *Princeton Theological Review* 4 (1906): 555–58.

———. Review of *Philosophia Ultima. Princeton Theological Review* 4 (1906): 541–42.

"Was Adam the First Man?" *Scribner's Magazine* 1 (1871): 578–89.

Watt, Hugh. *New College Edinburgh: A Centenary History.* Edinburgh: Oliver and Boyd, 1946.

Wayne, Alice. "After Babel." *Ladies' Repository: A Monthly Periodical, Devoted to Literature, Arts, and Religion* 4, no. 5 (1876): 401–7.

Weigel, Gustav. "Gleanings from the Commentaries on Human Generis." *Theological Studies* 12 (1951): 520–49.

Weiser, R. Review of *Pre-Adamite Man* by Griffith Lee. *Evangelical Quarterly Review* 17 (1866): 222–36.

Weisman, Charles A. *Not of One Blood: A Biblical, Historical and Scientific Evaluation of the Racial Equality Doctrine.* Apple Valley, Minn.: Weisman Publications, 2001.

[Whedon, Daniel]. Review of *Adam and the Adamite* by Dominick McCausland. *Methodist Quarterly Review* 53 (1871): 153–55.

———. Review of *Adamites and Preadamites* by Alexander Winchell. *Methodist Quarterly Review* 60 (1878): 567.

———. Review of *How the World Was Peopled* by Edward Fontaine. *Methodist Quarterly Review* 54 (1872): 521–23.

Wheeler, Roxann. *The Complexion of Race: Categories of Difference in Eighteenth-Century British Culture.* Philadelphia: University of Pennsylvania Press, 2000.

White, Andrew D. *A History of the Warfare of Science with Theology in Christendom.* 1894. Reprint. New York: George Braziller, 1955.

White, Anne Terry. *Men before Adam.* London: Hale, 1949.

White, Charles. *An Account of the Regular Gradation in Man, and in Different Animals and Vegetables; and from the Former to the Latter.* London: printed for C. Dilly, 1799.

White, Michael. *The Pope and the Heretic: A True Story of Courage and Murder at the Hands of the Inquisition.* London: Little, Brown, 2002.

Williams, George W. *History of the Negro in America from 1619 to 1880*. 2 vols. New York: Putnam, 1883.

Winchell, Alexander. *Adamites and Preadamites: or, A Popular Discussion Concerning the Remote Representatives of the Human Species and Their Relation to the Biblical Adam.* Syracuse: John T. Roberts, 1878.

———. *Creation, the Work of One Intelligence and Not the Product of Physical Forces.* Ann Arbor: Young Men's Literary Association, 1858.

———. *The Doctrine of Evolution: Its Data, Its Principles, Its Speculations, and Its Theistic Bearings.* New York: Harper and Bros., 1874.

———. "Huxley and Evolution." *Methodist Quarterly Review,* 4th ser., 59 (1877): 289–305.

———. "Preadamite." In *Cyclopaedia of Biblical, Theological, and Ecclesiastical Literature,* ed. John McClintock and James Strong, vol. 8. New York: Harper and Bros., 1877.

———. *Preadamites: Or a Demonstration of the Existence of Men before Adam; Together with a Study of Their Condition, Antiquity, Racial Affinities, and Progressive Dispersion over the Earth.* Chicago: A. C. Griggs and Co., 1880.

———. *Reconciliation of Science and Religion.* New York: Harper and Bros., 1877.

———. "Religious Ideas among Barbarous Tribes." *Methodist Quarterly Review* 57 (January 1875): 5–26.

———. "The Religious Nature of Savages." *Methodist Quarterly Review* 57 (July 1875): 357–78.

———. *Sketches of Creation: A Popular View of Some of the Grand Conclusions of the Sciences in Reference to the History of Matter and Life.* London: Sampson Low, Son, and Marston, 1870.

———. *Theologico-Geology, or, The Teaching of Scripture, Illustrated by the Conformation of the Earth's Crust.* Ann Arbor: Davis and Cole, 1857.

Winsor, Mary P. "Louis Agassiz and the Species Question." In *Studies in History of Biology,* ed. William Coleman and Camille Limoges, 89–117. Baltimore: Johns Hopkins University Press, 1979.

Wiseman, Nicholas Patrick. *Twelve Lectures on the Connection between Science and Revealed Religion.* 1836. Reprint. Dublin: James Duffy, 1866.

Withers, Charles W. J. "Geography, Enlightenment, and the Paradise Question." In *Geography and Enlightenment,* ed. David N. Livingstone and Charles W. J. Withers, 67–92. Chicago: University of Chicago Press, 1999.

———. *Placing the Enlightenment: Thinking Geographically about the Age of Reason.* Chicago: University of Chicago Press, 2007.

Wittkower, Rudolf. "Marvels of the East: A Study in the History of Monsters." *Journal of the Warburg and Courtault Institutes* 5 (1942): 159–97.

Wokler, Robert. "Apes and Races in the Scottish Enlightenment: Monboddo and Kames on the Nature of Man." In *Philosophy and Science in the Scottish Enlightenment,* ed. Peter Jones, 145–68. Edinburgh: John Donald Publishers, 1988.

Wood, Paul B. "The Science of Man." In *Culture of Natural History,* ed. N. Jardine,

J. A. Secord, and E. C. Spary, 197–210. Cambridge: Cambridge University Press, 1996.

Woodward, B. B. rev. David Huddleston, "McCausland, Dominick (1806–1873)." *Oxford Dictionary of National Biography.* Oxford: Oxford University Press, 2004–6.

Woodward, David. "Medieval *Mappaemundi*." In *The History of Cartography.* Vol. 1: *Cartography in Prehistoric, Ancient, and Medieval Europe and the Mediterranean,* ed. J. B. Harley and David Woodward, 286–370. Chicago: University of Chicago Press, 1987.

Wyatt-Brown, Bertram. *Southern Honor: Ethics and Behavior in the Old South.* New York: Oxford University Press, 1982.

Yardeni, Myriam. "La religion de La Peyrère et 'le rappel des Juifs.'" *Revue d'histoire et de philosophie religieuse* 51 (1971): 245–59.

Yates, Frances. *Giordano Bruno and the Hermetic Tradition.* Chicago: University of Chicago Press, 1964.

Yoder, John S., Jr. "Winchell, Alexander." *Dictionary of Scientific Biography,* 14:438–40. New York: Scribner's, 1981.

Young, Davis A. *The Biblical Flood: A Case Study of the Church's Response to Extrabiblical Evidence.* Grand Rapids, Mich.: Eerdmans, 1995.

Young, Robert Anderson. *The Negro: A Reply to Ariel. The Negro Belongs to the Genus Homo—He Is a Descendant of Adam and Eve—He Is the Offspring of Ham—He Is Not a Beast, but a Human Being—He Has an Immortal Soul—He May Be Civilized, Enlightened, and Converted to Christianity.* Nashville, Tenn.: J. W. M'Ferrin, 1867.

Zahm, J. A. *Evolution and Dogma.* Chicago: D. H. McBride and Co., 1896.

INDEX

abolitionism, 184–86
Aborigines Protection Society, 112, 121
Abraham, 33
acclimatization, 181
Account of the Regular Gradation in Man. See White, Charles.
Acosta, José de, 20
Adair, James, 67
Adam: and anthropology, 119–25; Augustine on, 14–15; birth of, 137, 138, 167–68; as black, 120; bloodline of, 180, 214–18; body of, 6–7, 138–39, 162, 167; as Caucasian, 101, 102, 103, 105, 156, 214–18, 242n87; colour of, 92, 126, 151, 198; creation of, 1, 205; as Cro-Magnon man, 205; and Darwinism, 154–67; descended from black forebears, 145, 150, 151, 153; as dolichocephalic, 118; ethnological advocates, 121–25; and evolution, 7, 139–40, 155–68, 209–10; father of the Jews, 35, 36, 54, 106, 132, 221; Hebrew meaning of, 153; intelligence of, 213; language of, 49, 116–17; and monstrous races, 13–14; as myth, 111, 154, 174; not first man, 6; as olive colour, 125; political significance of, 52, 54, 64–79, 221; predated by American Indians, 22, 23, 24; representation of, 2; as representative of humanity, 156; restricted location of, 90; Sabian

speculations about, 8; significance for conservatives, 201, 214; and slavery, 65; as stumbling block, 81; in tropical paradise, 2; views of Agassiz, 95; views of Bruno, 23–24; views of Fleming, 205; views of Lane, 101; views of La Peyrère, 26–27, 32–37; views of Macloskie, 158–59; views of Messenger, 165–67; views of Nemo, 106; views of Ponton, 135; views of Randolph, 111–12; views of Short, 160–61; views of Smyth, 180–81; views of Winchell, 145, 150, 153, 186–91; views of Wiseman, 128. *See also* co-adamites; monogenism; polygenism; pre-adamites
Adam and the Adamite, 103–5; reviews of 105–6. *See also* McCausland, Dominick
Adamites and Preadamites. See Winchell, Alexander
Adger, John, 182
African Methodist Episcopal Church, 185, 198
Africans: Hamitic origins of, 54, 182, 192, 198, 199, 200; humanity of, 65, 185; as inferior, 68–69, 73, 190; pre-adamite origins of, 215–16; as sub-human, 193, 197, 198–99
Agassiz, Louis, 99, 122, 126, 147, 156, 159, 180, 181, 184, 199, 208; admired by

Clarkson, Laurence, 24, 42

Clarkson, Thomas, 66

climate: influence of, 55–57, 60, 66, 110–11, 181; and morality, 64; views of Buffon, 56; views of Crawfurd 113; views of Duns, 127; views of Montesquieu, 55–56, 57, 58; views of Prichard, 120; views of Stanhope Smith, 74–75, 77; views of Wiseman, 129. *See also* environmentalism

Co-Adamitae, or an Essay to Prove the Two Following Paradoxes . . . , 45

co-adamites, 45–46, 60, 159, 163. *See also* pre-adamites

Colenso, John William, 103, 107, 111

Colman, Henry, 153

Columbus, Christopher, 16, 17

Comparate, Bernard, 216. *See also* Christian Identity Movement

Condé, Prince of, 26, 32

Conkin, Paul, 143

Cope, Edward Drinker, 143

Coverdale, Miles, 1

crania, measurements of, 174–76. *See also* anthropometry

Crania Americana, 174. *See also* American School of Ethnology; Morton, Samuel George

Crawfurd, John, 112–14

creation narrative, 1–6; as allegorical, 62; authorship of, 48, 231n91; challenges from chronology, 8–11, 22; challenges from geography, 8, 16–24; challenges from La Peyrère, 28–37; challenges from monstrosity, 8, 11–16, 22; two accounts of, 81, 88, 92, 100–101; as vision, 106

Credibility of the Gospel History, 44. *See also* Lardner, Nathaniel

Cro-Magnons, 147, 205, 207

Cuvier, Georges, 123

d'Abbeville, Nicolas Sanson, 27

Dabney, Robert, 183, 184

Dagg, J. L., 184

Dailey, Jane, 191

Dalgarno, George, 49

Dana, James, 202

Danish Chronicle, 29

Dannhauer, Johann, 40

Dante, 49

Darwin, Charles, 88, 90, 108, 109, 116, 120, 121, 136, 137, 138, 140, 142, 158, 171; and Farrar, 115; and Lamarckism, 120; and linguistic diversification, 113; as reconciling monogenism and polygenism, 188; and unity of human races, 129; and Winchell, 143; and Zahm, 162. *See also* Darwinism; human evolution

Darwinism, 80, 91–92, 124, 135, 140, 169, 199, 202; and pre-adamism, 154–68, 208–14; used to support monogenism, 185. *See also* Darwin, Charles; human evolution

Darwinism and Catholic Thought, 165. *See also* Dorlodot, Henry de

Darwinism and Its Critics, 206

Daston, Lorraine, 12

Dawson, Charles, 204

Death of Adam, The, 137

Defence of Galileo. See Campabella, Tommaso

degenerationism, 13–14, 150, 153

Delany, Martin, 185

d'Envieu, Jules Fabre, 164

Desmarets, Samuel, 40

Desmond, Adrian, 139–40

Dickson, William, 207

Diderot, Denis, 40

Dierckx, François, 246n96

Discovery of the Large, Rich and Beautiful Empire of Guiana, The, 17

dismal science, the, 170–71

Dobbs, Francis, 43–44

Doctrine of Evolution, The, 142. *See also* Winchell, Alexander dolichocephalic head form, 118. *See also* anthropometry; crania

d'Orbingy, Alcide, 88

Dorlodot, Henry de, 165, 166, 209

Douglass, Frederick, 184, 194

Dove, John, 23

Dubois, Eugène, 90, 204

Duncan, George, 90

Duncan, Isabelle, 87–91, 92, 101; anonymous author, 88; and biblical criticism, 88; and gap between pre- and post-adamic worlds, 90; reviews of, 91

Dunk, Hermann W. von der, 253n61

Dunn, Robert, 121–22

Duns, John, 126–28

Du rappel des Juifs, 32, 36–37. *See also* La Peyrère, Isaac

Ebstorf map, 14, 15

Echoes of the Universe, 87

Eco, Umberto, 10, 49

Eden: date of 3, 5; and fall as sexual transgression, 191–97; Garden of, 2–6, 33, 34, 63; hyper-fecundity in, 53; racial reading of, 216; restricted scope, 90; serpent in, 191–92, 195, 196; speech in, 50

Egyptian chronology, 8, 9, 35, 36, 53, 147–48. *See also* world chronology

Eiseley, Loren, 223

Elliston, John, 120

Elpis Israel, 86

Elzevier (publisher), 32

Emerson, Ralph Waldo, 173

environmentalism, 56–57, 121, 185, 189. *See also* climate

Ephrem, Saint, 165

Essay on the Causes of the Variety of Complexion, 74. *See also* Smith, Samuel Stanhope

Essay on the Treatment and Conversion of African Slaves in the British Sugar Colonies, 67. *See also* Ramsay, James

Essays and Reviews, 103

Ethnological Society of London, 112–14, 171. *See also* Anthropological Society of London

ethnology, 103–4; and Bendysche, 115–16; and Crawfurd, 112–14; and Farrar, 115; institutions of, 112; and language, 116–17; and Poole, 114–15; professional, 109. *See also* anthropology

ethno-theology, 186–200; and Ariel controversy, 192–97; and British Israelism, 214; challenged by Blackie, 194; challenged by Bolding, 198; challenged by Young, 193–94; and Christian Identity Movement, 214–18; and idea of Eve's seduction, 191; views of Campbell, 199–200; views of Carroll, 198–99; views of Cartwright, 191–92; views of Hasskarl, 197–98; views of Lester, 196–97; views of M.S., 194–95; views of Payne, 192–93; views of Winchell, 186–91

Europe, invention of, 16

Eusebius of Caesarea, 9

Eve, seduction of, 191–97

evolution. *See* Darwin, Charles; Darwinism; human evolution; Lamarckism

Evolution and Dogma, 162. *See also* Zahm, John

Evolution or Creation? 203. *See also* Fleming, Ambrose

Evolution Protest Movement, 202–3

Ewing, J. Franklin, 210

Exeter Hall, 171

Fabre, Abbé, 164

Farrar, Frederic W., 115

Field, John, 2

Fiering, Norman, 78

Fisher, Dick, 208

Fiske, John, 155

FitzRoy, Robert, 121
Fleischman, Albert, 204
Fleming, Ambrose, 202–8; criticised
  by Keith, 206–7; *Evolution or Crea-*
  tion? 203; on Cro-Magnons, 205; and
  Evolution Protest Movement, 202–3;
  and hominids, 204; on Neanderthal
  Man, 205; and pre-adamism, 204–7;
  on race mixing, 206; on spontane-
  ous generation, 251n9; *The Origin of*
  Mankind, 204
Fleming, John, 126
Fletcher, John, 2
Fontaine, Edward, 151
Forbes, Edward, 122, 156
Foucault, Michel, 135
Frederickson, George, 180, 192
French national identity, 37
Friedman, John Block, 16
Frothingam, N. L., 96–97
Fundamentals, The, 202

Galileo, 144
Gall, James, 157; and Darwinism,
  91–92; on map projections, 91, 101;
  and pre-adamism, 92; and two
  creation narratives, 92
gap theory, 83, 85
Garrigan, O. W., 164
Gassendi, Pierre, 27
Gay, Peter, 47
Gegenbaur, Karl, 143
Genesis of the Earth and of Man, 100.
  See also Lane, Edward William;
  Poole, Reginald Stuart
geographers, 34
geographical imagination, 3, 12–13, 16,
  17–18, 31
geography: challenges to traditional
  creation narrative, 16–24; moral,
  64–65; of northern regions, 27–31
geological time, 81–83, 84
geology, 81–85, 93, 111, 131–32, 133, 142,
  151

Gessner, Conrad, 49
Girardeau, John, 182
Gliddon, George, 125, 156, 159, 171, 173,
  181, 184, 211; approves of Lane, 101;
  opposes theology, 174; opposes unity
  of human race, 180. *See also* Ameri-
  can School of Ethnology
Gobineau, Arthur de, 159, 176
Godwyn, Morgan, 65
Goethe, Johann Wolfgang von, 72
Goldsmith, Oliver, 56–57, 64, 69
Good, John Mason, 93–94
Gore, Charles, 207
Gould, Stephen Jay, 90, 175
Graf, Karl Heinrich, 131
Grafton, Anthony: on early European
  perceptions of America, 19, 20; on
  La Peyrère, 34, 36, 37, 48, 51; on refu-
  tations of La Peyrère, 38
Green, William Henry, 125–26
Greene, John C., 137
Greenland, 27–28, 177
Gregory of Nyssa, Saint, 6, 7, 165; and
  evolution, 225n13
Grotius, Hugo, 22, 27, 30, 31, 32
Grynaeus, Somon, 17
Guide of the Perplexed. See Maimonides,
  Moses
Guyot, Arnold Henry, 214

Haeckel, Ernst, 154, 167, 198
Hale, Matthew, 22, 39, 66
Ham: as black, 54, 182, 192, 198; cursed,
  14, 182, 102; and monstrosity, 14; and
  slavery, 184, 192; views of Winchell,
  190–91. *See also* Noah, sons of
Hammond, James Henry, 180
Harriot, Thomas, 22
Harris, John, 82–85; reviewers of,
  84–85
Harris, Marvin, 74–75
Harvey, Paul, 199
Hasskarl, Gottlieb Christopher Henry,
  197–98

35, 51; on northern environments, 228n19; opposed by Good, 93–94; opposed by Hale, 39; opposed by Renaudot, 52; opposed by Ridgley, 53–54; opposed by Smyth, 180; opposed by Stillingfleet, 39; and polygenism, 50; and pre-adamite theory, 32–37; and Quatrefages, 122–23; recants, 26, 37–38; scholarship of, 48; and skepticism, 35–36, 51, 53; support of Amringe, 94; support of Bendysche, 115–16; support of Frothingham, 96; support of Philalethes, 116; supporters of, 42–44; and voyages of exploration, 34–35; and Winchell, 146, 151; work on northern geography, 27–31. *See also* pre-adamites
Laplace, Pierre-Simon, 82
Lardner, Nathaniel, 44–45
Large Collection of Ancient Jewish and Heathen Testimonies to the Truth of the Christian Religion, A, 44. *See also* Lardner, Nathaniel
Las Casas, Bartolomé de, 19–20, 21
Latham, R. G., 113, 122, 127
Lavater, Johann Kaspar, 61
Layfield, John, 3
Lecky, W. E. H., 43, 48
Lee, Griffith. *See* Randolph, Paschal Beverly, 110–12
Leo XIII, Pope, 165
Leroy, Dalmace, 162, 163, 165
Lesringant, Frank, 17
Lester, A. Hoyle, 196–97
Levy, David, 170–71
Lincoln, Abraham, 111
Linnaeus, Carolus, 61, 71
Locke, John, 37
Lodwick, Francis, 40–41
Lombard, Pierre, 2
London, Jack, 223–24
Long, Edward, 62, 68–69, 70, 71–72, 73; opposes Buffon, 68, 69
Lost Tribes of Israel, 214

Luther, Martin, 5
Lyell, Charles, 151

M.S., 194–96
Macloskie, George, 157; and evolution, 158–59
Madison, James, 78
Magne, Charles Lee, 216
Maimonides, Moses, 7–8, 40
Malcolm, Noel, 51
Mandeville's Travels, 11
Man in Genesis and Geology. See Thompson, Joseph Parish
mappaemundi, 2, 5–6, 11–12, 14–16
map projections, 91
Marconi, Guglielmo, 203
Maresius, Samuel. *See* Desmarets, Samuel
Marlowe, Christopher, 22
Martin, Ranald, 172
Martini, Martin, 9–10
Marvels of the East, 11
Marvin, E. M., 149
Mason, Peter, 12
Masoretic Bible, 38
Mass, A. J., 164
Maury, Louis Ferdinand Alfred, 176
McCausland, Dominick, 110, 131, 148, 151, 156, 164, 198, 204, 208, 218; *Adam and the Adamite*, 103–5; reviewers of, 105–6
McClintock, John, 146, 151
McCosh, James, 142, 157
McTyiere, Holland N., 143
Meigs, J. Aiken, 175–76. *See also* American School of Ethnology; anthropometry; crania
Men before Adam. See La Peyrère, Isaac; *Prae-Adamitae*; pre-adamites
Mercator, Gerard, 17
Mersenne, Marin, 32
Messenger, Ernest C., 165–67
Methodism, 142–46, 151
millennialism, 43–44

Miller, Hugh, 82, 89, 100, 106, 126, 214; on Adam as Caucasian, 242n87

miscegenation. *See* race mixing

Mitochondrial Eve, 218

Mivart, St. George Jackson, 139–40, 151, 156, 157, 163, 165, 167, 209

Modern Discovery and the Bible, 160–61. *See also* Short, A. Rendle

Monboddo, James Burnett, Lord, 70, 71, 74, 78; on language, 72–73

monkey chart, 176–79. *See also* Gliddon, George R.

monogenism: advocated by Prichard, 119–21; and African Episcopal Church, 185; and environment, 121–22; and Huxley, 188; and language, 102; pre-adamite version of, 138, 141–53; and Princeton, 125; support of Wiseman, 128–29; theological support for, 125–30; and Victorian Catholics, 128–31; views of Brock, 126; views of Brownson, 130–31; views of Dunn, 121; views of Duns, 126–28; views of Kennedy, 121; views of Miller, 126; views of Quatrefages, 122–23; views of Reddie, 124–25; views of Reusch, 129; views of Wagner, 123–24; views of Walworth, 129–30; views of Winchell, 141–53

monstrous births, 13

monstrous races, 11–16; in concentric circles, 226n34

Montaigne, Michel de, 10

Montesinos, Antonio de, 19

Montesquieu, M. de Secondat, Baron de, 60, 75, 78; and climate, 55–56, 57, 58; opposed by Kames, 57; opposed by King, 63

Moody, D. L., 202

Moore, G. E., 117

Moore, George, 117, 118

Moore, W. H., 96

morality: and cartography, 64–65, 177–78; and climate, 64; and history, 253n61

moral Newtonianism, 78

Morgan, Philip, 253n61

Morris, Herbert William, 85, 86

Morsels of Criticism, 81. *See also* King, Edward

Morton, Samuel George, 95, 116, 133, 145, 173–79, 181, 184, 185, 188, 199. *See also* American School of Ethnology; Nott, Josiah Clark

Mosaic Account of the Creation and Deluge, Illustrated by the Discoveries of Modern Science, The, 132. *See also* Smith, John Pye

Müller, Max, 102, 113, 116, 117

Munck, Jens, 28

Münster, Sebastian, 12, 17

Murchison, Roderick, 112

Murphy, Jeremiah, 140

Murphy, John Joseph, 139

Nabatean Agriculture, 7

Nachash, 192, 196

Natural History of the Human Species, The, 97. *See also* Smith, Charles Hamilton

natural slaves, 19–20

Naudé, Gabriel, 27

Navy-Surgeon The (Atkins), 61

Neanderthals, 147, 161, 164, 205, 206, 212, 246n90

Negro and the World Crisis, The, 216

Negro's and Indians Advocate, The (Godwyn), 65

neo-Lamarckism. *See* Lamarckism

Newman, John Henry, 129

Newton, Isaac, 43, 137

New York Public Theatre, 18

Nimrod, 14

Noah: sons of, 5–6; 10, 13–14, 19, 121, 184–85, 206; dispersion of sons of,

147–49; flood, 35, 38, 42, 51, 101, 107, 115, 132, 193; as progenitor of white races, 153

Noll, Mark, 76, 183

Nott, Josiah Clark, 125, 145, 156, 159, 171, 173, 174, 179, 180, 183, 184, 192, 199, 211. *See also* American School of Ethnology; Morton, Samuel George

Noyce, Elisha, 126

O'Brien, John, 209–10

Ogilby, John, 2

On the Emendation of Chronology, 9. *See also* Scaliger, Joseph John

On the Genesis of Species, 139, 140. *See also* Mivart, St. George Jackson

orang-utang, 69–73

Origen, 7, 130

Origines Sacrae, 39. *See also* Stillingfleet, Edward

Origin of Mankind, The, 204. *See also* Fleming, Ambrose

Orr, James, 160

Ortelius, Abraham, 17

Owen, Richard, 142, 194

pagan chronicles, 8–11. *See also* world chronology

paganism, idea of, 18–19

Palaeologus, Jacob, 22

Palmieri, Domenico, 164

Paracelsus, 23, 40, 159

Park, Katharine, 12

Parkinson, John, 2, 4

Payne, Buckner H. (Ariel), 192–94, 198, 199, 215, 217. *See also* Ariel controversy

Pearce, E. K. Victor, 213–14

Pedersen, Christiern, 29

Pember, George Hawkins, 93, 213

Perthes, Jacques Boucher de, 88, 116

Pesch, Christian, 163

Peschel, Oscar, 123

Peters, Arno, 92

Philadelphia Academy of Natural Sciences, 174

Philalethes, 116

Phillips, D. G., 196

Phillips, Wendell, 190

Philosophia Ultima, 155. *See also* Shields, Charles Woodruff

Pianciani, G. B., 161

Piltdown man, 204–5

Pithecanthropus erectus, 204

Pius IX, Pope, 140, 161

Pius XII, Pope, 209

Plato, 10

Plinian races, 11, 13. *See also* Pliny

Pliny, 10, 11, 17. *See also* Plinian races

plurality of worlds, 23–24, 149, 164

politics: Adamic, 52, 54, 64–79, 180–86, 221; and American School of Ethnology, 173–80; and modern ethnotheology, 186–200, 214–18; and polygeny, 170–73; of racial supremacy, 169–200, 214–18. *See also* slavery

polycentrism, 211

polygenism, 50, 58, 64, 92, 95, 99, 131–35, 169; advocates of, 109–17, 170–73, 173–80; and American School of Ethnology, 173–80; anthropological opposition to, 121–25; and Ariel controversy, 195; in Britain, 170–73; Catholic attitudes toward, 128, 129–31, 163–64, 209–12; and evolution, 140, 146; Protestant opposition toward, 125–28, 180–86; rejection of, 118–31; response of African-Americans to, 184–86; secular, 115. *See also* centers of creation; co-adamites; monogenism; pre-adamites

Ponton, Mungo, 134–35

Poole, Reginald Stuart, 100, 156, 164; attacks Bunsen's monogenism, 102; opposes use of polygenism in American South, 114–15. *See also* Lane, Edward William

Poole, William, 24, 40
Popkin, Richard, 27, 30, 37, 40, 43, 51
Pouchet, Georges, 172–73
Poulin, M., 120
Prae-Adamitae, 26, 27, 32–37, 39, 41, 51. *See also* La Peyrère, Isaac; pre-adamites
pre-adamite earth, 81–85
Pre-Adamite Earth, 82. *See also* Harris, John
Pre-Adamite Man (Duncan), 88–91. *See also* Duncan, Isabelle
Pre-Adamite Man (Randolph), review of, 110. *See also* Randolph, Paschal Beverly
Preadamites, 145. *See also* Winchell, Alexander
pre-adamites: as angels, 86, 90; 85–108; and anthropology, 114–19; and anti-evolutionism, 201–8; and anti-Semitism, 217; as apologetic device, 80–81, 85–108; Catholic stance on, 128–31, 161–67, 209; and conservative Protestants, 154–61; as conservative strategy, 91, 107, 154–55; as demons, 93; early speculations about, 6–8; and ethno-theology, 186–200, 214–18; and evolution theory, 137–68, 208–14; extinction of, 89–90, 92, 202, 210; and French national identity, 37; as gentiles, 36; as heresy, 65; hominization of, 211, 213; and *Humani Generis,* 209; illustrations of, 188, 189, 190; Mongolian, 197; and monogenism, 138, 141–52; move from heresy to orthodoxy, 220; as Neanderthals, 205; opposed by Stanhope Smith, 75–78; as potential harmonizing strategy, 131–35; and racial supremacy, 169–200, 214–18; revival of, 208–18; as sub-human, 193, 197–200; 20th-century legacy, 201–24; versatility of theory of, 219–24; views of

Alexander, 125; views of Atkins, 61; views of Bendysche, 115–16; views of Blount, 42; views of Brock, 126; views of Bruno, 23–24; views of Christmas, 87; views of Clarkson, 42; views of Crawfurd, 112–14; views of Dobbs, 43–44; views of Duncan, 87–91; views of Duns, 126–27; views of Farrar, 115; views of Fleming, 204–7; views of Harriot, 22; views of Harris, 117; views of Hodge, 126–27; views of Johnson, 163–64; views of Kames, 59–60; views of Kasujja, 211–12; views of Keith, 207; views of King, 62–63; views of Lane, 100–103; views of La Peyrère, 26–27, 32–37; views of Lardner, 44–45; views of Lodwick, 41–42; views of Macloskie, 158; views of McCausland, 103–6; views of Miller, 126; views of Moore, 117–18; views of Nemo, 106–7; views of Poole, 114–15; views of Rahner, 210; views of Raleigh, 22; views of Randolph, 111–12; views of Renaudot, 52; views of Short, 160–61; views of Thomas, 85–86; views of Voltaire, 42; views of Wake, 118; views of Winchell, 141–54, 188–91; views of Wiseman, 128–29; and Wagner, 123–24. *See also* co-adamites; monogenism; polygenism
Prichard, James Cowles, 84, 99, 113, 114, 121, 122, 127, 128, 129, 131, 151, 174, 180, 194; on Adam as black, 120, 241n55; and climate, 120; disgust at slavery, 173; opposes polygenism, 119
Primeval Man, 105. *See also* Argyll, George Douglas Campbell
Primeval Man Unveiled, 91, 157. *See also* Gall, James
Principia of Ethnology (Delany), 185
Prospero, 196
Purchas, Samuel, 73

skepticism: and geography, 27–31; influence of La Peyrère on, 35–36; Kames charged with, 76; and pre-adamism, 46, 80

Sketches of Creation, 142. *See also* Winchell, Alexander

Sketches of the History of Man. See Kames, Henry Home

Skreglinguer people, 29, 30

slavery, 65–69, 95, 114, 124, 171, 222; biblical support for, 182–84; black opposition to, 184–86; natural, 19–20; and science, 169–70, 179–80; and story of Ham, 184, 192

Sloan, James A., 184

Smith, Adam, 71, 170

Smith, Charles Hamilton, 97–99

Smith, John Pye, 89, 131–33, 158

Smith, Samuel Stanhope, 73–79, 120, 125, 180; anthropology of, 74–75; and climate, 74–75, 77; opposes Kames, 74, 76; opposes Monboddo, 73–74; and political philosophy, 78–79; rejects polygenism, 75–78

Smyth, Thomas, 99, 116, 182; and unity of human race, 180–81

Snobolen, Stephen, 90, 91

Sömmering, Samuel Thomas von, 61

soul: and Cartwright's racism, 192; Catholic debate on, 166–67, 209; and evolution of Adam, 139; inserted into human body, 138; and race in Hodge, 181; views of Mivart, 162; views of Origen, 7; views of Orr, 160; views of Wake, 118

Spencer, Herbert, 155

Spinoza, Baruch, 36, 48, 51

Stanton, William, 74

Stepan, Nancy, 173

Stewart, Dugald, 58, 72

Stillingfleet, Edward, 38, 39, 51

Stocking, George W., 58, 120, 140

Stokes, Mason, 191, 196

Strabo, 11

Strauss, Leo, 35

Strong, James, 146, 151

Suarez, Francisco, 139

Summers, T. O., 144

supremacism, 169; and American School of Ethnology, 173–80; in Britain, 170–73; and modern ethno-theology, 186–200. *See also* Christian Identity Movement; race hierarchy

Swift, Wesley, 216, 217. *See also* Christian Identity Movement

Tanner, Benjamin Tucker, 185

Teilhard de Chardin, Pierre, 205

Tenth Federalist, The, 78

Testimony of Modern Science to the Unity of Mankind (Cabell), 181

Theatrum Orbis Terrarum. See Ortelius, Abraham

Thevet, André, 17

Thomas, John, 85–86, 88

Thompson, Joseph Parish, 133–34

Thompson, W. O., 185

Thomson, A. S. 172

Thomson, Ann, 65

Thornwell, James Henley, 183

Tobin, James, 170

Topinard, Paul, 198

Torrey, Reuben A., 202, 213

Twelve Lectures on the Connection between Science and Revealed Religion (Wiseman), 128–29

Types of Mankind, 125, 173; review of, 180. *See also* Glidden, George R.; Nott, Josiah Clark

Tyson, Edward, 61, 73; and orang-utang, 70–71

Unity of the Human Race, 99, 126, 180. *See also* Smyth, Thomas

Universal History, 54

Ussher, James, 5

Valladolid debate, 19–20

Vanderbilt University, 141, 142, 143–45,